Control and Optimization

CHAPMAN & HALL MATHEMATICS SERIES

OTHER TITLES IN THE SERIES INCLUDE

Functions of Two Variables
S. Dineen

Network Optimization
V. K. Balakrishnan

Sets, Functions and Logic
A foundation course in mathematics
Second edition
K. Devlin

Algebraic Numbers and Algebraic Functions
P. M. Cohn

Dynamical Systems
Differential equations, maps and chaotic behaviour
D. K. Arrowsmith and C. M. Place

Elements of Linear Algebra
P. M. Cohn

Full information on the complete range of Chapman & Hall mathematics books is available from the publishers.

Control and Optimization

B. D. Craven

Reader in Mathematics
University of Melbourne
Australia

CHAPMAN & HALL
London · Glasgow · Weinheim · New York · Tokyo · Melbourne · Madras

Published by Chapman & Hall, 2–6 Boundary Row, London SE1 8HN, UK

Chapman & Hall, 2–6 Boundary Row, London SE1 8HN, UK

Blackie Academic & Professional, Wester Cleddens Road, Bishopbriggs, Glasgow G64 2NZ, UK

Chapman & Hall GmbH, Pappelallee 3, 69469 Weinheim, Germany

Chapman & Hall USA, 115 Fifth Avenue, New York, NY 10003, USA

Chapman & Hall Japan, ITP-Japan, Kyowa Building, 3F, 2-2-1 Hirakawacho, Chiyoda-ku, Tokyo 102, Japan

Chapman & Hall Australia, 102 Dodds Street, South Melbourne, Victoria 3205, Australia

Chapman & Hall India, R. Seshadri, 32 Second Main Road, CIT East, Madras 600 035, India

First edition 1995

Typeset in 10/12 pt Times by Thomson Press (India) Ltd, New Delhi, India
Printed in Great Britian by Hartnolls Ltd, Bodmin, Cornwall

ISBN 0 412 55890 4

A catalogue record for this book is available from the British Library

Library of Congress Catalog Card Number: 95-68504

∞ Printed on permanent acid-free text paper, manufactured in accordance with ANSI/NISO Z39.48-1992 and ANSI/NISO Z39.48-1984 (Permanence of Paper).

Contents

Preface

Many questions of optimization, and optimal control, arise in management, economics and engineering. An *optimum* (maximum or minimum) is sought for some function describing the system, subject to *constraints* on the values of the variables. Often the functions are nonlinear, so more specialized methods than linear programming are required. In optimal control, the variables in the problem become functions of time – state and control functions. A state function could be a path to be followed, or describe sales rate or earnings in a business model or water level in a water storage system, and the control function describes what can be controlled, within limits, in order to manage the system. This subject, as well as contributing to a large diversity of applications, has generated a great deal of mathematics and a variety of computational methods.

This book presents a systematic theory of optimal control, in relation to a general approach to optimization, applicable in other contexts as well. Chapter 1 introduces the subject, and summarizes various mathematical tools that are required. Chapter 2 sets up a diversity of optimal control models for the application areas mentioned and various others. Chapter 3 presents the underlying mathematics – some key words are *convex*, *linearization*, *Lagrangian conditions*, *duality*, *nonsmooth optimization* – and obtains general conditions describing an optimum. In Chapter 4, these are applied to optimal control problems, in discrete time and in continuous time, to obtain the Pontryagin principle, and also to analyse questions of sensitivity and stability when a system is perturbed. Chapter 5 is devoted to worked examples of optimal control problems. However, many practical problems in control, or otherwise in optimization, cannot be solved by formulae and require numerical computation. To this end, Chapter 6 describes the principles of some algorithms for control problems. These depend on approximating the control problem by some optimization problem in a finite number of variables so that it can be computed by one of various optimization algorithms. Sensitivity to parameter changes and the treatment of some nonsmooth terms in control problems are also discussed.

Chapter 7 presents proofs for Pontryagin's principle and other related results, as well as analysing questions of approximation and sensitivity. Lists of selected references may be found at the end of chapters.

This book includes recent results which may not be found elsewhere, concerning sensitivity and approximation, invex (generalized convex) functions in optimization models and methods for nonsmooth problems (when the functions do not always have derivatives).

The book is aimed at mathematics students at senior or graduate level at an American university, at second or third year of an honours course or postgraduate level in the UK and at third or fourth year or graduate level in an Australian university. For an indication of the mathematical background assumed, reference may be made to section 1.5, where the tools needed are summarized, including relevant topics in matrix algebra and normed spaces. An applied course will emphasize Chapters 2 (models), 5 (worked examples) and portions of 6 (algorithms). Students of pure mathematics will read Chapters 3 and 4, perhaps omitting the proofs in Chapter 7. Exercises are distributed through Chapter 3, and appear at the end of Chapters 2, 4, 5 and 6.

It thank Dr Barney Glover for useful suggestions and checking, and my final honours class for finding typing errors (but I am responsible for any remaining errors).

B. D. Craven
Melbourne
November 1994

1

Optimization – ideas and background

1.1 INTRODUCTION

Many questions in management and planning lead to mathematical models requiring *optimization*. Thus, some function of the variables that describe the problem must be maximized (or minimized) by a suitable choice of the variables within some permitted *feasible region*. For example, it may be required to calculate the conditions of operation of an industrial process which gives the maximum output or quality, or which gives the minimum cost. Such calculations are always subject to restrictions on the variables – not all combinations of the variables are practicable, there are restrictions on resources available, profit is to be maximized subject to some minimum level of quality being attained, and so on. Such restrictions, or *constraints*, define the feasible region for the variables of the problem.

A mathematical problem in which it is required to calculate the maximum or minimum (the word *optimum* includes both) of some *objective function*, usually of a vector variable, subject to constraints, is called a problem of *mathematical programming*. The numbers of variables and constraints may be finite or infinite – the same underlying theory applies. In the second case, the optimization is with respect to some *function*, for example a continuous function describing the state of the system being optimized.

In a large class of problems, both a *state function* and a *control function* are involved. Typically, the state function describes the state of the system as a function of time, whereas the control function describes some input to the system, which can be chosen, subject to some constraints, so as to optimize an objective function. Such a problem is called an *optimal control problem*. It may be noted that it differs from a problem of the calculus of variations in that the control function is subject to inequality constraints.

1.2 SIMPLE INVENTORY MODEL

To motivate this, consider the following inventory model. At time t, a firm holds a stock of its several product, represented by a vector $\mathbf{s}_t$; each component of $\mathbf{s}_t$ refers to a different product. Consider only integer times, $t = 0, 1, 2, \ldots$. In the tth time interval (thus between times t and $t+1$) an amount $\mathbf{m}_t$ is manufactured. This manufacture uses, as raw material, some of the existing stock; suppose that this amount is $\boldsymbol{G}\mathbf{m}_t$, where $\boldsymbol{G}$ is a suitable matrix. During $(t, t+1)$, a demand requirement $\mathbf{d}_t$ must be met. Then the stock vector at the end of this time interval is given by

$$\mathbf{s}_{t+1} = \mathbf{s}_t + \mathbf{m}_t - \boldsymbol{G}\mathbf{m}_t - \mathbf{d}_t \quad (t = 0, 1, 2, \ldots, n-1)$$

assuming a planning period of n time intervals. In a simple model, the objective function to be minimized could be a cost function:

$$f(\mathbf{s}, \mathbf{m}) = \sum (a_t^T \mathbf{m}_t + c_t^T (\mathbf{s}_t)_+ + k_t^T (\mathbf{s}_t)_-$$

Here $\mathbf{m} = (m_0, m_1, \ldots, m_{n-1})$ and $\mathbf{s} = (\mathbf{s}_0, \mathbf{s}_1, \ldots, \mathbf{s}_{n-1})$ describe the sequences of stocks and amounts made, $(\mathbf{s}_t)_+$ is obtained from $\mathbf{s}_t$ by replacing negative components by zero, $(\mathbf{s}_t)_-$ is obtained from $\mathbf{s}_t$ by replacing positive components by 0 then changing the sign of negative components, and the summations are from $t = 0$ to $n-1$. The unit cost vectors are a_t for amounts manufactured, c_t for holding stock (cost of holding one unit in stock for unit time) and k_t for running out of stock (penalty cost for not having a unit in stock for each time period in which it is demanded). All these quantities are written as column vectors (because usually there are many products to consider); superscript T denotes matrix transpose (see section 1.4). There may be bounds placed on inventory levels (thus $\mathbf{p}_t \le \mathbf{s}_t \le \mathbf{q}_t$) and an upper bound on the amounts manufactured (thus $0 \le \mathbf{m}_t \le \mathbf{r}_t$).

While this problem could be solved as a linear program – since all the functions in it are linear – that throws away the time-dependent structure of the problem. Instead, this inventory model can be considered as an *optimal control problem*, with $\mathbf{s}$ as the *state function* and $\mathbf{m}$ as the *control function* – appropriately, since the management can directly vary it.

This problem has an analogue in continuous time, say $0 \le t \le T$, instead of discrete time, $t = 0, 1, 2, \ldots, n$. The summations become integrals, thus the objective function takes the form

$$\int_0^T [\mathbf{a}(t)^T \mathbf{m}(t) + \mathbf{c}(t)^T \mathbf{s}_+(t) + \mathbf{k}(t)^T \mathbf{s}_-(t)]\, dt$$

where now $\mathbf{m}_t$ becomes a function $\mathbf{m}(t)$, and similarly for the other quantities. For inventory, such a continuous-time model is less useful than a discrete-time model. However, continuous-time models are useful in other contexts.

1.3 MATHEMATICAL FORMULATION – STAGE ONE

In each optimization problem considered, it is required to find an optimum (minimum or maximum) of a function $f(x)$, where x is a vector, or function, in some specified class V, and is subject to constraints – equations or inequalities – which may be symbolized by $x \in K$, where K is the *feasible set* (or *constraint set*). In particular, x may be a vector of n components (thus, $x \in \mathbf{R}^n$, the Euclidean space of n dimensions), or x may be a continuous function from an interval $I = [a, b]$ into $\mathbf{R}^n$ (denoted by $x \in C(I, \mathbf{R}^n)$). When x is in a space of functions, it is necessary to specify what kind of functions, for example $x \in V = C(I, \mathbf{R}^n)$.

Although V is not in general a vector space, it simplifies the presentation to make V into a vector space. Usually this is done by shifting the origin. For example, suppose that V consists of all real continuous functions x on the interval $[a, b]$, such that $x(a) = 0$ and $x(b) = c \neq 0$. Then the sum of two such functions x does not satisfy $x(b) = c$; so this V is not a vector space. However, the functions x can be expressed in terms of new functions y, defined by $y(t) = x(t) - c(t-a)/(t-b)$; then $y(a) = y(b) = 0$. So these functions y satisfy the vector space property: if $y^{(1)}$ and $y^{(2)}$ are two such functions, then so also is $\alpha y^{(1)} + \beta y^{(2)}$, if α and β are any constants. It is seldom necessary to calculate this shift of origin; it is enough to know that it can be done.

Consider the simple example (with $x \in \mathbf{R}$):

$$\text{Minimize}_{x \in \mathbf{R}}\ f(x) = x^2 \quad \text{subject to} \quad x \text{ in } K = [a, b]$$

where $0 < a < b$. The minimum obviously occurs when $x = a$, where f has gradient $2a \neq 0$. (Thus the constrained minimum point is not a *stationary* point of f, where the gradient vanishes.) The method of Lagrange multipliers, first given for minimization subject to equality constraints, extends to inequality constraints. The requirement $x \in K$ is equivalent to the pair of inequalities $x \geq a$, $x \leq b$; it is convenient to write both as ≤ 0 inequalities; thus by $a - x \leq 0$, $x - b \leq 0$. Then the *Lagrangian* function for this problem is

$$L(x) = x^2 + \lambda_1(a - x) + \lambda_2(x - b)$$

where λ_1 and λ_2 are *Lagrange multipliers*. The *Karush–Kuhn–Tucker theorem*, described later, shows that, if $\bar{x}$ is an optimum point, then the gradient of the Lagrangian vanishes there; thus

$$L'(\bar{x}) = 2\bar{x} - \lambda_1 + \lambda_2 = 0$$

and also $\lambda_1 \geq 0$, $\lambda_2 \geq 0$ (corresponding to the constraint inequalities $a - x \leq 0$ and $x - b \leq 0$), and $\lambda_1(a - \bar{x}) = 0$, $\lambda_2(\bar{x} - b) = 0$. These solve to $\bar{x} = a$, $\lambda_1 = 2a$, $\lambda_2 = 0$.

Constraints are usually expressed by equations or inequalities. A finite system of constraints often takes the form:

$$g_i(x) \leq 0 \quad (i = 1, 2, \ldots, m); \quad h_j(x) = 0 \quad (j = 1, 2, \ldots, r)$$

where the g_i and h_j are real functions. It is convenient to combine $g_1, g_2, \ldots, g_m$ into one column vector $\mathbf{g}$, and combine $h_1, h_2, \ldots, h_r$ into one column vector $\mathbf{h}$; then the constraints are expressed as

$$-\mathbf{g}(x) \in S \quad \mathbf{h}(x) = 0$$

where S is the nonnegative orthant in $\mathbf{R}^m$ (thus, $\mathbf{S}$ is the set of vectors in $\mathbf{R}^m$ for which all components v_i are nonnegative).

All these ideas apply equally to an optimal control problem. In that context, a constraint system $g_i(x) \le 0$ $(i = 1, 2, \ldots, m)$ may be replaced by

$$(\forall t \in [a, b]) \quad \mathbf{g}(x)(t) \le 0$$

Here $\mathbf{g}$ maps the vector (or function) x into a space of functions – thus $\mathbf{g}(x)$ is a function of a variable t in an interval $[a, b]$, replacing the index $i = 1, 2, \ldots, m$. So the constraint requires the function $g(x)$, evaluated at each point t in the interval, to take nonnegative values always. Suppose that the functions considered are the continuous real functions on $[a, b]$; thus, $\mathbf{g}(x) \in C[a, b]$. Denote by $\mathbf{S}$ the set of functions $\varphi \in C[a, b]$, such that $\varphi(t) \ge 0$ for each $t \in [a, b]$. Then the constraint may be written equivalently as $-\mathbf{g}(x) \in \mathbf{S}$.

The sets S and $\mathbf{S}$ just introduced are *convex cones*, so named for a geometric property that they share with both nonnegative orthants and ice-cream cones. It will appear that an optimal control problem can also be usefully expressed in a symbolic form:

$$\text{Minimize}_{z \in V} \; F(z) \quad \text{subject to} \quad -\mathbf{g}(z) \in \mathbf{S}$$

where V is a space of functions, z comprises both *state* and *control* functions (or variables) and $\mathbf{S}$ is some convex cone. (For the inventory model of section 1.2, z comprises both $\mathbf{s}$ and $\mathbf{m}$.)

Given such a problem, several questions arise:

(i) Are the constraints *consistent*? Thus, is there some $\hat{z} \in V$ for which $-\mathbf{g}(\hat{z}) \in \mathbf{S}$? (This is not always so. The constraints may be impossible to fulfil.) Assuming consistency, then:

(ii) What vectors z, satisfying the constraint, are *near* $\hat{z}$? In which directions can z move away from $\hat{z}$ by a short distance and still satisfy the constraints? (If $\hat{z}$ is a minimum point, then $F(z) \ge F(\hat{z})$ for such feasible points z.) *Implicit function theorems* and the concept of *local solvability* relate to this question.

(iii) Under what restrictive assumptions are conditions using Lagrange multipliers, such as the Karush–Kuhn–Tucker theorem, *necessary* for a minimum, or *sufficient* for a minimum? Such matters depend on *convex sets* and *convex functions* and on linearizing a nonlinear problem.

(iv) How can Lagrange multipliers be *represented*? (For optimal control problems, the required Lagrange multipliers belong to *dual spaces* of certain function spaces, and what is needed are integral formulae to describe them. This relates to descriptions using differential equations.)

Then also (importantly):

(v) As well as theory, what *algorithms* are available, and effective, to compute numerically an optimum for an optimal control problem?

These questions will all be discussed in the following chapters.

1.4 A ROCKET PROBLEM

Suppose that a spaceship is to be sent to Mars by rocket propulsion. At time t, the *state* of the spaceship will be described by a vector $x(t)$, whose elements are the components of position, velocity and angular velocity, relating to the spaceship at time t. The *control* is a function $u(t)$, specifying the programmed rate of burning the rocket fuel. Newton's law of motion describe acceleration in terms of force applied, which in this case is determined by the rocket thrust, depending on $u(t)$. Newton's second law leads to a second-order differential equation for the position. This reduces to a first-order differential equation in terms of $x(t)$, which includes the velocities as well. Thus, the dynamics of the spaceship can be described in the form:

$$\dot{x}(t) = m(x(t), u(t), t) \quad (t \geq 0); \quad x(0) = x_0$$

Here $\dot{x}(t) \equiv \mathrm{d}x(t)/\mathrm{d}t$, and an *initial condition* $x(0) = x_0$ must be specified. Note that the rate of burning fuel cannot be negative, and has some upper bound (else the motor would explode); thus there is an inequality constraint on the control, requiring $0 \leq u(t) \leq k$ for all times t, where k is some constant.

What is to be minimized? This is perhaps the time to reach Mars, or perhaps instead some combination of time to arrive with something depending on fuel consumption. Such an objective function may be expressed, in an integral form, as

$$F(x, u) = \int_0^T f(x(t), u(t), t)\,\mathrm{d}t$$

integrated over some time interval starting at zero (rocket launch) time. If $f(.) \equiv 1$, and the integration is from zero until the destination is reached, then $F(x, u)$ equals the time (T, say) to arrive. This gives a *time-optimal* control problem, in which T is variable. Or one may consider a fixed time T, and take $f(.)$, for example, in the form

$$f(x(t), u(t), t) = \alpha x(t)^2 + \beta u(t)^2 + \gamma$$

where α, β, γ are positive weights; thus a *fixed-time* optimal control problem arises.

Often the objective function also includes a term depending on the final state; thus

$$F(x, u) = \Phi(x(T)) + \int_0^T f(x(t), u(t), t)\,\mathrm{d}t$$

It will be convenient to write this objective in an equivalent form, using *Dirac's delta-function* $\delta(.)$ (see section 1.5.9), as

$$F(x, u) = \int_0^T [f(x(t), u(t), t) + \delta(t - T)\,\Phi(x(t))]\,\mathrm{d}t$$

This puts the problem with the extra objective term in Φ into the same shape as the previous case. It is useful to do so, for then a single theory will cover the several cases. It will be shown later that the time-optimal problem fits into this shapc also when $f(.)$ is suitably chosen.

The optimal control problem for the spaceship can then be expressed in the summary form:

$$\text{Minimize}_{x,u}\, F(x, u) \quad \text{subject to} \quad Dx = M(x, u),\, u \in \Gamma$$

Here the inequality on $u(t)$ for each t requires the control function $u(.)$ to lie in a certain bounded set; and the differential equation for $x_9(t)$, with initial condition, can be expressed by saying that the whole function (or graph) of the derivative $\dot{x}(.)$, symbolized by Dx, is a certain function M of the graph of $x(.)$ and the graph of $u(.)$.

1.5 SOME MATHEMATICAL BACKGROUND

To understand this book requires some proficiency in matrix algebra, calculus and normed vector spaces. Some of these essentials are summarized as follows – read now the parts that you need! Some additional concepts (function spaces, weak*closed, adjoint mapping, delta function, measure, representation of dual spaces) are here for later reference.

1.5.1 Matrix calculations

A *matrix* is a rectangular array of numbers, together with rules for algebraic operations (addition, multiplication etc.) on such arrays. An *m* × *n matrix* means a matrix with m rows and n columns; as special cases, an $m \times 1$ matrix is a *column vector*, and a $1 \times n$ matrix is a *row vector*. The rules for addition and multiplication are indicated by the following examples.

$$\boldsymbol{A} + \boldsymbol{B} = \begin{bmatrix} a_{11} & a_{12} & a_{13} \\ a_{21} & a_{22} & a_{23} \\ a_{31} & a_{32} & a_{33} \end{bmatrix} + \begin{bmatrix} b_{11} & b_{12} & b_{13} \\ b_{21} & b_{22} & b_{23} \\ b_{31} & b_{32} & b_{33} \end{bmatrix} = \begin{bmatrix} a_{11}+b_{11} & a_{12}+b_{12} & a_{13}+b_{13} \\ a_{21}+b_{21} & a_{22}+b_{22} & a_{23}+b_{23} \\ a_{31}+b_{31} & a_{32}+b_{32} & a_{33}+b_{33} \end{bmatrix}$$

$$\boldsymbol{DH} = \left[\begin{array}{cc|c} d_{11} & d_{12} & d_{13} \\ \hline d_{21} & d_{22} & d_{23} \end{array}\right] \left[\begin{array}{cc|c} h_{11} & h_{12} & h_{13} \\ h_{21} & h_{22} & h_{24} \\ \hline h_{31} & h_{32} & h_{34} \end{array}\right] = \left[\begin{array}{cc|c} c_{11} & c_{12} & c_{13} \\ \hline c_{21} & c_{22} & c_{23} \end{array}\right]$$

where $c_{ij} = \Sigma d_{ik} h_{kj}$, summing over $k = 1, 2, 3$. Note that matrices of the same size and shape can be added (and then the corresponding matrix elements are added); and an $m \times n$ matrix can multiply an $n \times p$ matrix, obtaining a product which is an $m \times p$ matrix. (The horizontal and vertical bars shown in $\boldsymbol{DH}$ relate to *partitioned multiplication*; see below.)

In these operations, each matrix is treated as a single entity (a sort of 'super number'). However, not all properties of numbers extend to matrices; observe that $\boldsymbol{AB}$ is not usually equal to $\boldsymbol{BA}$, and that $\boldsymbol{AB} = 0$ and $\boldsymbol{A} \neq 0$ do not imply $\boldsymbol{B} = 0$. For example:

$$\begin{bmatrix} 1 & 0 \\ 0 & 0 \end{bmatrix} \begin{bmatrix} 0 & 0 \\ 0 & 1 \end{bmatrix} = \begin{bmatrix} 0 & 0 \\ 0 & 0 \end{bmatrix}$$

If $\boldsymbol{A}$ is the matrix with elements a_{ij}, and λ is a real number, then $\lambda\boldsymbol{A}$ is the matrix with elements λa_{ij}. Matrices satisfy distributive laws: $\boldsymbol{A}(\boldsymbol{B} + \boldsymbol{C}) = \boldsymbol{AB} + \boldsymbol{AC}$ and $(\boldsymbol{P} + \boldsymbol{Q})\boldsymbol{R} = \boldsymbol{PR} + \boldsymbol{QR}$, provided that all the matrix expressions make sense.

The *transpose* of the $m \times n$ matrix $\boldsymbol{A}$ is obtained from $\boldsymbol{A}$ by interchanging rows with columns. Thus, the *transpose* $\boldsymbol{A}^{\mathrm{T}}$ is an $n \times m$ matrix, $\boldsymbol{B}$ say, with $b_{ij} = a_{ji}$.

Let $\mathbf{v}$ be a column vector with n components; thus $\mathbf{v}$ is a matrix with n rows and 1 column, or an $n \times 1$ matrix for short. The transpose $\mathbf{v}^{\mathrm{T}}$ of $\mathbf{v}$ is then a $1 \times n$ matrix. Note that $\mathbf{v}^{\mathrm{T}}\mathbf{v}$ is a 1×1 matrix, which may be identified with a scalar (thus, a number), whereas $\mathbf{v}\mathbf{v}^{\mathrm{T}}$ is an $n \times n$ matrix.

A *square matrix* $\boldsymbol{A}$ (thus, number of rows = number of columns) may, but need not, have an *inverse* $\boldsymbol{A}^{-1}$, defined by $\boldsymbol{AA}^{-1} = \boldsymbol{A}^{-1}\boldsymbol{A} = \boldsymbol{I}$, where $\boldsymbol{I}$ is the *unit matrix*

$$\begin{bmatrix} 1 & 0 & \cdots & 0 \\ 0 & 1 & & 0 \\ \vdots & & \ddots & \\ 0 & 0 & & 1 \end{bmatrix}$$

A theorem states that the following are equivalent for a square matrix $\boldsymbol{A}$:

(i) $\boldsymbol{A}$ is *invertible* (or *nonsingular*), meaning that $\boldsymbol{A}$ has an inverse.
(ii) The rows of $\boldsymbol{A}$ are *linearly independent*. (If the rows of $\boldsymbol{A}$ are denoted by $a_{1.}, a_{2.}, \ldots, a_{m.}$, then these vectors are *linearly independent* if $\alpha_1 a_{1.} + \alpha_2 a_{2.} + \ldots + \alpha_m a_{m.} = 0$, which implies that $(\alpha_1 = \alpha_2 = \ldots = \alpha_m = 0.)$

(iii) The columns of $\boldsymbol{A}$ are linearly independent.
(iv) The determinant of $\boldsymbol{A}$ is nonzero.

If the matrix product $\boldsymbol{AB}$ is defined, then the transpose $(\boldsymbol{AB})^{\mathrm{T}} = \boldsymbol{B}^{\mathrm{T}}\boldsymbol{A}^{\mathrm{T}}$. If $\boldsymbol{A}$ and $\boldsymbol{B}$ are invertible (square) matrices, then the inverse $(\boldsymbol{AB})^{-1} = \boldsymbol{B}^{-1}\boldsymbol{A}^{-1}$.

In the matrix multiplication formula given above, the matrix elements a_{ij} and b_{ij} may themselves be smaller matrices (called *submatrices*); then matrix multiplication is still valid, provided that all matrix multiplications and additions involved make sense. An example of partitioning is shown above, with partitions shown by horizontal and vertical bars; $\boldsymbol{D}$ is partitioned into four and

$\boldsymbol{H}$ into four submatrices; then $\boldsymbol{DH}$ is partitioned into four submatrices. The partitioning of the columns of $\boldsymbol{D}$ must agree with the partitioning of the rows of $\boldsymbol{H}$ (in the example, 2 + 2 in each case). If the submatrices of $\boldsymbol{D}$ and $\boldsymbol{H}$ are labelled by capital letters, then the following multiplication is correct when the submatrices in the partitioned matrix on the right are multiplied out.

$$\boldsymbol{DH} = \begin{bmatrix} \boldsymbol{D}_{11} & \boldsymbol{D}_{12} \\ \boldsymbol{D}_{21} & \boldsymbol{D}_{22} \end{bmatrix} \begin{bmatrix} \boldsymbol{H}_{11} & \boldsymbol{H}_{12} \\ \boldsymbol{H}_{21} & \boldsymbol{H}_{22} \end{bmatrix} = \begin{bmatrix} \boldsymbol{D}_{11}\boldsymbol{H}_{11} + \boldsymbol{D}_{12}\boldsymbol{H}_{21} & \boldsymbol{D}_{11}\boldsymbol{H}_{12} + \boldsymbol{D}_{12}\boldsymbol{H}_{22} \\ \boldsymbol{D}_{21}\boldsymbol{H}_{11} + \boldsymbol{D}_{22}\boldsymbol{H}_{21} & \boldsymbol{D}_{21}\boldsymbol{H}_{12} + \boldsymbol{D}_{22}\boldsymbol{H}_{22} \end{bmatrix}$$

For a partitioned square ($n \times n$) matrix $\boldsymbol{A}$, which is invertible, there are formulae (see section 6.2.7) for the inverse of $\boldsymbol{A}$ in terms of submatrices, and also (see section 6.2.8) for the change in the inverse of $\boldsymbol{A}$ when certain changes are made to $\boldsymbol{A}$ – for example, changing one column of $\boldsymbol{A}$.

The *rank* of an $m \times n$ matrix $\boldsymbol{A}$ is the largest number of linearly independent rows of $\boldsymbol{A}$ (or, equivalently, columns of $\boldsymbol{A}$). The matrix $\boldsymbol{A}$ has *full rank* if $\boldsymbol{A}$ has the largest rank allowed by m and n, namely the smaller of m and n.

Consider an $m \times n$ matrix $\boldsymbol{A}$, with $m < n$. An *elementary row operation* on $\boldsymbol{A}$ is one of:

- Interchanging two rows of $\boldsymbol{A}$.
- Multiplying a row by a nonzero constant.
- Adding a constant multiple of one row to another row.

Each such operation can be described by multiplying $\boldsymbol{A}$ on the left by a suitable (and simple) invertible matrix, $\boldsymbol{M}$ say; and this does not change the rank of $\boldsymbol{A}$. (*Exercise*: show that $\boldsymbol{M}$, to add a constant multiple of one row to another row, differs from a unit matrix in just one row and column.) If $\boldsymbol{A}$ has full rank, then $\boldsymbol{A}$ can be reduced by a number of elementary row operations to a standard form, shown by the partitioned matrix

$$\boldsymbol{EA} = [\boldsymbol{I} \,|\, \boldsymbol{C}\}$$

where $\boldsymbol{E}$ is the product of the matrices describing the row operations, $\boldsymbol{I}$ is an $m \times m$ unit matrix, $\boldsymbol{B}$ is some $m \times (n - m)$ matrix, and the columns of $\boldsymbol{A}$ have been reordered to bring the columns of $\boldsymbol{I}$ into standard order. Then *elementary column operations* (= elementary row operations applied to $\boldsymbol{A}^{\mathrm{T}}$) can produce a further reduction to $[\boldsymbol{I} \,|\, \boldsymbol{0}]$, where $\boldsymbol{0}$ stands for a submatrix of zeros.

1.5.2 Vector spaces and linear mappings

The *Euclidean space* $\mathbf{R}^n$ consists of all column vectors of n (real) elements. If such a vector has elements $a_1, a_2, \ldots, a_n$, this vector can be written as the *linear combination* $\alpha_1\mathbf{e}_1 + \alpha_2\mathbf{e}_2 + \ldots + \alpha_n\mathbf{e}_n$, where

$$\mathbf{e}_1^{\mathrm{T}} = (1, 0, 0, \ldots, 0), \quad \mathbf{e}_2^{\mathrm{T}} = (0, 1, 0, \ldots, 0), \ldots, \quad \mathbf{e}_n^{\mathrm{T}} = (0, 0, 0, \ldots, 1)$$

are unit vectors, forming a *basis* for $\mathbf{R}^n$ (thus, each vector in $\mathbf{R}^n$ is a linear combination of vectors in the basis and the basis vectors are linearly independent).

A mapping M from $\mathbf{R}^n$ into $\mathbf{R}^n$ (denoted $A: \mathbf{R}^n \to \mathbf{R}^n$) is *linear* if

$$M(\alpha\mathbf{x} + \beta\mathbf{y}) = \alpha M\mathbf{x} + \beta M\mathbf{y}$$

holds for all vectors $\mathbf{x}$, $\mathbf{y}$ and numbers α, β. Choose a basis of unit vectors $\mathbf{e}_j$ for $\mathbf{R}^m$ and $\mathbf{e}'_j$ for $\mathbf{R}^n$ (this amounts to defining coordinates). Since M is linear, $M\mathbf{e}_j = \sum_i a_{ij}\mathbf{e}'_i$ for some coefficients a_{ij}. If $\mathbf{x} = \sum x_j\mathbf{e}_j$ then

$$M\mathbf{x} = \sum_j x_j M(\mathbf{e}_j) = \sum_i \left(\sum_j a_{ij}x_j\right)\mathbf{e}'_i$$

Thus the linear mapping M is represented by the $m \times n$ matrix A with elements a_{ij}. (Often one does not need to use a different symbol for the mapping and the matrix that represents it, given coordinate systems.)

For example,

$$A = \begin{bmatrix} 2 & 3 \\ 4 & 5 \end{bmatrix}$$

maps the basis vectors

$$\begin{bmatrix} 1 \\ 0 \end{bmatrix} \quad \text{and} \quad \begin{bmatrix} 0 \\ 1 \end{bmatrix}$$

to the vectors

$$\begin{bmatrix} 2 \\ 4 \end{bmatrix} \quad \text{and} \quad \begin{bmatrix} 3 \\ 5 \end{bmatrix}$$

The reduction of an $m \times n$ matrix A (of full rank, with $m < n$) to a standard form $[I \mid \mathbf{0}]$ by row and column operations can also be considered as changing the bases (thus, changing the coordinate systems) in the domain space $\mathbf{R}^n$ and the range space $\mathbf{R}^m$.

If two linear mappings A and B are composed (thus if $\mathbf{z} = A\mathbf{x}$ and $\mathbf{w} = B\mathbf{z}$, when $\mathbf{w} = B(A\mathbf{x})$), then the composition mapping, written BA or $B \circ A$, is represented by the product of the matrices representing B and A. (The definition of matrix product is arranged to do exactly this.)

Let $\mathbf{R}_+ := [0, \infty)$, and $\mathbf{R}^n_+ := \{x \in \mathbf{R}^n : \mathbf{x}_i \geq 0\ (i = 1, 2, \ldots, n)$.

1.5.3 Norms and function spaces

A *vector space*, of which $\mathbf{R}^n$ is the prototype, has the property that, if $\mathbf{x}$ and $\mathbf{y}$ are any two vectors in the space, and α and β are any two numbers (or *scalars*), then also $\alpha\mathbf{x} + \beta\mathbf{y}$ is a vector in the space. There are algebraic rules for manipulating vectors (such as $\alpha(\mathbf{x} + \mathbf{y}) = \alpha\mathbf{x} + \alpha\mathbf{y}$) which are not detailed here, since they follow automatically, for all the spaces of functions considered, from the properties of numbers.

The (Euclidean) length of a vector $\mathbf{x} \in \mathbf{R}^n$ is $\|\mathbf{x}\| = (\sum x_j^2)^{1/2}$. The symbol $\|\mathbf{x}\|$ is called the *norm* of $\mathbf{x}$, and it has the properties:

$$\|\mathbf{x}\| \geq 0; \quad \|\mathbf{x} + \mathbf{y}\| \leq \|\mathbf{x}\| + \|\mathbf{y}\|; \quad \|\alpha\mathbf{x}\| = |\alpha|\,\|\mathbf{x}\|; \quad \|\mathbf{x}\| = 0 \Leftrightarrow \mathbf{x} = \mathbf{0}$$

for all vectors $\mathbf{x}$ and $\mathbf{y}$ and constant numbers α. The *triangle inequality* for

$\|\mathbf{x}+\mathbf{y}\|$ means geometrically that the length of one side of a triangle is not greater than the sum of the lengths of the other two sides. This norm will now be written as $\|\mathbf{x}\|_2$, to distinguish it from other norms satisfying the same list of properties. One such is $\|\mathbf{x}\|_1 = \Sigma|x_j|$.

A *norm* can often be defined for a space of functions. For an interval $I = [a, b]$, denote by $C(I)$ the vector space of real-valued continuous functions defined on I. Addition of functions, and multiplication of a function by a constant, are defined by

$$(\mathbf{x} + \mathbf{y})(t) = \mathbf{x}(t) + \mathbf{y}(t); \quad (\alpha\mathbf{x})(t) = \alpha\mathbf{x}(t)$$

This ensures that $C(I)$ satisfies the defining property of a *vector space*: if $\mathbf{x} \in C(I)$ and $\mathbf{y} \in C(I)$, and α and β are constants, then $\alpha\mathbf{x} + \beta\mathbf{y} \in C(I)$, using the theorem that the sum of two continuous functions is a continuous function. The algebraic properties for a vector space (associative, distributive etc.) need not be detailed here, since they follow immediately from the similar properties of real numbers. The usual norm used for $C(I)$ is the *uniform norm*:

$$\|\mathbf{x}\|_\infty = \max_{t \in I} |\mathbf{x}(t)|$$

Sometimes the L^2-norm $\|\mathbf{x}\|_2 = [\int_I |\mathbf{x}(t)|^2 \, dt]^{1/2}$ is also used for $C(I)$. (It also satisfies the stated defining properties for a norm.)

Consider also the space $C(I, \mathbf{R}^m)$ of continuous functions from the interval I into $\mathbf{R}^m$. The uniform and L^2-norms for $C(I, \mathbf{R}^m)$ are defined as above, but now interpreting $|\mathbf{x}(t)|$ as the Euclidean norm of the vector $\mathbf{x}(t)$ in $\mathbf{R}^m$. One must distinguish between the function evaluation $\mathbf{x}(t) \in \mathbf{R}^m$ and the function $\mathbf{x} \in C(I, \mathbf{R}^m)$, also written $\mathbf{x}(.) \in C(I, \mathbf{R}^m)$. (Although many authors write $\mathbf{x}$ as a contraction for $\mathbf{x}(t)$, it is misleading to thus contract until well accustomed to the ideas involved.)

Denote by $L^1(I)$ the vector space of functions $\mathbf{x} : I \to \mathbf{R}$ whose absolute values have finite integrals; the norm here is $\|\mathbf{x}\|_1 = \int_I |\mathbf{x}(t)| \, dt$. Similarly, $L^2(I)$ is the vector space of functions $\mathbf{x} : I \to \mathbf{R}$ whose squares have finite integrals; the norm here is $\|\mathbf{x}\|_2 = [\int_I |\mathbf{x}(t)|^2 \, dt]^{1/2}$. These two vector spaces consist of those functions for which the stated norm is finite. Obviously, $C(I)$ with the L^2-norm is a subspace of $L^2(I)$.

1.5.4 Continuous linear mappings

Let X and Y be normed vector spaces. A mapping $M: X \to Y$ is *continuous* if $\|Mu - Mv\|_Y$ can be made arbitrarily small by making $\|u - v\|_X$ sufficiently small. The notation $M: X \to Y$ means that every $x \in X$ has an image Mx; it does not imply that every $y \in Y$ is the image of some $x \in X$. If it happens that every $y \in Y$ has a corresponding $x \in X$ (or more than one) such that $y = Mx$, then M is called *surjective*, or *onto* Y.

Suppose now that M is *linear*. It then follows that M is *continuous* exactly when, for some (finite) constant κ, and each $x \in X$, $\|Mx\|_Y \leq \kappa\|x\|_X$. The minimal such κ is then called the *norm* of M, and denoted by $\|M\|$. (Of course, this norm depends on the chosen norms $\|.\|_X$ and $\|.\|_Y$.) If X and Y are finite dimensional, then a linear mapping $M: X \to Y$ is automatically continuous; however, a linear mapping from a function space need not be continuous. It is usually unnecessary to distinguish in notation between $\|.\|_X$ and $\|.\|_Y$; it is enough to write $\|.\|$ for both. (It is, however, necessary to distinguish between different norms on the same vector space.) The vector space of all continuous linear mappings from the normed vector space X into the normed vector space Y is denoted by $\mathbf{L}(X, Y)$.

If X and Y have finite dimensions, then M is represented by a matrix (section 1.5.2). For $n \times n$ matrices M, define also the *Frobenius norm*

$$\|M\|_{\text{Frob}} := \left[\sum_i \sum_j |M_{ij}|^2 \right]^{-1/2}$$

If $\|M\|$ is the matrix norm constructed from $X = Y = \mathbf{R}^n$ with Euclidean norms, then $\|M\| \leq \|M\|_{\text{Frob}}$. Even when X and Y have infinite dimensions, it is convenient to visualize $M \in \mathbf{L}(X, Y)$ as if it were a matrix, because formulae for calculating with M (e.g. section 1.5.7) have the same form as for matrices.

Given a function, or mapping, $f: X \to Y$, and $E \subset X$, denote

$$f(E) = \{f(x) : x \in E\}$$

The function f is *onto* if $f(X) = Y$. Given $U \subset Y$, let

$$f^{-1}(U) = \{\mathbf{x} \in X : f(\mathbf{x}) \in U\}$$

This is a conventional notation, and does not imply that f is one-to-one.

A *subspace* K of the vector space X is a subset such that

$$\mathbf{x}, \mathbf{y} \in K \quad \text{and} \quad \alpha, \beta \leq \mathbf{R} \Rightarrow \alpha\mathbf{x} + \beta\mathbf{y} \in K$$

Lines and planes through the origin are examples. Note that $\mathbf{0} \in K$ for each subspace K. A displaced subspace $V = \mathbf{c} + K = \{\mathbf{c} + \mathbf{x} : \mathbf{x} \in K\}$, where $\mathbf{c}$ is some constant vector, will be called here an *affine variety*. (The usual name is *linear variety*, but that suggests that $\mathbf{0} \in V$.) The *affine hull* of a set $E \subset X$ is the set aff E of all affine varieties containing E.

Mappings from X into $\mathbf{R}$ are called *functionals* on X. The vector space of all *continuous linear functionals* on X is called the *dual space* of X. In this book, the dual space of X will be denoted by X'. (The alternative notation X^* is needed here for another concept, *dual cone*.) In finite dimensions, each vector $\mathbf{x} \in X$ is represented by a column vector, and each vector $\mathbf{f} \in X'$ is represented by a row vector, so that

$$\mathbf{f}(\mathbf{x}) = \begin{bmatrix} f_1 & f_2 & f_3 & \cdots & f_n \end{bmatrix} \begin{bmatrix} x_1 \\ x_2 \\ \vdots \\ x_n \end{bmatrix} \in \mathbf{R}$$

Let $0 \neq \mathbf{f} \in X'$; thus, $\mathbf{f}$ is a nonzero continuous linear functional on X. Let $H = \mathbf{f}^{-1}(0)$, If $\mathbf{x}, \mathbf{y} \in H$ and $\alpha, \beta \in \mathbf{R}$, then $\mathbf{f}(\alpha\mathbf{x} + \beta\mathbf{y}) = \alpha\mathbf{f}(\mathbf{x}) + \beta\mathbf{f}(\mathbf{y}) = 0$ since $\mathbf{f}$ is linear. If $\{\mathbf{x}_n\}$ is a sequence in H, and $\{\mathbf{x}_n\} \to \mathbf{x}_0$, then $0 = \mathbf{f}(\mathbf{x}_n) \to \mathbf{f}(\mathbf{x}_0)$, since $\mathbf{f}$ is continuous; so $\mathbf{x}_0 \in H$. Thus H is a closed subspace of X. Since $\mathbf{f} \neq 0$, $\mathbf{f}(\mathbf{b}) \neq 0$ for some $\mathbf{b} \in X$. Given $\mathbf{x} \in X$, let $\mathbf{z} = \mathbf{x} - [\mathbf{f}(\mathbf{x})/\mathbf{f}(\mathbf{b})]\mathbf{b}$; then $\mathbf{f}(\mathbf{z}) = 0$, so $\mathbf{z} \in H$. Thus $X = H + U$, where $U = \{\alpha\mathbf{b} : \alpha \in \mathbf{R}\}$ is the one-dimensional subspace generated by $\mathbf{b}$. Intuitively, H is 'one dimension down' from X. The only subspace properly containing H is X itself. Such a subspace H is called a *hyperplane*. A set $\{\mathbf{x} \in X : \mathbf{f}(\mathbf{x}) \geq \beta\}$, where β is constant, is called a *closed halfspace* of X. Its boundary is the displaced hyperplane $\mathbf{c} + \mathbf{f}^{-1}(0)$.

1.5.5 Differentiation of vector functions

This is best approached by an example. Consider the function of two real variables $f(x, y) = x^2 - 3xy - y^3$ near the point $(x, y) = (2, -1)$. To do so, set $x = 2 + \xi$ and $y = -1 + \eta$. Then, expanding the expression,

$$f(2 + \xi, -1 + \eta) = 11 + (7\xi - 9\eta) + (\xi^2 - 3\xi\eta + 3\eta^2 + \eta^3)$$
$$= f(2, -1) + [7 -9]\begin{bmatrix} \xi \\ \eta \end{bmatrix} + r(\xi, \eta)$$

where the remainder term $r(\xi, \eta) = \xi^2 - 3\xi\eta + 3\eta^2 + \eta^3$ satisfies

$$|r(\xi, \eta)|/(\xi^2 + \eta^2)^{1/2} \to 0 \quad \text{as} \quad \|\mathbf{v}\| = (\xi^2 + \eta^2)^{1/2} \to 0, \quad \text{where} \quad \mathbf{v}^{\mathrm{T}} = [\xi, \eta]$$

Denote also $\mathbf{c}^{\mathrm{T}} = [2, -1]$ and $\mathbf{z}^{\mathrm{T}} = [x, y]$. Then

$$f(\mathbf{c} + \mathbf{v}) - f(\mathbf{c}) = f'(\mathbf{c})\mathbf{v} + \mathbf{r}(\mathbf{v})$$

where $f'(\mathbf{c}) = [7, -9]$, so that $f'(\mathbf{c})\mathbf{v}$ is a *linear* function of $\mathbf{v}$, and $\mathbf{r}(\mathbf{v}) = o(\|\mathbf{v}\|)$, meaning that $\mathbf{r}(\mathbf{v})/\|\mathbf{v}\| \to 0$ as $\|\mathbf{v}\| \to 0$.

Consider now a vector function $F(x)$ with two components, $f(x)$ as above, and $g(x) = x^3 - y^3$. A similar calculation shows that

$$F(\mathbf{c} + \mathbf{v}) - F(\mathbf{c}) = F'(\mathbf{c})\mathbf{v} + \mathbf{r}(\mathbf{v})$$

where

$$F'(\mathbf{v}) = \begin{bmatrix} 7 & -9 \\ 12 & -3 \end{bmatrix}$$

and $\mathbf{r}(\mathbf{v}) = o(\|\mathbf{v}\|)$ as $\|\mathbf{v}\| \to 0$. This means that $F(\mathbf{c} + \mathbf{v}) - F(\mathbf{c})$ is approximated, when $\|\mathbf{v}\|$ is small, by a *linear function* $F'(\mathbf{c})\mathbf{v}$, with an error that tends to zero faster than any linear term.

By holding $\mathbf{v}_2 = 0$, it follows that $7 = \partial f/\partial x$ and $12 = \partial g/\partial x$, evaluating the partial derivatives at $\mathbf{c} = (2, -1)$. Similarly, $12 = \partial g/\partial x$ and $-3 = \partial g/\partial y$ at $\mathbf{c}$. However, the basic idea is linear approximation of the function; the partial derivatives are merely a means to calculate the terms.

A definition is now given for a function $F : X \to Y$, where X and Y are two normed vector spaces. The function f is *Fréchet differentiable* at $\mathbf{c} \in X$ if there is a continuous linear mapping, denoted $F'(\mathbf{c})$, such that

$$F(\mathbf{x}) - F(\mathbf{c}) = F'(\mathbf{c})\,(\mathbf{x} - \mathbf{c}) + \mathbf{r}(\mathbf{x} - \mathbf{c})$$

where $\mathbf{r}(\mathbf{x} - \mathbf{c}) = o(\|\mathbf{x} - \mathbf{c}\|)$, meaning that $\|\mathbf{r}(\mathbf{x} - \mathbf{c})\|/\|\mathbf{x} - \mathbf{c}\| \to 0$ as $\|\mathbf{x} - \mathbf{c}\| \to 0$. The mapping $F'(\mathbf{c})$ (or the matrix representing it) is called the *Fréchet derivative* of F at $\mathbf{c}$. The word *continuous* is needed for function spaces; it happens automatically in finite dimensions.

Example

Define a quadratic function $f: \mathbf{R}^n \to \mathbf{R}$ by $f(\mathbf{x}) := \mathbf{x}^{\mathrm{T}} A \mathbf{x}$, where A is a symmetric $n \times n$ matrix (with real elements). (If A is not symmetric, then

$$\mathbf{x}^{\mathrm{T}} A \mathbf{x} = (\mathbf{x}^{\mathrm{T}} A \mathbf{x})^{\mathrm{T}} = \mathbf{x}^{\mathrm{T}} A^{\mathrm{T}} \mathbf{x}$$

so

$$\mathbf{x}^{\mathrm{T}} A \mathbf{x} = \mathbf{x}^{\mathrm{T}} A^{\mathrm{T}} \mathbf{x} = \mathbf{x}^{\mathrm{T}} \tilde{A}\,\mathbf{x}$$

where $\tilde{A} = \tfrac{1}{2}(A + A^{\mathrm{T}})$ is symmetric. So A can be assumed symmetric.)

Then

$$f(\mathbf{x} + \mathbf{v}) - f(\mathbf{x}) = 2\mathbf{x}^{\mathrm{T}} A \mathbf{v} + \mathbf{v}^{\mathrm{T}} A \mathbf{v} = f'(\mathbf{x})\mathbf{v} + o(\mathbf{v})$$

In order to have F *Fréchet differentiable*, it is not enough just to have partial derivatives. It suffices if those partial derivatives are continuous functions of $\mathbf{x}$; F is then called *continuously differentiable*.

Suppose now that a differentiable function $\mathbf{f}$ maps the vector $\mathbf{w}$ to the vector $\mathbf{s}$, and that the differentiable function $\mathbf{g}$ maps the vector $\mathbf{s}$ to the vector $\mathbf{t}$. Diagrammatically,

$$\mathbf{w} \xrightarrow{\mathbf{f}} \mathbf{s} \xrightarrow{\mathbf{g}} \mathbf{t}$$

Then the composition $\mathbf{h} = \mathbf{g} \circ \mathbf{f}$ of the functions $\mathbf{f}$ and $\mathbf{g}$ maps $\mathbf{w}$ to $\mathbf{t}$. Then

$$\mathbf{f}(\mathbf{w}) - \mathbf{f}(\mathbf{c}) = A(\mathbf{w} - \mathbf{c}) + \theta \quad \text{and} \quad \mathbf{g}(\mathbf{f}(\mathbf{w})) - \mathbf{g}(\mathbf{f}(\mathbf{c})) = B(\mathbf{f}(\mathbf{w}) - \mathbf{f}(\mathbf{c})) + \rho$$

where A and B are suitable matrices (or continuous linear mappings), and θ and ρ are 'small' terms. This suggests the approximation

$$\mathbf{g}(\mathbf{f}(\mathbf{w})) - \mathbf{g}(\mathbf{f}(\mathbf{c})) \approx BA(\mathbf{w} - \mathbf{c})$$

It can be shown that, if $\mathbf{f}$ and $\mathbf{g}$ are differentiable, then so also is $\mathbf{h} = \mathbf{g} \circ \mathbf{f}$, and the Fréchet derivative $\mathbf{h}'(\mathbf{c}) = BA$, which is calculated as the ordinary matrix product. This *chain rule* formula can also be written as

$$(\mathbf{g} \circ \mathbf{f})'(\mathbf{c}) = \mathbf{g}'(\mathbf{f}(\mathbf{c})) \circ \mathbf{f}'(\mathbf{c})$$

which includes the usual chain rule formulae in calculus books as special cases.

If $\mathbf{a} \in \mathbf{R}^n$ and $\mathbf{f}$ maps $\mathbf{R}^n$ into $\mathbf{R}^m$, then the Fréchet derivative $\mathbf{f}'(\mathbf{a})$, being a linear mapping, is represented (section 1.5.2) by an $m \times n$ matrix. In particular, if $m = 1$ the matrix is $1 \times n$, and thus a *row vector*. (Some authors use, instead, its transpose, $\mathbf{f}'(\mathbf{a})^{\mathrm{T}}$, and sometimes denote it by $\mathbf{f}'(\mathbf{a})$ and sometimes by $\nabla \mathbf{f}(\mathbf{a})$.)

If $\mathbf{f}: X \rightarrow Y$ is $\mathbf{f}'(\mathbf{a})$ is Fréchet differentiable at each point $\mathbf{x}$, and if also $\mathbf{f}'(.)$ is continuous (in other words, the derivative $\mathbf{f}'(\mathbf{x})$ is a continuous function of $\mathbf{x}$), then the following stronger property results (e.g. Craven, 1981). For each $\varepsilon > 0$, there exists $\delta(\varepsilon) > 0$, such that

$$\mathbf{f}(\mathbf{x} + \mathbf{w}) - \mathbf{f}(\mathbf{x}) = \mathbf{f}'(\mathbf{a})\mathbf{w} + \xi(\mathbf{x}, \mathbf{w})$$

where $\|\xi(\mathbf{x}, \mathbf{w})\| < \varepsilon\|\mathbf{w}\|$ whenever both $\|\mathbf{x} - \mathbf{a}\| < \delta(\varepsilon)$ and $\|\mathbf{w}\| < \delta(\varepsilon)$. In this case, the continuous linear mapping $\mathbf{f}'(\mathbf{a})$ leads to a local linear approximation to $\mathbf{f}(.)$, which is uniform over all $\mathbf{x}$ close enough to the point $\mathbf{a}$.

Fréchet derivatives are not always available. The mapping $\mathbf{g}: X \rightarrow Y$ is *linearly Gâteaux differentiable* at $\mathbf{a}$ if there is a continuous linear mapping $\mathbf{g}'(\mathbf{a}): X \rightarrow Y$ such that, for each (fixed) $\mathbf{x} \in X$,

$$\alpha^{-1}[\mathbf{g}(\mathbf{a} + \alpha\mathbf{x}) - \mathbf{g}(\mathbf{a}) - \alpha\mathbf{g}'(\mathbf{a})\mathbf{x}] \rightarrow 0 \quad \text{as} \quad \alpha \downarrow 0 \text{ in } \mathbf{R}_+$$

Obviously, $\mathbf{g}'(\mathbf{a})\mathbf{x}$ is the *directional derivative* of $\mathbf{g}$ at $\mathbf{a}$ in direction $\mathbf{x}$; but here we have also that $\mathbf{g}'(\mathbf{a})$ is a linear mapping. Note that Fréchet differentiability implies linear Gâteaux differentiability.

1.5.6 Balls, open sets, neighbourhoods, convergence

In a normed vector space X, the *closed ball* with centre $\mathbf{c}$ and radius $\mathbf{r}$ is the set $B(\mathbf{c}, r)$ of vectors $\mathbf{x}$ for which $\|\mathbf{x} - \mathbf{c}\| \leq r$. A set E, contained in X, is called *open* if for each point $\mathbf{x} \in \mathbf{E}$ there is some ball $B(\mathbf{x}, r(\mathbf{x}))$ contained in E. (This ball is called a *neighbourhood* of $\mathbf{x}$.) The set of points $x \in E$, for each of which there is some ball $B(\mathbf{x}, r(\mathbf{x})) \subset E$, is called the *interior* of E, denoted int E. Note that int E can be empty. A set F is *closed* if its *complement* $X \backslash F$ (the set of points in X but not in F) is open. Do not write $X - F$ for complement of F; the notation $X - F$ is needed for set difference. Thus

$$X - F = \{x - f : x \in X, f \in F\}; \quad X \backslash F = \{x \in X : x \notin F\}$$

It is often convenient to refer to *points* $\mathbf{x}$, instead of vectors $\mathbf{x}$. (While the ideas are distinct, it is often not worth the trouble to separate them. A vector may be pictured as an arrow starting at the origin $\mathbf{0}$, with the point $\mathbf{x}$ at the tip of the arrow.)

A sequence of vectors $\mathbf{x}_1, \mathbf{x}_2, \mathbf{x}_3, \ldots$ *converges* to a vector $\mathbf{x}$, called the *limit* of the sequence, if $\|\mathbf{x}_n - \mathbf{x}\| \rightarrow 0$ as $n \rightarrow \infty$. Equivalently, each ball $B(\mathbf{x}, r)$ contains all the sequences from a certain stage (depending on r) onwards. This sort of convergence is called *strong convergence* to distinguish it from other kinds. It follows that a set is closed exactly when it contains all limits of sequences in the set.

In a space of functions, strong convergence does not always have the properties required. Consider first the space $C(I)$ of real continuous functions on an interval I and vectors $\mathbf{x}_1, \mathbf{x}_2, \mathbf{x}_3, \ldots$ and $\mathbf{x}$ in $C(I)$. *Pointwise convergence* of the sequence $\mathbf{x}_1, \mathbf{x}_2, \mathbf{x}_3, \ldots$ to $\mathbf{x}$ means that, for each fixed $t \in I$, $\mathbf{x}_n(t) \rightarrow \mathbf{x}(t)$ as

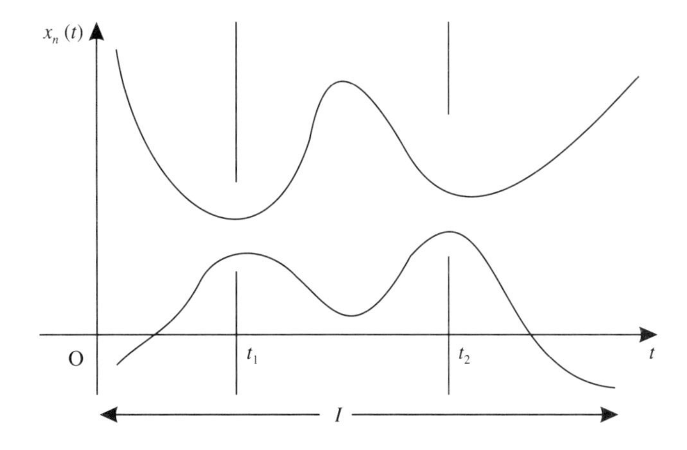

Figure 1.1 Pointwise convergence.

$n \to \infty$. In contrast, *uniform convergence* of $\mathbf{x}_1, \mathbf{x}_2, \mathbf{x}_3, \ldots$ to $\mathbf{x}$ means that $\|\mathbf{x}_n - \mathbf{x}\| \to 0$ as $n \to \infty$. Consequently, uniform convergence implies pointwise convergence, but pointwise convergence does not imply uniform convergence.

Figure 1.1 illustrates pointwise convergence. At two values $t_1, t_2 \in I$ (or any finite number), the graphs for $\mathbf{x}_n(t_1)$ come close together when n is large and the graphs for $\mathbf{x}_n(t_2)$ come close together when n is large, but the graphs for $\mathbf{x}_n(t)$ can still be far apart for other values $t \in I$, as shown by the graphs passing through 'windows' at t_1 and t_2. The role of the ball $B(\mathbf{x}, \mathbf{r})$ in describing convergence is taken over by sets of functions passing through such windows.

For the *dual space* X' of a normed space X, a different notion of *open set* is useful. A *weak∗neighbourhood* of $\mathbf{p} \in X'$ is any set

$$N(\mathbf{p}) = \{\mathbf{y} \in X' : |y(\mathbf{x}_i) - \mathbf{p}(\mathbf{x}_i)| < r \quad (i = 1, 2, \ldots, k)\}$$

specified by finitely many points $x_1, x_2, \ldots, x_k$ in X and a positive number r. These sets $N(p)$ take the place now of balls $B(\mathbf{p}, r)$, for defining convergence. A subset $F \subset X'$ is *weak∗closed* if $X' \backslash F$ is *weak∗open*, i.e. if every point $\mathbf{p} \in X' \backslash F$ has a weak∗neighbourhood $N(\mathbf{p})$ that does not meet F. (In $\mathbf{R}^n$, weak∗closed coincides with closed.) A *topology* (such as *weak∗topology*) means the collection of all the open sets.

Weak∗neighbourhoods may be visualized by their close analogy with pointwise convergence. Since $\mathbf{y}(\mathbf{x})$ must be close to $\mathbf{p}(\mathbf{x})$ at a finite number only of points $\mathbf{x}$, the graph of $\mathbf{y}(\mathbf{x})$ must pass through a finite number of *windows* (two are shown in Figure 1.2), but is otherwise not restricted. (In contrast, uniform convergence would require $\mathbf{y}(\mathbf{x})$ to be close to $\mathbf{p}(\mathbf{x})$ for all values of $\mathbf{x}$. Of course, the diagram has to illustrate the vector space X by a single co-ordinate axis.)

The dual space of X', denoted by X'', contains X, but often X'' is strictly larger than X. If X' is given the weak∗open sets, instead of the open sets described

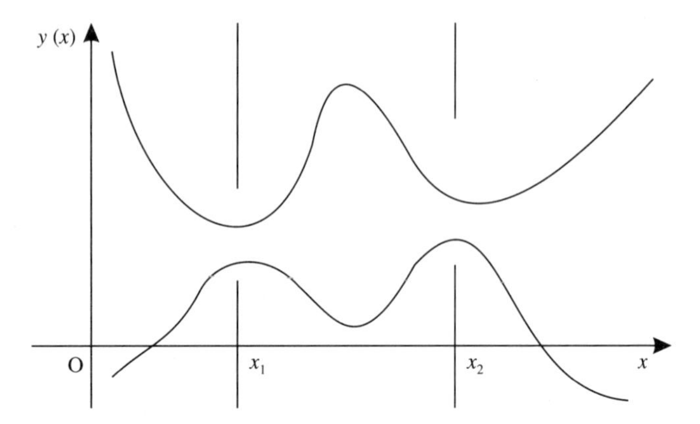

Figure 1.2 Weak∗convergence.

by the dual norm on X':

$$\|x'\| := \sup \{|x'(x)| : x \in X, \|x\| \leq 1\}$$

then X'' can be identified with X. This is needed for using the *separation theorem* (e.g. Simmons (1973) and section 3.1.2).

The *closure* of a set E denoted cl E, is the intersection of all closed sets containing E. This means that cl E consists of E together with all limits of points in E. Here, *closed* may mean norm-closed, or weak∗closed, depending on the context. (Unfortunately the two are different, except in finite dimensional spaces where they agree.)

A normed vector space is called a *Banach space* if a sequence $\mathbf{x}_1, \mathbf{x}_2, \mathbf{x}_3, \ldots$ always has a strong limit whenever it has the *Cauchy property* that $\|\mathbf{x}_m - \mathbf{x}_n\| \to 0$ as m and $n \to \infty$. In particular, the spaces $\mathbf{R}^n$, $C(I)$ and $L^p(I)$ defined above are Banach spaces.

When can an equation $F(x, y) = 0$ be solved for y? Assume that F is a continuously Fréchet differentiable mapping of $X \times Y$ into Y, and that the partial Fréchet derivative $F_y(\bar{x}, \bar{y}): Y \to Y$ is *invertible*. An *implicit function theorem* states that there is a differentiable function $\varphi : N \to X$, where N is some ball with centre $\bar{x}$, and $F(x, \varphi(x)) = 0$ identically on N. Thus the solution exists locally, near $(\bar{x}, \bar{y})$.

A subset E of a normed space X is called *compact* if, whenever some family of open balls covers E, there are finitely many of these balls that also cover E. From this abstract definition, it follows that compact sets share some properties of single points. If E is compact, and $f : E \to \mathbf{R}$ is continuous, then f is bounded on E, and reaches its maximum and minimum at some points of E. If E (in a normed space) is compact, then any sequence in E has some subsequence convergent to some point of E. If $X = \mathbf{R}^n$, then bounded closed sets are compact; in particular, the unit (closed) ball $\{\mathbf{x} \in \mathbf{R}^n : \|\mathbf{x}\| \leq 1)$ and the unit sphere $(\mathbf{x} \in \mathbf{R}^n : \|\mathbf{x}\| = 1\}$. If X has infinite dimensions, then closed bounded sets are *not* usually compact; indeed, compact sets are then often hard to find. Now let Y be a normed space and Y' its dual. *Weak*compact* subsets of Y' are defined by replacing open balls by (open) weak*neighbourhoods in the definition. The ball $\{x' \in X' : \|x'\| \leq 1\}$ is weak*compact, by Alaoglu's theorem (Simmons, 1973).

1.5.7 Algebra on sets of vectors

Let A and B be subsets of a vector space X; let $\alpha \in \mathbf{R}$. Define the sets $A + B$, $A - B$ and αB by

$$A + B = \{a + b : a \in A, b \in B\}; \quad A - B = \{a - b : a \in A, b \in B\}; \quad \alpha a = \{\alpha a : a \in A\}$$

Observe that $A + B$ is not the same as the union $A \cup B$ and $A - B = A + (-B) \neq A \backslash B$.

Suppose now that $X = X_1 \oplus X_2$; this *direct sum* means that $X = X_1 + X_2$ and $X_1 \cap X_2 = \{0\}$, so that each $x \in X$ is uniquely expressible as $x = x_1 + x_2$ with $x_1 \in X_1$ and $x_2 \in X_2$, and also that the projection that takes each x to its corresponding x_1 is a *continuous* linear mapping. (This last requirement is a nontrivial assumption for normed vector spaces of infinite dimension.) Then $M \in \mathbf{L}(X, Y)$ can be represented by $M\mathbf{x} = M_1\mathbf{x}_1 + M_2\mathbf{x}_2$,where (for $i = 1, 2$) M_i is the restriction of M to X_i. This equation is conveniently written in partitioned matrix notation as

$$M\mathbf{x} = [M_1 \quad M_2] \begin{bmatrix} \mathbf{x}_1 \\ \mathbf{x}_2 \end{bmatrix}$$

If also $Y - Y_1 \oplus Y_2$, then $M\mathbf{x} \in Y$ can be similarly partitioned, giving a further partitioned matrix representation

$$M\mathbf{x} = \begin{bmatrix} \mathbf{y}_1 \\ \mathbf{y}_2 \end{bmatrix} = \begin{bmatrix} M_{11} & M_{12} \\ M_{21} & M_{22} \end{bmatrix} \begin{bmatrix} \mathbf{x}_1 \\ \mathbf{x}_2 \end{bmatrix}$$

in terms of suitable continuous linear mappings M_{ij}. Note that the multiplications and additions here are just as for partitioned matrices, even when the vector spaces have infinite dimensions; these are valid provided that X and Y are described as above in terms of direct sums.

The linear mapping $M \in \mathbf{L}(X, Y)$ has *full rank* if $X = X_1 \oplus X_2$, where X_2 is the *nullspace* $M_1^{-1}(0)$ of M_1 (thus $M_1^{-1}(0) = (\mathbf{x}_1 : M_1\mathbf{x}_1 = 0\})$, and M_2 maps X_1 *one-to-one* onto (all of) Y. Since M_1 is continuous, the *open mapping theorem* of functional analysis shows then that the inverse M_1^{-1} is also continuous. In finite dimensions, consider M as an $m \times n$ matrix with $m < n$; then M has full rank by this definition exactly when the matrix M has its full rank of m by the definition in section 1.5.2. These ideas are needed for an *implicit function theorem* (Craven (1981) section 3.7).

1.5.8 Transpose and adjoint

An element of a dual space is both a function (thus, a linear mapping) and a vector. Neither function notation (as $f(\mathbf{x})$ nor matrix–vector notation fulfils

Table 1.1 Comparison of function notation and matrix notation

Object	*Function notation*	*Object*	*Matrix notation*
Vector	$\mathbf{x} \in X$	Column vector	$\mathbf{x}$
Dual vectors	$\mathbf{f} \in X', \mathbf{g} \in Y'$	Row vectors	$\mathbf{f}^T, \mathbf{g}^T$
Evaluation	$f(\mathbf{x})$	Inner product	$\mathbf{f}^T\mathbf{x}$
Linear mapping	$A: X \to Y$	$(m \times n)$ matrix	A
	$\mathbf{g}(A\mathbf{x})$		$\mathbf{g}^T(A\mathbf{x})$
Adjoint mapping	A^T	Transpose matrix	A^T
	$\mathbf{g}(A\mathbf{x}) = (A^T\mathbf{g})(\mathbf{x})$		$\mathbf{g}^T(A\mathbf{x}) = (A^T\mathbf{g})^T\mathbf{x}$
(since $\mathbf{g}(A\mathbf{x})$ defines a linear mapping from $\mathbf{x}$ into $\mathbf{R}$, and that mapping is itself linear in $\mathbf{g}$, denote it by $A^T\mathbf{g}$. It can be shown that $\lVert A^T \rVert = \lVert A \rVert$)		(corresponding matrix expressions in terms of transposes)	

every purpose. Table 1.1 compares the two notations. Here, consider first the spaces $X = \mathbf{R}^n$ and $Y = \mathbf{R}^m$. The matrix notation uses column vectors for vectors in the given space and row vectors for vectors in the dual space (thus, $\mathbf{f}^T$ if $\mathbf{f}$ is a column vector).

It is convenient to use the same notation A^T for *adjoint linear mapping* and *transpose matrix*, since their behaviour is similar. Moreover, these manipulations need not be restricted to finite dimensions; they work just the same in (infinite dimensional) function spaces.

1.5.9 Delta function

The *Dirac delta function* $\delta(.)$ is described by $\delta(.) \geq 0$, $\delta(t) = 0$ when $t \neq 0$ in $\mathbf{R}$, and $\int_{\mathbf{R}} \delta(t)\,dt = 1$. To obtain these properties, the concept of *function* has to be enlarged. This can be done rigorously in a number of ways, which will not be detailed here. The following is a visualizable description of one such way – a rigorous account is given in Craven (1983).

Denote by ε an *infinitesimal*, with the assumed property that $0 < \varepsilon < r$ for every $r \in \mathbf{R}$ with $r > 0$. The usual arithmetic operations applied to $\mathbf{R}$ with ε adjoined lead to elements x of the form

$$x = c_{-r}\varepsilon^{-r} + c_{-r+1}\varepsilon^{-r+1} + \ldots + c_0 + c_1\varepsilon + c_2\varepsilon^2 + \ldots + c_n\varepsilon^n + \ldots$$

Denote by $\mathbf{R}^\$$ the vector space of such *generalized real numbers*. Algebraic operations on the elements of $\mathbf{R}^\$$ follow the rules for polynomials, treating ε just as a symbol. Limit operations, including integrals, are done by considering separately the terms in each power of ε. Note that ε^{-1} is an *infinite value*; $r < \varepsilon^{-1}$ for each $r \in \mathbf{R}$.

Suppose that a control problem is defined on a time interval $[0, T] \subset \mathbf{R}$, and that a delta function, denoted by $\delta(. - T)$, is to be located at T, but considered as falling inside the interval $[0, T]$. To construct this delta function, let

$$\varphi(t) = c \exp(-t^{-1} - t^2) \quad \text{for} \quad t > 0, \quad \varphi(t) = 0 \quad \text{for} \quad t \leq 0$$

with c chosen so that $\int_{\mathbf{R}} \varphi(t) \mathrm{d}t = 1$. Note that $\varphi(.)$ has derivatives of all orders, and they all vanish at 0. Cut the real line $\mathbf{R}$ between $[0, T)$ and T, paste $-\varepsilon\mathbf{R}_+ = \{-\varepsilon r : r \in \mathbf{R}, r \geq 0\}$ in this gap, and define $\delta(t - T)$ to be

$$\varepsilon^{-1}\varphi(r) \quad \text{for} \quad -\varepsilon r \in -\varepsilon\mathbf{R}_+$$

and zero elsewhere. Then $\delta(. - T)$ maps $\mathbf{R}^{\$}$ into $\mathbf{R}^{\$}$, and

$$\int_{R^{\$}} \delta(t - T)\, \mathrm{d}t = \int_{-\varepsilon\mathbf{R}+} \varepsilon^{-1}\varphi(r)\, (\varepsilon\, \mathrm{d}r) = 1$$

so that the desired properties of the delta function are obtained. Since $\delta(t - T) = \varepsilon^{-1}\varphi(-\varepsilon^{-1}t)$, defining $\varphi(.) = 0$ for arguments outside $\mathbf{R}$, there follows

$$\delta'(t - T) = \varepsilon^{-2}\varphi'(-\varepsilon^{-1}(t - T))$$

and similarly for higher derivatives. Thus, for example, $\delta(.)$ is the derivative of the *Heaviside unit function* $\pi(.)$, where $\pi(t) = 1$ for $t \geq 0$, $\pi(t) = 0$ for $t < 0$.

As one consequence,

$$\int_0^T f(x(t), u(t), t)\, \mathrm{d}t + h(x(T)) = \int_0^T [f(x(t), u(t), t) + h(x(t))\delta(t - T)]\, \mathrm{d}t$$

Thus an 'end-point term' $h(x(T))$ in a control problem can be included in the integrand, and so is not an additional case to be treated separately. Note that integrals involving delta functions and their derivatives can be manipulated in the usual way; in particular integration by parts works as usual. Thus, for example, if f is continuously differentiable, then

$$\int_{-1}^{1} f(t)\delta'(t)\, \mathrm{d}t = [f(t)\delta(t)]_{-1}^{1} - \int_{-1}^{1} f'(t)\delta(t)\, \mathrm{d}t = 0 - f'(0)$$

1.5.10 Measures and representations

In a conventional integral, $\int f(t)\, \mathrm{d}t$, taken over $\mathbf{R}$ or a subset of $\mathbf{R}$, the symbol $\mathrm{d}t$ refers to length, or a generalization of length which applies to more complicated sets. Similarly, in a double integral $\iint F(x, y)\, \mathrm{d}x\, \mathrm{d}y$, the symbol $\mathrm{d}x\, \mathrm{d}y$ relates to area. However, the construction of an integral, and many of its properties (namely those that do *not* relate to differentiation) do not depend specifically on length or area, but on certain properties that length or area have in common. It is convenient to state these properties in a generalized form, applicable to a large class of sets. A *measure* μ on a set S (where S could be $\mathbf{R}$ or $\mathbf{R}^2$ or, say, an interval $I = [a, b] \subset \mathbf{R}$) is a function from a family of subsets of S (called *measurable sets*) into $\mathbf{R}_+ = [0, \infty)$, having the property

$$\mu(E_1 \cup E_2 \cup \ldots \cup E_n \cup \ldots) = \mu(E_1) + \mu(E_2) + \ldots + \mu(E_n) + \ldots$$

whenever $E_1, E_2, \ldots, E_n, \ldots$ is any sequence of non-overlapping measurable sets in S. The measure μ is a *probability measure* on S if also $\mu(S) = 1$. The measure on $\mathbf{R}$ which agrees with ordinary length on intervals, the measure on $\mathbf{R}^2$ which agrees with ordinary area on rectangles, and their analogues in $\mathbf{R}^n$, are called *Lebesgue measures* (on $\mathbf{R}$, $\mathbf{R}^2$, $\mathbf{R}^n$).

Note that any union, or intersection, of a sequence of measurable sets (whether or not overlapping) is also measurable. The *Borel sets* are obtained by starting with bounded closed sets in $\mathbf{R}$ (or $\mathbf{R}^n$), and taking all the sets obtained by countably many operations of set union, intersection and set difference (in any order). Borel sets are always measurable. In order to be integrated, a function f must be such that $f^{-1}(E)$ is measurable whenever E is measurable. However, while sets that are not Borel, or that are not measurable, exist, they are so exotic that they can be disregarded in applications (and in this book). A set of zero measure is called a *null set*; this does *not* mean an empty set. Changing the value of an integrand on a null set does not affect the value of the integral.

As one example, the *Dirac measure* $\boldsymbol{\delta}$ is defined by $\boldsymbol{\delta}(E) = 1$ if $0 \in E$, $\boldsymbol{\delta}(E) = 0$ if $0 \notin E$. If $f: \mathbf{R} \to \mathbf{R}$ is continuous then, in terms of the delta function, denoted $\delta(.)$,

$$\int_{\mathbf{R}} f(t)\delta(t)\,\mathrm{d}t = \int_{\mathbf{R}} f\,\mathrm{d}\boldsymbol{\delta} = f(0)$$

Consider now a normed space X and its dual space X'. Elements of X' arise (as Lagrange multipliers), and it is necessary to describe them. Often they can be *represented* by elements of some space of functions, or of *measures*. Some important instances are as follows.

If $X = L^2(I)$, where $I = [a, b]$ or $I = \mathbf{R}$, and if $\bar{\lambda} \in X'$, then there is a function $\lambda \in L^2(I)$ which *represents* $\bar{\lambda}$, meaning that

$$(\forall x \in L^2(I)) \quad \bar{\lambda}(x) = \int_I \lambda(t)x(t)\,\mathrm{d}t$$

The function $\lambda(.)$ is not unique, since changing its value on any null set does not change the value of the integral. Conversely, each function $\lambda \in L^2(I)$ defines an element $\bar{\lambda}$ in the dual space.

If $X = C(I)$, where $I = [a, b]$ is some finite interval, then the elements of the dual space X' are *represented* by the *Borel measures* on I. Thus, if $\bar{\mu} \in X'$, there exists a Borel measure μ, such that

$$(\forall x \in C(I)) \quad \bar{\mu}(x) = \int_I x\,\mathrm{d}\mu$$

In an important special case,

$$\int_I x\,\mathrm{d}\mu = \int_I x(t)[\theta(t) + \sum c_i\delta(t - t_i)]\,\mathrm{d}t$$

Here, $\theta(.)$ is *integrable* (thus, $\int_I |\theta(t)|\,\mathrm{d}t$ is finite), and the summation Σ is finite, or countable. (A measure can have also a *singular* component, not described by the last formula, but that happens seldom in control problems.)

1.6 REFERENCES

Craven, B. D. (1981) *Functions of Several Variables*, Chapman & Hall, London.

Craven, B. D. (1983) Generalized functions for applications, *J. Austral. Math. Soc., Ser. B.*, **26** 362–74.

Simmons, G. F. (1973) *Introduction to Topology and Modern Analysis*, McGraw-Hill/Kogakusha, Tokyo.

2

Optimal control models

2.1 INTRODUCTION

Some of the optimal control models in this chapter are set up in *continuous time*, with an *objective function* of the general form

$$F(x,u) = \Phi(x(T)) + \int_0^T f(x(t), u(t), t)\,\mathrm{d}t$$

subject to a differential equation (with initial conditions)

$$\dot{x}(t) = m(x(t), u(t), t)\ (t \geq 0); \quad x(0) = x_0$$

relating the *state function* $x(.)$ to the *control function* $u(.)$ (so the objective function can also be expressed as $J(u)$), and subject to inequality constraints on the control, say $(\forall t)\ u(t) \in [\alpha, \beta]$, and perhaps also on the state. The objective function can also be written, using a Dirac *delta function* $\delta(.)$, as

$$F(x,u) = \int_0^T [f(x(t), u(t), t) + \Phi(x(t))\,\delta(t-T)]\,\mathrm{d}t$$

Other models are in discrete time ($t = 0, 1, 2, \ldots$), with an *objective function*

$$J(u) \equiv F(x,u) = \sum_t \psi^{(t)}(x_t, u_t)$$

(summed over $t = 0, 1, 2, \ldots, N$) subject to a difference equation

$$\Delta x_t \equiv x_{t+1} - x_t = \varphi^{(t)}(x_t, u_t) \quad (t = 0, 1, 2, \ldots, N-1); \quad x_0 = c$$

and to bounds on the u_t, such as $\alpha \leq u_t \leq \beta\ (t = 0, 1, 2, \ldots, N)$. Here the state x and control u are functions of an integer variable $t = 0, 1, 2, \ldots, N$, so they can be expressed as vectors $\mathbf{x} = (x_0, x_1, \ldots, x_N)$ and $\mathbf{u} = (u_0, u_1, u_2, \ldots, u_N)$. Note that any of the $u(t)$, $x(t)$, u_t and x_t can themselves be real-valued or vector-valued. The different functions $\varphi^{(t)}$ and $\psi^{(t)}$ for different values of t are labelled by superscripts; subscripts are needed for partial derivatives. Thus $\varphi_x^{(t)}(x_t, u_t)$ will mean $(\partial/\partial x_t)\,\varphi^{(t)}(x_t, u_t)$.

In either case, the theory, and some of the computation, depends on a *Hamiltonian* function. This is defined for continuous time by

$$h(x(t), u(t), t; \lambda(t)) = f(x(t), u(t), t) + \Phi(x(t))\,\delta(t - T) + \lambda(t) m(x(t), u(t), t)$$

and for discrete time by

$$h^{(t)}(x_t, u_t; \lambda_t) = \psi^{(t)}(x_t, u_t) + \lambda_t\, \varphi^{(t)}(x_t, u_t)$$

The *costate* function, $\lambda(t)$ or $(\lambda_0, \lambda_1, \lambda_2, \ldots, \lambda_N)$, relates to Lagrange multipliers; and the Hamiltonian is closely related to the Lagrangian (see section 1.2 for an introduction). The theory will be given in sections 4.2 and 7.2; various worked examples in Chapter 5 show how optimality conditions for control problems can be derived from the Hamiltonian.

The bounds on the control function can depend on t; thus α and β can be given functions of t. If $u(t)$ is vector-valued, suitable sets must be specified. For example, if $u(t) = (u_1(t), u_2(t))$, then a possible constraint is given by (Figure 2.1)

$$u_1(t) \geq 0, \quad u_2(t) \geq 0, \quad u_1(t) + u_2(t) \leq 1$$

Observe that these constraints on the control place restrictions on $u(t)$ for each t separately. Thus, for example, there is no constraint involving

$$\sum (u_{t+1} - u_t)^2$$

That (or a similar term in the objective function) would present a difficulty for the theory, even though such a constraint might be relevant for the problem being modelled.

If $\mathbf{v}$ is a vector, denote by $\mathbf{v}_+$ the vector obtained from $\mathbf{v}$ by replacing all negative components of $\mathbf{v}$ by zero.

2.2 AN ADVERTISING MODEL

A model for sales response to advertising was formulated by Vidale and Wolfe (1957). Optimizing this model leads to optimal control problems. While this

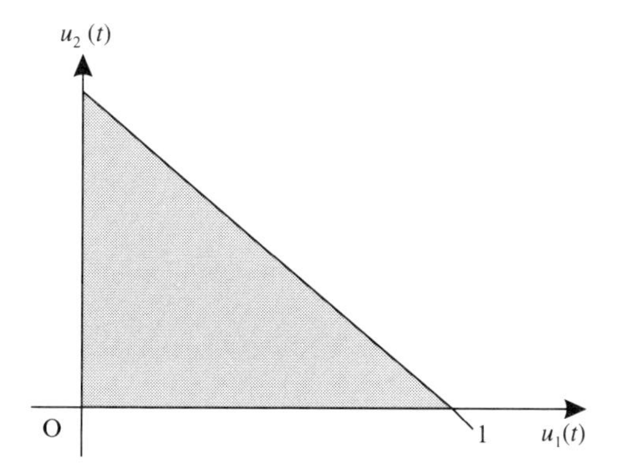

Figure 2.1 Constraints on the control $u(t)$.

model necessarily simplifies the real-world problem, it is based on observed data.

If sales promotion ceases, the sales are observed to decrease with time. If monthly sales of various products are graphed, on a logarithmic scale, against time (years) on a linear scale, the data for various products is approximately described by a straight line with negative slope. Superposed on this line are some random fluctuations and also periodic seasonal effects in some cases. If the latter are large, an average over each year may be considered. The straight line represents an exponential decay of sales with time.

Other observations show that, if the advertising is continued over a period of time the sales increase, but then tend to level off. There appears to be a saturation effect.

Combining these observations, the following mathematical model was formulated. Denote by $S(t)$ the rate of sales (thus, sales per unit time) at time t, and by $A(t)$ the rate of expenditure on advertising. The model is described by the differential equation

$$\mathrm{d}S(t)/\mathrm{d}t = rA(t)[(M - S(t))/M] - \beta S(t) \tag{2.1}$$

Here r is a scale factor, describing the sales initially generated per dollar spent on advertising, β is the decay constant for the exponential decay and M is the saturation level for sales.

In order to handle this equation, it is helpful to scale the variables, so as to put the equation in a dimensionless form. This has the desirable result of reducing the number of parameters in the problem, and thereby minimizing the number of cases that must be studied. Define then $x(t) = S(t)/M$, $u(t) = A(t)/A_0$, where A_0 is some fixed rate of expenditure, taken as a reference, and $\rho = rA_0/M$. The differential equation then becomes

$$\mathrm{d}x(t)/\mathrm{d}t = \rho u(t)\,[1 - x(t)] - \beta x(t) \tag{2.2}$$

There are now just two parameters, ρ being independent of scales of measurement and β depending on the time unit (year). For a control formulation, $x(t)$ is the state function, whereas $u(t)$, representing the rate of expenditure that the company or advertising agency can choose to vary, is the control function. Note that $0 \le u(t) \le c$ for some upper bound c.

This model can have some other features added quite simply. If the saturation level for sales increases with time (Sethi, 1973, 1974) (thus if constant M becomes a given function $M(t)$) then $x(t)$ is now taken as $S(t)/M(t)$. Since

$$(\mathrm{d}/\mathrm{d}t)]\, S(t)/M(t)] = M(t)^{-1}\,\mathrm{d}S(t)/\mathrm{d}t - \theta(t)S(t)/M(t)$$

where $\theta(t) = M(t)^{-1}\,\mathrm{d}M(t)/\mathrm{d}t$, substitution into (2.2) gives the new differential equation

$$\mathrm{d}x(t)/\mathrm{d}t = \rho(t)u(t)\,[1 - x(t)] - (\beta + \theta(t))\,x(t) \tag{2.3}$$

where now $\rho(t) = rA_0/M(t)$. If $M(t)$ grows exponentially, then $\theta(t)$ is constant. Note that (2.3) is very similar to (2.2), so some results for (2.2) carry over easily to (2.3).

Another generalization adds a term $\mu x(t)\,[1 - x(t)]$ to the right-hand side of (2.3). This describes some increase of sales when new purchasers tend to follow the example of previous purchasers.

Consider now optimizing this model, so as to maximize an objective function:

$$J(u) = \int_0^T e^{-\delta t}\,[px(t) - u(t)]\,dt$$

Here p represents unit profit from sales; the cost of advertising, represented by $u(t)$, must be subtracted from the profit. The *discount factor* $e^{-\delta t}$ reduces all sums of money to equivalent amounts, as if they were spent or received at zero time. Here δ is an interest rate ($100\delta\%$), representing some comparison rate of earning, such as a bank rate of interest. (If the money is subject to inflation, then δ must be increased by some amount comparable to the inflation rate.)

The Hamiltonian (for the model using (2.2)) is then

$$-e^{-\delta t}\,[px(t) - u(t)] + \lambda(t)\,[\rho u(t)\,[1 - x(t)] - \beta x(t)]$$

Here, a minus sign is placed before the objective to produce an equivalent minimization problem, and the costate function is $\lambda(t)$.

2.3 SOME OTHER ADVERTISING MODELS

This model by Vidale and Wolfe is closely tied to observations of what happens to sales as time passes or as changes are made in the advertising. Some other models of advertising have been proposed (Sasieni, 1971; Sethi, 1977), based rather on ideas from economic theory. Nerlove and Arrow (1962) assumed that the level of sales depends on a stock of *goodwill*, built up by advertising in the past. Consider first a single company in a monopolistic position (but then, do they need to advertise?) and a single product. Denote by $A(t)$ the amount of goodwill at time t, and by $u(t)$ the rate of advertising expenditure at time t. Nerlove and Arrow assumed that $A(t)$ was described by the differential equation

$$\dot{A}(t) = u(t) - \beta A(t), \quad A(0) = A_0$$

where $\dot{A}(t) \equiv dA(t)/dt$ and β is a decay constant, analogous to β in the Vidale–Wolfe model. The rate of sales, $S(t)$, at time t is taken as a function $S(p(t), A(t), Z(t))$, where $p(t)$ denotes price and $Z(t)$ describes external factors, such as consumer income and population. An optimal control model maximizes, subject to this differential equation, a profit function:

$$J(u, p) = \int_0^\infty e^{-\delta t}\,[p(t)S(p(t), A(t), Z(t)) - h(S(p(t), A(t), Z(t))) - c_0 u(t)]\,dt$$

where c_0 denotes a unit cost for advertising expenditure and the term $h(S(p(t), A(t), Z(t)))$ represents costs involved, other than advertising. With this model one may maximize initially with respect to $p(t)$, since $p(t)$ is supposed (surprisingly) not to affect $A(t)$, giving an objective function to be maximized of the form:

$$J(u) = \int_0^\infty e^{-\delta t} P(u(t), A(t), Z(t))\, dt$$

Here $u(t)$ is the control function and $A(t)$ is the state function.

Jacquemin (see Sethi, 1977) extended this model, supposing that S also depends on $U(t)$, representing the advertising expenditure by other firms. However, he assumes that $U(t)$ is a function of $u(t)$; this very restrictive assumption does simplify the subsequent calculations. His sales function then has the form

$$S(p(t), u(t), U(t), A(t), t)$$

with assumptions on partial derivatives (shown here by suffixes) that

$$S_p < 0, \quad S_u < 0, \quad S_{uu} < 0, \quad S_U < 0, \quad S_A > 0, \quad S_{AA} < 0, \quad S_{Au} = 0$$

(Such qualitative information is often available in economic models – is a function increasing or decreasing, is the graph curving up (convex) or curving down (concave)? – even if numerical information is hard to get.)

Tsurumi and Tsurumi (1971) modified the above model so as to maximize sales under a profit constraint; thus

$$\text{Maximize} \int_0^\infty e^{-\delta t} p(t) S(p(t), A(t), Z(t))\, dt$$

subject to $J(u, p)$ (as above) taking a given value. One might instead assume $J(u, p) \geq k$ for a given positive k; that would be easier to handle.

Bensoussan, Bultez and Naert (1973) considered a number of firms, or competing products, and made some more specific assumptions about the functions in the model. For product i, assume a goodwill function of the form

$$A_i(t) = \int_0^t u_i(t) w_i(t - \tau)\, d\tau$$

i.e. a summation over effects of past times, with a weighting function

$$w_i(t - \tau) = \Gamma(t - \tau \mid \alpha_i, \gamma_i) = \gamma_i^{-\alpha_i} (\Gamma(\alpha_i))^{-1} (t - \tau)^{\alpha_i - 1} e^{-(t - \tau)/\gamma_i}$$

containing two parameters to adjust for position and scale. For n products (or firms), define price, goodwill and advertising vectors as

$$\mathbf{p}(t) = (p_1(t), p_2(t), \ldots, p_n(t))$$

$$\mathbf{A}(t) = (A_1(t), A_2(t), \ldots, A_n(t))$$

$$\mathbf{u}(t) = (u_1(t), u_2(t), \ldots, u_n(t))$$

and assuming a sales vector **S** to be a function of these three. Defining cost functions $h(.)$ and $\mathbf{c}_0$ analogously to the Nerlove–Arrow model, the optimal control problem becomes

$$\text{Maximize} \int_0^\infty e^{-\delta t} [\mathbf{p}(t)^\mathrm{T}\mathbf{S}(\mathbf{p}(t), \mathbf{u}(t), \mathbf{A}(t)) - h(\mathbf{S}(\mathbf{p}(t), \mathbf{u}(t), \mathbf{A}(t))) - \mathbf{c}_0^\mathrm{T}\mathbf{u}(t)]\,\mathrm{d}t$$

subject to a differential equation for $\mathbf{A}(t)$ derived from the assumed form for $A_i(t)$. This leads to a first-order matrix differential equation of the form:

$$\dot{\mathbf{A}}(t) = -C_1\mathbf{A}(t) + C_2 \int_0^t M(t-\tau)\mathbf{u}(\tau)\,\mathrm{d}t$$

in which M is a diagonal matrix with diagonal terms $\Gamma(t-\tau) | \alpha_i - 1, \gamma_i)$ and C_1 and C_2 are matrices calculable from the γ_i and α_i. The model may include an additional component with $i = 0$, which is a weighted sum of the form

$$\mathbf{A}_0(t) = \sum_{i=1}^{n} q_i\mathbf{A}_i(t) + q_0 \int_0^t u_0(\tau)w_0(t-\tau)\,\mathrm{d}\tau$$

with weights q_i summing to 1, representing *institutional goodwill*, an assumed goodwill for the whole industry, or range of products, built up by the advertising of each of them.

The above discussion is *not* intended as a state-of-the-art survey of the (many and complicated) advertising models in the literature, but rather as a guide to how such a model may be constructed. (One can always add on complications – if the data are available to determine the parameters. It can be very misleading to have a great number of parameters in a model without sufficient data to estimate their values.)

2.4 AN INVESTMENT MODEL

A public utility company must decide what proportion of its earnings to retain – to the advantage of future earnings at the expense of present dividends – and also what new stock issues should be made. In the model of Davis and Elzinga (1972), $P(t)$ is the market price of a share of stock at time t, $E(t)$ is the equity per share of outstanding common stock (that is, net worth of the company's assets divided by number of shares), $u_r(t)$ is the retention rate and $u_s(t)$ is the stock financing rate (to be described). This system is described by two differential equations:

$$\mathrm{d}P(t)/\mathrm{d}t = c[1 - u_r(t)]rE(t) - \rho P(t)$$

$$\mathrm{d}E(t)/\mathrm{d}t = rE(t)\,[u_r(t) + u_s(t)\,(1 - \tau E(t)/P(t))]$$

Here c is some constant describing how rapidly the share price changes when earnings and dividends change; since a fraction $u_r(t)$ of earnings is retained for increasing the company's capital assets, fraction $1 - u_r(t)$ is available for dividends; a rate of return of r on equity is assumed (the rate r may be limited (in effect, set) by legislation). The *discount rate* is ρ, i.e. a dollar at time t has a *present value* (equivalent value at time 0) of $e^{-\rho t}$). The second differential equation describes the effect of retained earnings, and also of new stock issues on the growth of the value of the company. An issue of new stock dilutes the value of existing shares, described by the factor $1 - \tau E(t)/P(t)$; $\tau < 1$, since the cost of floating a stock issue absorbs a fraction of the money invested.

It is required to maximize an objective function:

$$P(T)e^{-\rho T} + \int_0^T e^{-\rho t}[1 - u_r(t)]rE(t)\,dt$$

representing the present value (discounted to time 0) of the shareholders' assets in the company. This is subject to the two differential equations, with initial conditions at time 0, and also to inequalities on the two fractions $u_r(t)$ and $u_s(t)$, namely

$$u_r(t) \geq 0, \quad u_s(t) \geq 0, \quad u_r(t) + u_s(t) \leq k/r < 1 \quad (0 \leq t \leq T)$$

where k is an assumed maximum investment rate (at what maximum rate is it productive to expand the facilities of the company?).

In this problem, $u_r(t)$ and $u_s(t)$ are control functions, whereas $P(t)$ and $E(t)$ are state functions. A *planning horizon* of T is assumed. Because of the discount factor $e^{-\rho t}$, the exact choice of T may not be too critical, since $e^{-\rho T}$ is small when T is fairly large. Because the company is a utility (e.g. it supplies electricity or water), a steady demand for its output may be assumed, so that questions of forecasting future demand or influencing it by advertising need not enter the present model.

The Hamiltonian function is

$$-[e^{-\rho t}[1 - u_r(t)]rE(t) + P(T)e^{-\rho T}\delta(t-T)] + \lambda_1(t)[c[1 - u_r(t)]rE(t) - \rho P(t)]$$
$$+ \lambda_2(t)[rE(t)[u_r(t) + u_s(t)(1 - \tau E(t)/P(t))]]$$

2.5 PRODUCTION AND INVENTORY MODELS

A simple inventory model was given in section 1.2. The following is a more developed model, describing the interactions between a number of factories (or machines, or industries) over a number of time periods (here called years, though the same formulation applies also to shorter periods).

The *inputs* for the model comprise the (annual) quantities of raw materials used for each factory (or industry), including electric power but not including labour, which is dealt with elsewhere in the model. For year t, denote by $\mathbf{v}_t$ the (column) vector of these *inputs*. Denote by $\mathbf{z}_t$ the corresponding vector of *outputs*, namely the (annual) quantities of processed materials or products. Denote by $\mathbf{i}_t$ the vector of inventories held at the end of period t of each of the output quantities. For present simplicity, assume that no inventories are kept of inputs. Denote by $\mathbf{f}_t$ the vector those amounts of output that are delivered to customers (outside the system being modelled) in period t. Then

$$\mathbf{z}_t = M\mathbf{v}_t$$

with M some matrix representing the amounts of each input to make each output and

$$\mathbf{v}_t \leq \mathbf{r}_t$$

where $\mathbf{r}_t$ represents the amount of resources available; the inventory and production are related by

$$\mathbf{i}_{t+1} = [\mathbf{i}_t + \mathbf{z}_t - Q\mathbf{z}_t - \mathbf{f}_t]_+ \tag{2.4}$$

where $Q\mathbf{z}_t$ for a suitable matrix Q, represents those amounts of some products that are needed as inputs to make other products, and $[\ ]_+$ is required since inventory must not run negative. The demand for outputs is represented by a vector $\mathbf{d}_t$; since it is *not* assumed that all demands are met, denote by $\mathbf{e}_t \leq \mathbf{d}_t$ a vector of essential demands. Then

$$\mathbf{f}_t + \mathbf{u}_t = \mathbf{d}_t$$

where $\mathbf{u}_t$ is a slack variable, satisfying $\mathbf{u}_t \leq \mathbf{d}_t - \mathbf{e}_t$.

Denote by $\mathbf{y}_t$ the column vector, giving the number of people employed to produce output $\mathbf{z}_t$. (One could, instead, relate a labour vector $\mathbf{y}_t$ to input $\mathbf{v}_t$; and often several kinds of labour would be considered – here just one kind is assumed, for simplicity.) Then $\mathbf{y}_t = R\mathbf{z}_t$ for some matrix R.

In setting up a cost (or profit) functional for this economic model, it is essential to distinguish between *marginal costs* and *overheads*. To show this in its simplest form, consider just one product (Figure 2.2). Typically, there is a *setup cost* OS, as well as a cost component (shown by the line SP) that is proportional to the level of production. Suppose that P is the current *operating point* (where the process is running now). The *marginal cost* is the slope of the line SP, namely PQ/SQ; it is this cost that must enter any optimization model which contemplates changing the operating point P. (The setup cost enters separately – see below.) The marginal cost must not be confused with the 'cost per unit' at

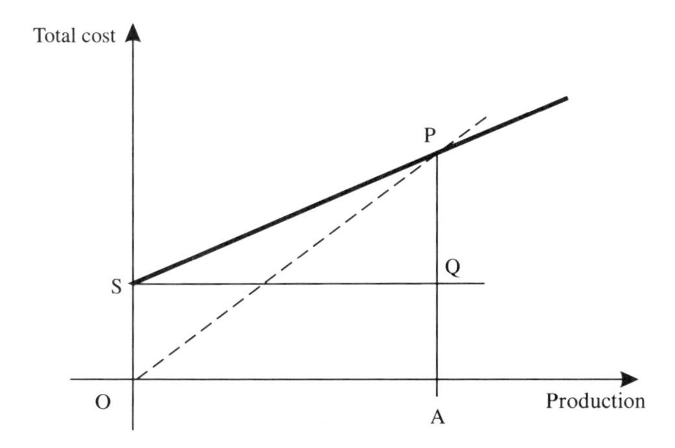

Figure 2.2 Distinguishing between marginal costs and overheads.

the current operating point (thus PA/OA), which is often recorded for accounting purposes. For (usually) many products, a set proportion of overheads (comprising setup costs, together with other costs not directly related to level of production) is commonly combined with a marginal cost to give a 'cost per unit'. However, an optimization model still requires instead a marginal cost for each product.

Assume then cost coefficient (row) vectors as follows:

$\boldsymbol{\beta}_t$ for $\mathbf{r}_t$	marginal costs for raw materials and energy
γ_t for $Q\mathbf{z}_t$	transport costs, but not material costs, which have already been included in $\mathbf{r}_t$
$\boldsymbol{\delta}_t$ for $\mathbf{z}_t$	maintenance costs, but not energy costs
$\boldsymbol{\alpha}_t$ for $\mathbf{y}_t$	wages and 'fringe benefits' less unemployment payments
$\boldsymbol{\theta}_t$ for $\mathbf{u}_t$	penalty cost for demand unfulfilled

Here maintenance means, more generally, the cost of keeping the process running, *except* for material costs, which have been already charged against production. It is assumed that the community will not let its unemployed people starve; hence the cost component $\boldsymbol{\alpha}_t\mathbf{y}_t$ represents the excess of actual wages (plus health insurance, superannuation etc.) over what the community pays to an unemployed person. This aspect of the present cost model is therefore from the viewpoint of the community as a whole; an individual firm may have a different criterion.

Taxation would be taken as proportional to the throughputs $\mathbf{v}_t$, $\mathbf{z}_t$ and $\mathbf{r}_t$, appropriate to a 'value added' tax system, and thus built into the marginal costs. (The complexities of an actual tax system would probably complicate the model to an unmanageable extent.) Assumed profit margins could be similarly included. However, the present model is a cost model rather than a profit model. Consequently, it is not affected by transfer prices between sections of a large organization.

The total marginal cost is obtained as

$$C_{\text{marg}} = \sum_t \rho^t[\boldsymbol{\beta}_t\mathbf{r}_t + \gamma_t Q\mathbf{z}_t + \boldsymbol{\delta}_t\mathbf{z}_t + \boldsymbol{\alpha}_t\mathbf{y}_t + \boldsymbol{\theta}_t u_t]$$

summed over the time periods t. The coefficient ρ^t is a discount factor to convert all costs to equivalent *net present values* at time zero; this is achieved by taking $\rho = 1/(1 + \mathrm{e})$, where 100e is an appropriate percentage interest rate. The problem of minimizing C_{marg}, subject to the constraints stated above and non-negative variables, may be regarded as an optimal control problem in discrete time $t = 0, 1, 2, \ldots$, with $\mathbf{v}_t$ (resources used) and $\mathbf{z}_t$ (production) as control variables, and with $\mathbf{i}_t$ (inventory) and $\mathbf{y}_t$ (labour) as state variables.

However, C_{marg} does not allow for important factors. These include setup costs when manufacture of some product is begun or when it is stopped, as well as other overhead costs. Note that the setup costs may include substantial components associated with changes in the level of the labour vector $\mathbf{y}_t$. If a new product is to be made, and the machinery and other facilities to do so do not already exist, then a large component of capital cost would be added to the setup costs. If these components of cost, additional to marginal costs, total (over all time periods) to S, then $S + C_{\text{marg}}$ should be minimized. But S depends on the products which are started or stopped and the times when these events happen.

In order to approximate this fairly complicated situation, introduce a further variable $\mathbf{j}_t$ to represent the *capital stock*, namely the inventory of productive equipment held at the end of period t. Then $\mathbf{j}_{t+1}$ is related to $\mathbf{j}_t$ by an inventory equation of similar form to (2.4). Here $\mathbf{j}_t$ becomes a further state variable.

For these models, unlike linear programming models, an optimum solution may have more nonzero variables than there are constraints in the model. When the setup costs are included, it is no longer optimal to start or stop an activity as readily as an analysis of marginal costs alone would predict. The setup costs could be added to C_{marg} by introducing *zero–one variables*. Each setup cost would be multiplied by a variable that equals 1 for the period in which the setup cost in incurred (thus, when the activity starts), and is otherwise 0. However, such variables would required combinatorial (*branch and bound*) methods to compute, which would make the control model much more difficult. Alternatively, a number of different scenarios could be optimized under different assumptions on when activities start, and the total costs compared.

2.6 WATER MANAGEMENT MODEL

Consider a network of reservoirs to store water, streams and channels, with input from rainfall and streamflow and output used for irrigation, town water and generation of hydroelectric power. Since the demand for these outputs generally exceeds what is available, there is an optimization problem here – how best to use the the limited resources? Consider then a *network*, consisting of *nodes* (points) and *arcs* (lines, which need not be straight, connecting various of the nodes). The nodes will represent reservoirs and hydroelectric power stations, and the arcs will represent streams and channels and pipelines.

The network diagram of Figure 2.3 shows a very simplified example, with two reservoirs and one hydroelectric power station. For this simple model, the *state functions* may be taken as the height of the water level above sea level, for each reservoir and at the input to the power station. The inputs are the two inflow rates from catchment areas. The control functions are the amounts (flow rates) of water taken from reservoir B for town water and for irrigation, and the amount (flow rate) of water taken for the power station. The demands are the flow rates required (if not always satisfied) for town water and irrigation.

The equation describing a reservoir is:

$$Q_{\text{in}}(t) - Q_{\text{out}}(t) = A(h(t))\dot{h}(t)$$

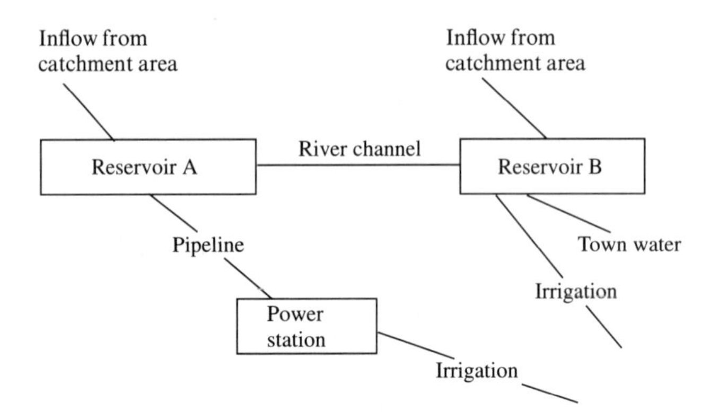

Figure 2.3 A simple network diagram.

where $Q_{in}(t)$ and $Q_{out}(t)$ are total inflow and outflow rates as functions of time t, $\dot{h}(t)$ is the rate of change of water level and $A(h)$ is the cross-sectional area of the reservoir at height h. For reservoir B, $Q_{in}(t)$ is the sum of two inflow rates, one from the river channel and one from a catchment area. The outflow rate $Q_{out}(t)$ is the sum of water over the spillway and water sent through a pipeline, say for hydroelectricity or for town water. If an output Q_{spill} from a reservoir flows over a spillway into a channel, then outflow rate Q_{out} and change in water level Δh are related by $Q_{spill} = cL(\Delta h)^{3/2}$, where c is a constant and L is the length of the spillway. The rate of flow Q_{pipe} of water along a pipe, or channel, is related to the change in water level (Δh) by $Q_{pipe} = k(\Delta h)^{1/2}$, where k is a constant depending on the length and cross-section of the pipe or channel. The electric power $P(t)$ generated by the power station from a flow rate Q_{power} and a change of height Δh is given by $k'(\Delta h)Q_{power}$.

Denote by $Q_{town}(t)$ and $Q_{irrig}(t)$ the flow rates delivered for town water and for irrigation, and by $P(t)$ the rate of electric power generated. The coefficients p_1, p_2 and p_3 represent the profit rate for each of these three uses for the water. (In fact, they are *net profits* – profit minus running costs.) An optimal control model can then be set up as follows, to maximize present value of net profit, assuming a discount factor $e^{-\delta t}$:

$$\text{Maximize} \int_0^T e^{-\delta t} [p_1 Q_{town}(t) + p_2 Q_{irrig}(t) + p_3 P(t)] \, dt$$

subject to equations describing the network (as described above), and to lower bounds (representing essential requirements) and upper bounds on $Q_{town}(t)$ and $Q_{irrig}(t)$. It is assumed here that the towns and farms depend on this water supply network as their only supply of water, but that electric power can also come from other sources, so that it is appropriate to measure the value of the electric power just by its sale price.

Although various linear programming models have been used for water resource planning, particularly for balancing supply and demand for water, optimal control models do not seem yet to have been applied in this area.

2.7 THE FISH MODEL

This model was given by Clark (1976) in relation to fish populations, but applies equally well to other biological populations. At time t, suppose that the population (of a species of fish, or whatever) is $z(t)$. Assume (with some simplification) that the rate of increase of this population equals $F(z(t))$, where $F(z)$ is approximately proportional to z when z is small and positive, but that $F(x)$ decreases when z is larger, if only because food supply will run out. Such a function is the *logistic function*

$$F(z) = kz(M - z)$$

where $k > 0$ is constant and M represents a maximum possible population. Of course, a more complicated function than this $F(x)$ might be appropriate – provided that observational data were available to determine the function.

The population differential equation when there is no exploitation of the fish is

$$\dot{z}(t) = kz(t)\,(M - z(t)), \quad \dot{z}(0) = z_0$$

There are two equilibrium points, $z = 0$ and $z = M$, at which $\dot{z}(t) = 0$. If $0 < z_0 < M$, then $\dot{z}(0) > 0$, so that $z(t)$ moves away from the unstable equilibrium point $z = 0$ to the stable equilibrium point $z = M$. The change of variables to $\tau = kMt$, $x(\tau) = z(t)/M$ transforms the differential equation to dimensionless form:

$$\mathrm{d}x(\tau)/\mathrm{d}\tau = x(\tau)\,(1 - x(\tau)), \quad x(0) = x_0 = z_0/M$$

This solves to

$$x(\tau) = 1/[1 + b\mathrm{e}^{-\tau}]$$

where $b = x_0/(1 - x_0) = z_0/(M - z_0)$. Hence $x(\tau) \to x(\infty) \equiv 1$ as $\tau \to \infty$ (or $t \to \infty$).

Suppose now that there is exploitation of the fish (or other biological) population, resulting in catching fish at rate $q(t)z(t)$. Here $q(t)$ (where $0 < q(t) < q_{max}$) is some measure of the effort put into catching fish (with an upper bound q_{max}), and the rate of catching fish is assumed also proportional to the population of fish currently in the sea. The population equation then becomes

$$\dot{z}(t) = kz(t)\,(M - z(t)) - q(t)z(t), \quad z(0) = z_0$$

The dimensionless form is then:

$$\mathrm{d}x(\tau)/\mathrm{d}\tau = x(\tau)\,(1 - x(\tau)) - r(\tau)x(\tau), \quad x(0) = x_0 \qquad (2.5)$$

where $r(\tau) \equiv q(t)$.

Consider the following profit function, representing (to some approximation) the present value of net profit from fishing with an effort $q(t)$ over a time period of T, and denoting by $v(t) = q(t)z(t)$ the rate at which fish are caught:

$$\int_0^T e^{-\delta t} [p_1 v(t) - p_2 v(t)^2 - c_1 q(t)]\, dt$$

Here the coefficient p_1 represents rate of profit from the sale of fish, when the amount $x(t)$ is not too large. The term $-p_2 v(t)^2$ describes a situation of *diminishing returns* when there is a large amount of fish to sell, and the sale price consequently falls. The term $c_1 q(t)$ represents the cost of the fishing operation, supposed to be proportional to the effort $q(t)$. In terms of the dimensionless variables, the net profit function becomes

$$J(u) = (kM)^{-1} \int_0^{T'} e^{-\delta' \tau} [p_1 \varphi(u(\tau)) - m(x(\tau))u(\tau)] d\tau$$

where $u(\tau) = v(t)$ is now the control function, representing the rate at which fish are caught, $\varphi(u(\tau)) = u(\tau) - (p_2/p_1)u(\tau)^2$, $m(x(\tau)) = c_1/x(\tau)$ and $T' = T/(kM)$. Thus $J(u)$ is to be maximized, subject to the differential equation (2.5) and to some bound on the fishing effort, represented here by $0 \le u(\tau) \le u_{\max}$. It is also obviously necessary that $x(\tau) > 0$; however, the differential equation will ensure that, given an initial condition x_0 in $0 < x_0 < M$. There could also be further restrictions on $u(\tau)$, perhaps limiting how often $u(\tau)$ may jump between 0 and $u_{\max}$.

For this model the Hamiltonian is then

$$-e^{-\delta' \tau} [p_1 \varphi(u(\tau)) - m(x(\tau))u(\tau)] + \lambda(\tau) [x(\tau)(1 - x(\tau)) - r(\tau)x(\tau)]$$

where $\lambda(\tau)$ is the costate function. A minus sign has been inserted to convert from this maximizing problem to the corresponding minimizing problem. It is convenient to extract the (positive) discount factor $e^{-\delta' t}$ by defining the *present value costate function* to be

$$\Lambda(\tau) = e^{\delta' \tau} \lambda(\tau)$$

Then the Hamiltonian equals

$$e^{-\delta' \tau} \{ -[p_1 \varphi(u(\tau)) - m(x(\tau))u(\tau)] + \Lambda(\tau) [x(\tau)(1 - x(\tau)) - r(\tau)x(\tau)]\}$$

It will appear (section 5.8) that $\Lambda(t)$ may attain a steady state as $t \to \infty$, whereas $\lambda(t)$ cannot do this.

2.8 EPIDEMIC MODELS

Sethi (1970) and Sethi and Sethi (1978) proposed models for the optimum deployment of limited medical resources in order to deal with an epidemic. At time t, suppose that the *state* $I(t)$ represents the number of *infectives* (those with the disease) in the population of N people, and that the *control* $v(t)$ represents the *intensity of medicare*, namely the rate at which the limited medical resources are being expended at time t. Clearly $0 \le I(t) \le N$, and $0 \le v(t) \le V$, where V is some maximum practicable level.

Some model must be assumed for the rate of increase (or decrease) of the number of infectives. Such biological models are necessarily rough (they cannot hope for the precision of a law of physics), and they must not contain more than a few parameters, or there is no hope of estimating good values of those parameters from observed data. One such model is expressed by the differential equation:

$$\dot{I}(t) = bI(t)\,[N - I(t)] - v(t)I(t), \quad I(0) = I_0, \quad I(T) = I_T$$

Observe that the logistic function is again used, and that initial and final levels are required for $I(.)$.

Some objective function must be chosen to measure the effectiveness of the medical efforts. Here consider the cost function (to be minimized):

$$J(v) = \int_0^T e^{-\delta t}\,[kv(t) + KI(t)]\,dt$$

in which $e^{-\delta t}$ is a discount factor, and unit costs of k and K are attached respectively to the use of medical resources and to the numbers of people affected by the epidemic. The latter term must at least account for the work not done by sick people, though it is far from obvious how human pain and suffering can thus be quantified.

The resulting Hamiltonian is then:

$$e^{-\delta t}\,[kv(t) + KI(t)] + \lambda(t)\,\{bI(t)\,[N - I(t)] - v(t)I(t)\}$$
$$= e^{-\delta t}\,\{[kv(t) + KI(t)] + \Lambda(t)\,[bI(t)\,[N - I(t)] - v(t)I(t)]\}$$

where $\Lambda(t) = e^{\delta t}$ is the *present value costate* function.

This model attaches no cost to *changing* the level of medical services – an assumption difficult to believe. A cost term involving $v(t)$ could be put into the model; that would disturb the theory, but an optimum could still be computed.

Another of Sethi's models considers the effect of inoculating some of the people against the disease. At time t, suppose that there are $I(t)$ *infectives*, $S(t)$ *susceptibles* (who may get the disease), and $E(t) = N - S(t)$ people who have not been inoculated. The assumed differential equations are

$$\dot{I}(t) = bS(t)I(t); \quad I(0) = I_0 > 0; \quad \dot{S}(t) = -bS(t)I(t) - u(t); \quad S(0) = S_0 > 0$$

Here the rate of increase of the epidemic is assumed proportional to the number of contacts between infectives and susceptibles; this is plausible if not too many get the disease, but if a large proportion of the population get the disease, then some term like $N - I(t)$, as in the logistic function, may be needed to get a less artificial model. Each person who gets the disease ceases to be a susceptible and becomes an infective. It is further assumed that a level $u(t)$ of medical services, here just for inoculating people against the disease, will reduce the rate of growth of the epidemic (see the second differential equation). Again, an idealized cost function is assumed:

$$J(v) = \int_0^T e^{-\delta t}\,[cu(t) + KI(t)]\,dt$$

leading to a Hamiltonian:

$$\begin{aligned} &e^{-\delta t}\,[cu(t) + KI(t)] + \lambda(t)bS(t)I(t) + \mu(t)\,[-bS(t)I(t) - u(t)] \\ &\quad = e^{-\delta t}\,\{[cu(t) + KI(t)] + \Lambda(t)bS(t)I(t) + M(t)\,[-bS(t)I(t) - u(t)]\} \end{aligned}$$

in which $\lambda(t)$ and $\mu(t)$ are the two components of the costate function, and $\Lambda(t) = e^{\delta t}\,\lambda(t)$ and $M(t) = e^{\delta t}\,\mu(t)$ are the two components of the present value costate function.

Many other biological models, concerning the control of cancer, have been given by Martin and Teo (1994). One such model describes the tumour cell population $N(t)$ in relation to the drug concentration $v(t)$ by

$$\dot{N}(t) = \lambda N(t)\log[\theta/N(t)] - k[v(t) - c]_+ N(t), \quad N(0) = N_0, \quad N(t) \le N_{\max}$$

$$v(t) = \sum_i u_i \exp[-\gamma(t - t_i)]\pi(t - t_i), \quad 0 \le v(t) \le v_{\max}, \quad \int_0^T v(t)\,dt \le v_{\text{cum}}$$

Here λ, θ, k, c, t_i, T, N_0, $N_{\max}$, $v_{\max}$ and v_{cum} are parameters, and $N(T)$ is to be maximized with respect to the control variables u_i, describing drug doses. This control model does not fit well into the theory, but can be computed.

2.9 STABILITY?

Any mathematical model must simplify the real-world situation in order to make analysis and computation practicable. There is always a matter of judgement as to which real-world features must be retained in the model. Optimal control models are quite often inadequate concerning *stability*.

The following parable about a computer-controlled car illustrates the importance of *stability against small perturbations*. The occupants of the computer-controlled car are looking always at their computer screen, but never look out to see where the car is going. They wonder why the car has stopped. In fact, the car has run into a tree. The computer control has been programmed with some *control function*, set down in advance, and this does not and cannot allow for small disturbances which may change the path of the car – in this example, cause the car to run off the road. This is an example of *open-loop control*, where there is no feedback of current information about the position of the car.

With an optimal control model, theory or computation often provides an optimal control function $u_{\text{opt}}(t)$. If it happens that, for some function Ψ,

$$u_{\text{opt}}(t) = \Psi(x(t), \dot{x}(t))$$

i.e. if, for each time t, $u_{\text{opt}}(t)$ depends on the values of $x(t)$ and its derivative $\dot{x}(t)$ at that time, then the deficiency of open-loop control may be mended. Note that such a function Ψ does not always exist; $u_{\text{opt}}(t)$ could depend on $x(s)$ and $\dot{x}(s)$ for all previous times $s \leq t$. However, if such a function Ψ exists, then $u_{\text{opt}}(t)$ may be replaced by $\Psi(x(t), \dot{x}(t))$ in the differential equation for $\dot{x}(t)$. Consequently, any small disturbance in $x(t)$, although *not* part of the original model, can be compensated by the *feedback* of current information about the actual position of $x(t)$ and $\dot{x}(t)$.

The feedback idea is illustrated by Figure 2.4, applicable to a linear system. The *black box* is so labelled because what matters about it is the relation between its input and output, rather than any details of what is inside the box. Here, the output equals input multiplied (in some sense) by $-A$. But the input to the black box consists of the given input $x(t)$ plus some part $\beta y(t)$ of the output, representing feedback of information from output back to input. Hence

$$y(t) = -A(x(t) + \beta y(t)) + p(t)$$

so that

$$\begin{aligned} y(t) &= -(A^{-1} + \beta)^{-1} x(t) + (1 + \beta A)^{-1} p(t) \\ &\approx -\beta^{-1} x(t) + (1 + \beta A)^{-1} p(t) \end{aligned}$$

if A is large positive. Thus the relation between $x(t)$ and $y(t)$ depends mainly on the feedback path β, and the effect of the disturbance $p(t)$ is reduced by the factor $(1 + \beta A)^{-1}$. These qualitative conclusions stand, although generally A and β are linear operators rather than just numbers. (They can be handled by Laplace transform methods.) The minus sign, appearing here in $-A$, is important; if it is reversed, then the feedback will produce oscillations, and thus an unstable system, instead of tending to stability.

However, time lags in a feedback system can easily produce instability. This may be illustrated by the differential equation

$$\ddot{x}(t + \varepsilon) + x(t) = 0$$

where ε is a positive time lag. If $\mathbf{I}$ denotes the integral operator $\int_0^t \ldots \mathrm{d}t$, then this differential equation describes the application of $\mathbf{I}$ twice to $x(t)$, then feeding

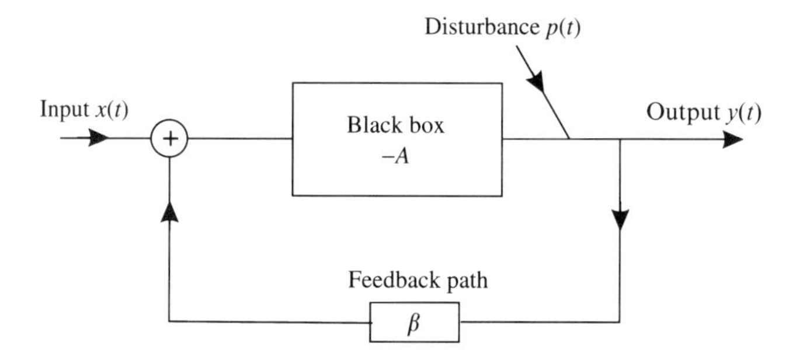

Figure 2.4 Stability of a positive feedback system.

back the result (with a minus sign) to $x(t)$, but with a time lag of ε. If $\varepsilon = 0$, the differential equation has solutions $ce^{\lambda t}$ with $\lambda = \pm\,\mathrm{i}$, giving periodic sine wave solutions. Denote by $\mathbf{D}$ the differential operator $\mathrm{d}/\mathrm{d}t$. If $x(t)$ is expansible in a Taylor series, then

$$e^{\varepsilon\mathbf{D}}x(t) = \sum_n (\varepsilon\mathbf{D})^n x(t)/n! = x(t+\varepsilon)$$

For ε small, this may be approximated by the first two terms of the series, thus by $(1+\varepsilon\mathbf{D})x(t)$. To this approximation, the differential equation becomes

$$(\varepsilon\mathbf{D}^3 + \mathbf{D}^2 + 1)x(t) = 0$$

This differential equation has a solution $x(t) = e^{\lambda t}$ if $\varepsilon\lambda^3 + \lambda^2 + 1 = 0$. Try a solution $\lambda = -\mathrm{i} + \theta$, and calculate up to linear terms in θ, since θ is small if ε is small. This leads to

$$\varepsilon(\mathrm{i} - 3\theta) + (-1 - 2\mathrm{i}\theta) + 1 \approx 0$$

so that $\theta \approx \varepsilon/2$. So $e^{\lambda t} \approx e^{(-\mathrm{i} + \varepsilon/2)t}$, which is unstable, since it grows at an exponential rate.

Unless the differential (or difference) equations for an optimal control model describe adequately the time lags in the system, a calculated optimum may represent an unstable system.

There is a large literature on feedback control; Rosenbrock (1970) is one of many. In the black box, A must represent time lags as well as amplification, and the analysis usually does this by using Laplace transforms to describe the linear differential equations involved. A nonlinear system may be analysed by *linearizing* it, thus approximating it by a linear system in a restricted region.

2.10 EXERCISES

2.1 Discuss in more detail the formulation of the water management model of section 2.6. Thus, for the system shown in Figure 2.3, list the variables and write down suitable constraint equations or inequalities.

2.2 Set up a tentative model for social policy concerning smoking and lung cancer, including costs for hospitalization of lung cancer patients, loss of their productive capacity and the costs of persuading smokers to stop smoking, with tax implications. Consider also a time lag between the smoking and the onset of lung cancer. (This is a difficult and open-ended question, and it is not clear what cost to put on human suffering.)

2.3 Consider a linearized model for production and demand, described by a *demand curve* $p_t = -aq_t$ and a *supply curve* $q_t = bp_{t-1} + cp_{t-2}$, where $t = 1, 2, \ldots$ is the period, p_t is the deviation of the demand in period t from a fixed level and q_t is the deviation of the quantity available for sale in period t from a fixed value. (This model is linearized, because the curves would not in fact be straight lines. Note that the amount produced for sale depends on a weighted average of prices in two earlier years.)

Eliminate q_t from the equations to obtain a linear difference equation for p_t of the form

$$p_t + 2\alpha p_{t-1} + \beta p_{t-2} = 0$$

Show by substitution that the solution has the form $p_t = A\xi^t + B\eta^t \, (t = 1, 2, \ldots)$, where A and B depend on initial conditions, and ξ and η are roots of a quadratic equation. Discuss in what region of the positive quadrant of the (α, β) plane the solutions are *stable*, which requires that $|\xi| < 1$ and $|\eta| < 1$ (then $p_t \to 0$). (This is another instance where time lags can cause instability.)

2.11 REFERENCES

Bensoussan, A., Bultez, A. and Naert, P. (1973) A generalization of the Nerlove–Arrow optimality condition, *Working paper 73-35*, European Inst. of Adv. Studies in Management, Brussels.

Clark, C. W. (1976) *Mathematical Bioeconomics: The Optimal Management of Renewable Resources*, Wiley, New York.

Davis, B. E. and Elzinga, D. J. (1972) The solution of an optimal control problem in financial modeling, *Operations Research*, **19** 1419–73.

Martin, R. and Teo, K. K. (1994) *Optimal Control of Drug Administration in Cancer Chemotherapy*, World Scientific, Singapore.

Nerlove, M. and Arrow, K. G. (1962) Optimal advertising policy under dynamic conditions, *Economica*, **39** 129–42.

Rosenbrock, H. H. (1970) *State-space and Multivariable Theory*, Nelson, London.

Sasieni, M. W. (1971) Optimal advertising expenditure, *Management Science*, **18**(4) part II, 64–72.

Sethi, S. P. (1970) Quantitative guidelines for communicable disease control programs, *Biometrics*, **30** 681–91.

Sethi, S. P. (1973) Optimal control of the Vidale–Wolfe advertising model, *Operations Research*, **21** 998–1013.

Sethi, S. P. (1974) Some explanatory remarks on the optimal control of the Vidale–Wolfe advertising model, *Operations Research*, **22** 1013–14.

Sethi, S. P. (1977) Dynamic optimal control studies in advertising: a survey, *SIAM Review*, **19** 685–725.

Sethi, S. P. and Sethi, P. W. (1978) Optimal control of some simple deterministic epidemic models, *J. Operational Research Society*, **29** 129–36.

Tsurumi, H. and Tsurumi, Y. (1971) Simultaneous determination of market share and advertising expenditure under dynamic conditions: the case of a firm within the Japanese pharmaceutical industry, *The Economic Studies Quarterly*, **22** 1–23.

Vidale, M. L. and Wolfe, H. B. (1957) An operations research study of sales response to advertising, *Operations Research*, **5** 370–81.

3

Convexity, linearization and multipliers

3.1 CONVEXITY

Many optimization questions depend on a geometric property called *convexity* of various sets of vectors involved. This property is relevant to Lagrange multipliers, and also to questions of sensitivity and stability.

3.1.1 Convex sets

A set E in a vector space X is *convex* if E contains the line segment joining each two of its points. If $\mathbf{a}, \mathbf{b} \in X$, denote by $[\mathbf{a}, \mathbf{b}]$ the line segment joining $\mathbf{a}$ to $\mathbf{b}$. Then $E \subset X$ is convex if

$$\mathbf{a}, \mathbf{b} \in E \Rightarrow [\mathbf{a}, \mathbf{b}] \in E$$

Figure 3.1 illustrates this in two dimensions.

Exercise

Show that $\mathbf{R}_+^n$ is convex.

A *linear combination* $\alpha_1 \mathbf{x}_1 + \alpha_2 \mathbf{x}_2 + \ldots + \alpha_r \mathbf{x}_r$ of the vectors $\mathbf{x}_1, \mathbf{x}_2, \ldots, \mathbf{x}_r$ is called a *convex combination* of these vectors if the real coefficients α_i satisfy $(\forall i)\alpha_i \geq 0$ and $\Sigma_i \alpha_i = 1$.The *convex hull* of a set $E \subset X$ is the set co E of all convex combinations of finite sets of points in E. In particular, co$\{\mathbf{x}, \mathbf{y}\}$ is the straight line segment $[\mathbf{x}, \mathbf{y}]$ joining the points $\mathbf{x}$ and $\mathbf{y}$. It follows that E is convex exactly when E = co E. If the vector space X is normed, the closure of co E is called the *closed convex hull* of E, and is denoted by $\overline{\text{co}}\ E$. (The same notation is used in a context where weak∗closure replaces norm closure.)

Figure 3.2 illustrates the fact that the convex hull of three points in a plane is the triangular area with the three points as vertices. (Observe that $\mathbf{q} = \alpha\mathbf{u} + (1-\alpha)\mathbf{p}$ and $\mathbf{p} = \beta\mathbf{v} + (1-\beta)\mathbf{w}$ for some $\alpha, \beta \in [0, 1]$. Combining these, $\mathbf{q} = \alpha\mathbf{u} + (1-\alpha)\beta\mathbf{v} + (1-\alpha)(1-\beta)\mathbf{w}$, where the coefficients are nonnegative with sum 1.

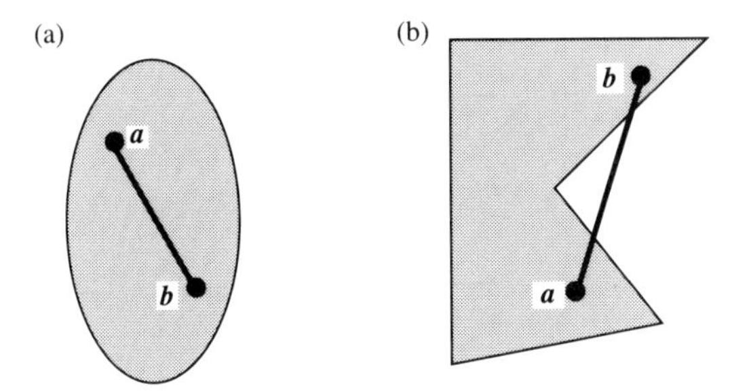

Figure 3.1 (a) A convex set; (b) a nonconvex set.

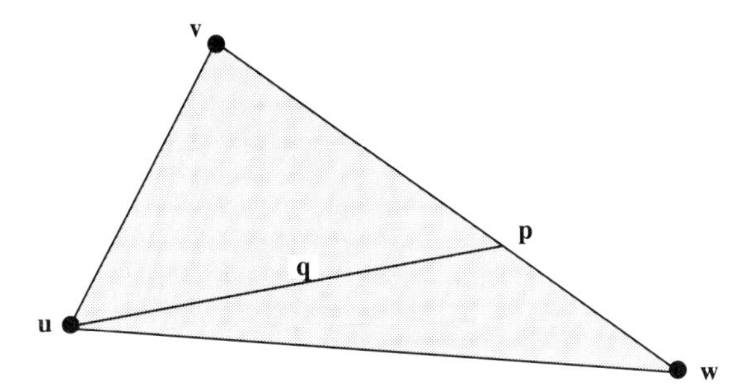

Figure 3.2 The convex hull of three points in a plane.

Conversely, any such combination **q** is a point of the triangular area.) There is an obvious extension to four or more points.

A point P is an *extreme point* of a set *E* if a line segment [A, B], containing P and contained in *E*, must have P at one end (thus either P = A or P = B). For the semicircular region shown in Figure 3.3, the boundary point P is an extreme point, but the boundary point P′ is not an extreme point. So the extreme points for this set form the semicircular part of the boundary. The vertices of a polygonal area are its extreme points; the present definition extends the idea to other sets.

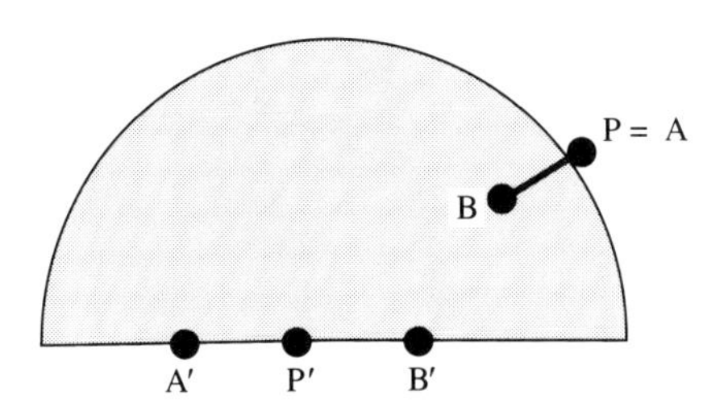

Figure 3.3 An extreme point.

Exercise
If E and F are convex sets, show that $E \cap F$ and $E + F$ are also convex sets, but that $E \cup F$ is *not* generally convex.

Exercise
If $A \subset G$ and G is convex, show that $A \subset \text{co}\, A \subset G$. Denote by H the intersection of all (infinitely many) convex sets containing A. Show that H is convex, $A \subset \text{co}\, A$ and $\text{co}\, A \subset H$; hence $H = \text{co}\, A$.

3.1.2 Separation theorem

Proofs that Lagrange multipliers exist for optimization problems are based ultimately on the *separation theorem* for disjoint convex sets. This is a version of the *Hahn–Banach theorem* in functional analysis (see e.g. Schaefer (1966)), here stated without proof. Figure 3.4 illustrates the property. Figure 3.4(a) shows two convex sets, with no common point, and there is a line with one convex set on each side of it. Figure 3.4(b) reduces one of the convex sets to a single point. Figure 3.5 shows one convex set, the open upper half-plane ($A = \{(x_1, x_2) : x_2 > 0\}$, shown shaded), and a second convex set, the x_1-axis ($B = \{(x_1, x_2) : x_2 = 0\}$. Here the separating line coincides with B; A is on one side of B, and B can be considered as on the other side of itself. Precise statements follow.

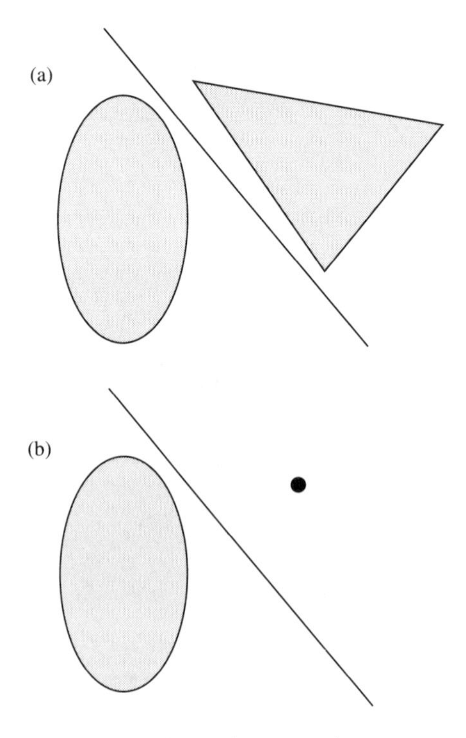

Figure 3.4 Strict separation.

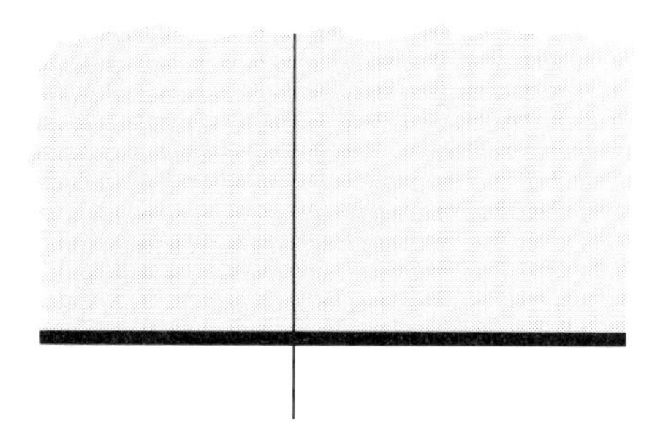

Figure 3.5 Nonstrict separation.

Let K and M be two disjoint convex sets in a space X. (*Disjoint* means that $K \cap M = \varnothing$.)

Case 1: If K is closed, and M consists of a single point $\mathbf{b}$, then there exists $\alpha \in \mathbf{R}$ and nonzero $g \in X'$ such that

$$\mathbf{g}(\mathbf{b}) < \alpha; \quad (\forall \mathbf{x} \in K)\, \mathbf{g}(x) \geq \alpha$$

Case 2: If K is open, there exists $\alpha \in \mathbf{R}$ and nonzero $\mathbf{g} \in X'$ such that

$$(\forall \mathbf{x} \in M)\, \mathbf{g}(\mathbf{x}) \leq \alpha; \quad (\forall \mathbf{x} \in K)\, \mathbf{g}(\mathbf{x}) \geq \alpha$$

The two cases are illustrated by Figures 3.4 and 3.5. They relate respectively to a *strict* inequality (<), and a *nonstrict* inequality (≤). The *affine variety* $V = \mathbf{g}^{-1}(\alpha)$ (section 1.5.3) has K on one side and M on the other.

Suppose now, instead, that K and M are disjoint convex subsets in a dual space X'. The above cases would find some $\mathbf{g}$ in the dual of X'. In order to find a suitable $\mathbf{g}$ in X, weak∗open and closed are substituted for open and closed in terms of the norm for X' (section 1.5.5). This makes no difference in finite dimensions.

Case 1′: If K is weak∗closed, and M consists of a single point $\mathbf{b}$, then there exists $\alpha \in \mathbf{R}$ and nonzero $\mathbf{g} \in X$ such that

$$\mathbf{b}(\mathbf{g}) < \alpha; \quad (\forall \mathbf{x} \in K)\, \mathbf{x}(\mathbf{g}) \geq \alpha$$

Case 2′: If K is weak∗open, there exists $\alpha \in \mathbf{R}$ and nonzero $\mathbf{g} \in X$ such that

$$(\forall \mathbf{x} \in M)\, \mathbf{x}(\mathbf{g}) \leq \alpha; \quad (\forall \mathbf{x} \in K)\, \mathbf{x}(\mathbf{g}) \geq \alpha$$

3.1.3 Convex cones

A set S in a vector space X is a *convex cone* if

$$S + S \subset S \quad \text{and} \quad (\forall \alpha \in \mathbf{R}_{+})\, \alpha S \subset S$$

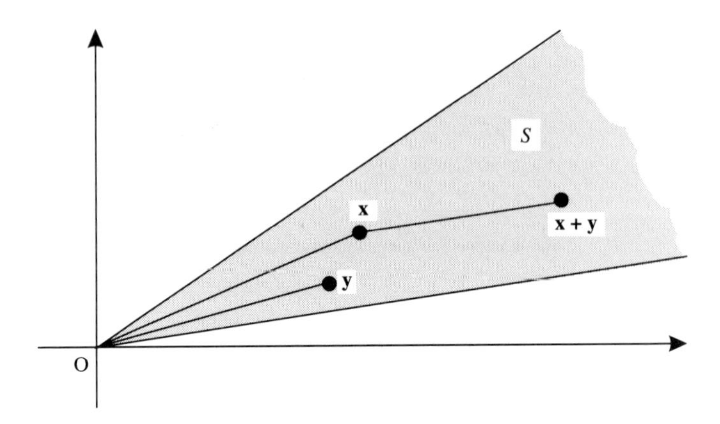

Figure 3.6 Definition of a convex cone.

These properties mean that if $\mathbf{x} \in S$ and $\mathbf{y} \in S$, then it follows that $\mathbf{x} + \mathbf{y} \in S$, and also that $\alpha\mathbf{x} \in S$ whenever $\alpha \geq 0$.

Figure 3.6 illustrates this definition for a convex cone S in $\mathbf{R}^2$, which is necessarily a sector. The vertex is the origin. The definition of convex cone is motivated by the prototype cone, $\mathbf{R}^n_+$.

A convex cone is a convex set (**Exercise**: prove this). A convex cone S defines a relation $\geq_S$ by

$$\mathbf{x} \geq_S y \Leftrightarrow x - y \in S$$

This relation has the properties

$$\mathbf{x} \geq_S \mathbf{x}; \quad (\mathbf{x} \geq_S \mathbf{y} \quad \text{and} \quad \mathbf{y} \geq_S \mathbf{z}) \Rightarrow \mathbf{x} \geq_S \mathbf{z}; \quad \mathbf{x} \geq_S \mathbf{y} \Rightarrow \mathbf{x} + \mathbf{u} \geq_S \mathbf{y} + \mathbf{u}$$

The first two of these make $\geq_S$ a *preorder*. It becomes a *partial order* (requiring also that $(\mathbf{x} \geq_S \mathbf{y}$ and $\mathbf{y} \geq_S \mathbf{x}) \rightarrow \mathbf{x} = \mathbf{y})$ if S is *pointed* $(S \cap (-S) = \{0\})$ (Figure 3.7(a)). (In the nonpointed example (Figure 3.7(b)), $S \cap (-S)$ is all the vertical axis.) If $S = \mathbf{R}^n_+$ then $\mathbf{x} \geq_S \mathbf{0} \Leftrightarrow (x_1 \geq 0, x_2 \geq 0, \ldots, x_n \geq 0)$.

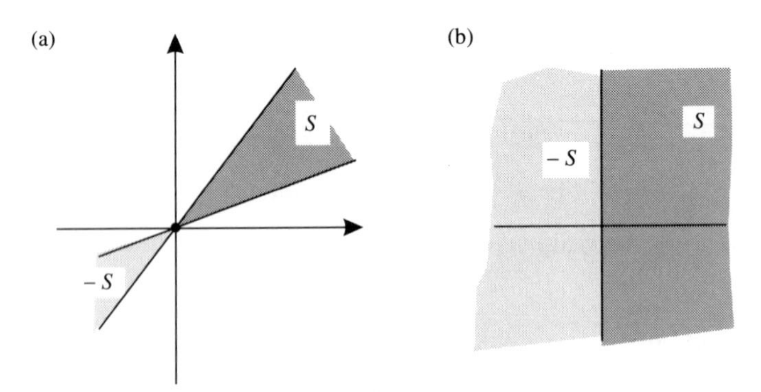

Figure 3.7 (a) Pointed and (b) nonpointed convex sets.

The *dual cone* of a set $E \subset X$ is the convex cone

$$E^* = \{\mathbf{y} \in X' : (\forall \mathbf{x} \in E)\ \mathbf{y}(\mathbf{x}) \geq 0\}$$

Another notation for the dual cone is E^+. (Note that some books define a dual cone with $\leq$ in place of $\geq$; their dual cone is the negative of E^* defined here. Often E is itself a convex cone S.)

Exercise
Show that E^* is a (weak$*$)closed convex cone. (Use

$$E^* = \cap_{\mathbf{x} \in E} \{\mathbf{y} \in X' : \mathbf{y}(\mathbf{x}) \geq 0\}$$

the intersection of (weak$*$)closed half-spaces, each of which is convex.)

Exercise
Denote by S the sector in $\mathbf{R}^2$ specified by $\mathbf{x}_1 \geq 0$ and $|\mathbf{x}_2| \leq c|\mathbf{x}_1|$, where c is a constant. Show that the dual cone S^* is also a sector. (*Hint*: A vector $\mathbf{y} \in \mathbf{R}^2$ is in S^* when $\mathbf{y}(\mathbf{x}) \geq 0$ for every $\mathbf{x} \in S$. Considering particular values of $\mathbf{x}$, such as $(1,0)^{\mathrm{T}}$ and $(1, \pm c)^{\mathrm{T}}$, leads to conditions which $\mathbf{y}$ must satisfy. Then $\mathbf{y}(\mathbf{x}) \geq 0$ must be verified for *all* $\mathbf{x} \in S$.)

The dual cone of a closed convex cone $S \subset X$ possesses the following separation property. If $\mathbf{b} \notin S$, then there exists $\mathbf{g} \in S^*$ such that $\mathbf{g}(\mathbf{b}) < 0$. In fact, from the separation theorem (case 1) there exist $\mathbf{g} \in X'$ and $\alpha \in \mathbf{R}$ such that $\mathbf{g}(\mathbf{s}) \geq \alpha$ for each $\mathbf{s} \in S$, and $\mathbf{g}(\mathbf{b}) < \alpha$. Since $0 \in S$, $\alpha \leq 0$. If $\mathbf{g}(\mathbf{s}_0) < 0$ for some $\mathbf{s}_0 \in S$; then, since $\lambda\mathbf{s} \in S$ for each $\lambda > 0$, there is λ such that $\mathbf{g}(\lambda\mathbf{s}_0) = \lambda\mathbf{g}(\mathbf{s}_0) < \alpha$, a contradiction. Hence $\mathbf{g}(S) \subset \mathbf{R}_+$, so $\mathbf{g} \in S^*$; and $\mathbf{g}(\mathbf{b}) < \alpha \leq 0$. Since $\mathbf{g} \neq 0$, S^* does not consist only of the zero vector.

Example
Let $S \subset X$ be a closed convex cone; then

$$\mathbf{x} \in S \Leftrightarrow (\forall \mathbf{y} \in S^*)\ \mathbf{y}(x) \geq 0$$

(One direction is given by the definition of S^*; the other follows from the separation property just proved.)

If X and Y are normed vector spaces, $S \subset X$ is a closed convex cone and $P: X \to Y$ is a continuous linear mapping, then the set $P(X)$ is also a convex cone.

Exercise
Prove this, from the definition of a convex cone.

Remark
The image cone $P(X)$ is *not* always closed. Ben-Israel (1969) has given a counterexample. Let S be a closed circular cone in $\mathbf{R}^3$, with axis vector $(1, 0, 1)^{\mathrm{T}}$ and generators at angle $\pi/4$ from this axis; thus S consists of all vectors whose angle

with $(1, 0, 1)^T$ is $\leq \pi/4$. Denote by P the matrix

$$\begin{bmatrix} 0 & 1 & 0 \\ 0 & 0 & 1 \end{bmatrix}$$

Then P is the orthogonal projection of $\mathbf{R}^3$ onto the $\mathbf{x}_2$, $\mathbf{x}_3$ plane. Calculation shows that $P(S)$ consists of the open upper half-plane, together with the origin; thus,

$$P(S) = (0, 0)^T \cup \{(\mathbf{x}_2, \mathbf{x}_3)^T : \mathbf{x}_3 > 0\}$$

In polar coordinates r, θ, thus with $x_2 = r \cos \theta$, $x_3 = r \sin \theta$,

$$P(S) = \{(r, \theta) : 0 < \theta < \pi, 0 \leq r\}$$

This convex cone is *not* closed, since it lacks its boundary (the rays for $\theta = 0$ and $\theta = \pi$). Intuitively, those points of the cone have been 'sent to infinity' by the projection. This does not happen with (closed) *polyhedral* cones, defined as follows.

Remark

If B is an $n \times k$ matrix, then $B(\mathbf{R}_+^k)$ is a closed convex cone (Ben-Israel, 1969). A *polyhedral cone* S is the intersection of a finite number of closed half-spaces; thus S has the form $\cap_j \{\mathbf{x} \in \mathbf{R}^n : \mathbf{c}_j^T \mathbf{x} \geq 0\}$, for a finite number of vectors $\mathbf{c}_j$. Since this S is expressible also as $B(\mathbf{R}_+^k)$, it follows that the image of a polyhedral cone by a matrix is another polyhedral cone, hence closed.

Example

A system of constraints $\mathbf{g}_i(\mathbf{x}) \leq 0$ ($i = 1, 2, \ldots m$), $\mathbf{h}_j(\mathbf{x}) = 0$ ($j = 1, 2, \ldots, r$) may be expressed as $-\mathbf{G}(\mathbf{x}) \in S$, where $\mathbf{G}(\mathbf{x})$ is the vector function with components $\mathbf{g}_1(\mathbf{x}), \ldots, \mathbf{g}_m(\mathbf{x}), \mathbf{h}_1(\mathbf{x}), \ldots, \mathbf{h}_r(\mathbf{x})$ and S is the convex cone

$$\mathbf{R}_+^m \times \{0\} = \{\mathbf{y} \in \mathbf{R}^{m+r} : \mathbf{y}_1 \geq 0, \ldots, \mathbf{y}_m \geq 0, \mathbf{y}_{m+1} = 0, \ldots, \mathbf{y}_{m+r} = 0]$$

where here $\{\mathbf{0}\}$ means the zero cone in $\mathbf{R}^r$.

Consider now a modified constraint system, given by $-M\mathbf{G}(\mathbf{x}) \in S$, where M is some $k \times (m + r)$ matrix. These modified constraints represent inequalities and equalities applied to certain linear combinations of the components of $\mathbf{G}(\mathbf{x})$, and so may describe a situation where some mutual compensation is possible between different requirements. The new constraints may be written $-\mathbf{G}(\mathbf{x}) \in U$, where $U = M^{-1}(S)$ is a polyhedral cone.

Exercise

Let X and Y be vector spaces, $A : X \to Y$ a linear mapping and $S \subset X$ a convex cone. Show (from the definition of convex cone) that the image $A(S) = \{Ax : X \in S\}$ of S is also a convex cone.

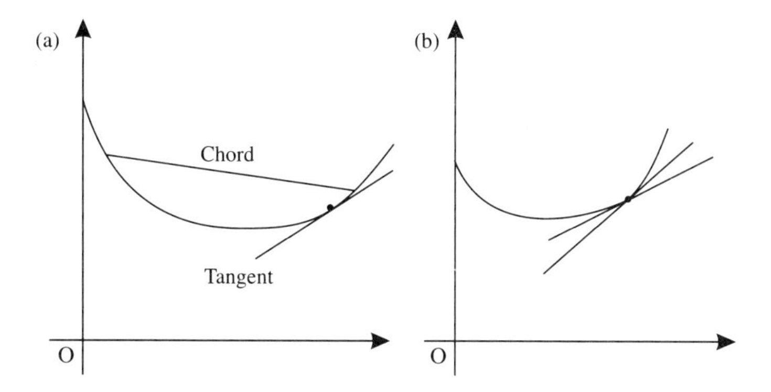

Figure 3.8 Examples of convex functions: (a) differentiable, showing a tangent and a chord; (b) nondifferentiable, showing two subdifferentials.

3.2 CONVEX FUNCTIONS

Let X and Y be normed vector spaces; let Γ be a convex subset of X; let $S \subset Y$ be a convex cone.

3.2.1 Definition

A function $f: \Gamma \to \mathbf{R}$ is *convex* if, for all $\mathbf{x}, \mathbf{z} \in \Gamma$ and all α in $0 < \alpha < 1$,

$$\alpha f(\mathbf{x}) + (1 - \alpha) f(\mathbf{z}) - f(\alpha \mathbf{x} + (1 - \alpha)\mathbf{z}) \in \mathbf{R}_+$$

Note that, since Γ is convex, $\alpha\mathbf{x} + (1 - \alpha)\mathbf{z} \in \Gamma$. The function f has not been assumed differentiable. Figure 3.8 illustrates this definition.

Geometrically, each chord drawn on the graph of f lies *above* the graph; and tangents lie *below* the graph – as proved below. The defining property holds trivially when $\alpha = 0$ and when $\alpha = 1$, so these cases need not be included in the definition.

Exercise

Show that $f: \Gamma \to \mathbf{R}$ is convex if and only if its *epigraph*

$$\operatorname{epi} f = \{(\mathbf{x}, y) : \mathbf{x} \in \Gamma, \mathbf{y} \geq f(\mathbf{x})\}$$

is a convex set.

This convex definition generalizes as follows to vector functions.

3.2.2 Definition

A function $f : \Gamma \to Y$ is *S-convex*, where $S \subset Y$ is a convex cone, if for all $\mathbf{x}, \mathbf{z} \in \Gamma$ and all α in $0 < \alpha < 1$,

$$\alpha f(\mathbf{x}) + (1 - \alpha) f(\mathbf{z}) - f(\alpha \mathbf{x} + (1 - \alpha)\mathbf{z}) \in S$$

Exercise

If $Y = \mathbf{R}^m$ and $S = \mathbf{R}^m_+$, show that $\mathbf{g}$ is S-convex if and only if each component of G is convex ($\equiv \mathbf{R}_+$-convex).

Exercise
Show (from the definition) that the sum of two convex functions (defined on some convex domain) is also a convex function. Prove a similar result for S-convex functions.

Exercise
Show that the set of all S-convex functions on a given convex domain D forms a convex conc.

Exercise
If $\Gamma = \{\mathbf{x} \in \mathbf{R}^n : \mathbf{x}^T A \mathbf{x} \leq 1\}$ and $f(\mathbf{x}) = \mathbf{c}^T\mathbf{x} + (\mathbf{x}^T M \mathbf{x})^{1/2}$, where A and M are positive semidefinite (real symmetric $n \times n$) matrices and $\mathbf{c}$ is a constant vector, show that Γ is a convex set and f is a convex function.

(*Remark*: If also $\mathbf{g}$ is a linear function and S is a convex cone, then

$$\text{Minimize } f(\mathbf{x}) \quad \text{subject to} \quad -\mathbf{g}(x) \leq S, \mathbf{x} \in \Gamma$$

is a convex problem.)

Hint: The *generalized Schwarz inequality*, valid when M is positive semidefinite, states that

$$(\forall \mathbf{w}, \mathbf{x} \in \mathbf{R}^n)\ \mathbf{w}^T M \mathbf{x} \leq (\mathbf{x}^T M \mathbf{x})^{1/2} (\mathbf{w}^T M \mathbf{w})^{1/2}$$

with equality exactly when $\mathbf{x}$ and $\mathbf{w}$ have the same direction (thus, when $\mathbf{w} = \beta\mathbf{x}$ for some constant number β). To prove it, note that the quadratic function $(\mathbf{x} + \beta\mathbf{w})^T M(\mathbf{x} + \beta\mathbf{w}) \geq 0$ for each real β; and (if $a > 0$)

$$(\forall \beta)\ 0 \leq a\beta^2 + 2b\beta + c = a(\beta - b/a)^2 + a^{-1}(ca - b^2) \Rightarrow b^2 \leq ca$$

3.2.3 Global minimization lemma

A local minimum of a convex function on a convex set is also a global minimum.

Proof
Let Γ be convex and $f : \Gamma \to \mathbf{R}$ reach a local minimum at $\mathbf{p} \in \Gamma$; this means that $f(\mathbf{x}) \geq f(\mathbf{p})$ for all $\mathbf{x} \in \Gamma \cap (\mathbf{p} + N)$, where N is some ball with centre $\mathbf{0}$. Suppose, if possible, that $\mathbf{p}$ is not a global minimum; then, for some $\mathbf{z} \in \Gamma, f(\mathbf{z}) < f(\mathbf{p})$. Since Γ is convex, $\alpha\mathbf{z} + (1 - \alpha)\mathbf{p} \in \Gamma$ whenever $0 < \alpha < 1$. Since f is convex,

$$\begin{aligned} f(\mathbf{p} + \alpha(\mathbf{z} - \mathbf{p})) - f(\mathbf{p}) &= f(\alpha\mathbf{z} + (1 - \alpha)\mathbf{p}) - f(\mathbf{p}) \\ &\leq \alpha f(\mathbf{z}) + (1 - \alpha) f(\mathbf{p}) - f(\mathbf{p}) \\ &= -\alpha[f(p) - f(\mathbf{z})] \\ &< 0 \end{aligned}$$

for arbitrarily small $\alpha > 0$, contradicting the local minimum. □

Example

A *nonconvex* function may have several local minima, which are not global minima, for example the function $f(x) = x^2 + 4 \sin x$, $x \in \mathbf{R}$.

Exercise

Generalize this result to a vector function $f: \Gamma \to Y$, with an order cone $Q \subset Y$, where int $Q \neq \varnothing$. A *local minimum* at $\mathbf{p}$ now means that

$$(\forall \mathbf{x} \in \Gamma \cap (\mathbf{p} + N))\, f(\mathbf{x}) - f(\mathbf{p}) \notin -\operatorname{int} Q$$

where N is some ball with centre, 0, and a global minimum means that

$$(\forall \mathbf{x} \in \Gamma)\, f(\mathbf{x}) - f(\mathbf{p}) \notin -\operatorname{int} Q$$

Thus $\leq$ is replaced here by $\leq_Q$, and <0 is replaced by $\in -\operatorname{int} Q$. This kind of minimum for a vector function is called a *weak minimum* (section 3.4.4).

3.2.4 Continuous convex function

Let $f: \Gamma \to \mathbf{R}$ be convex, with Γ a convex set, let $\mathbf{p} + N \subset \Gamma$ for some ball N, and let f be bounded above on $\mathbf{p} + N$. Then f is continuous at $\mathbf{p}$.

Proof

Choose $\mathbf{h} \neq \mathbf{0}$ so that $\pm\mathbf{h} \in N$. By assumption, $f(\mathbf{x}) \leq \kappa$ when $\mathbf{x} \in \mathbf{p} + N$ for some constant κ. Let $0 < \alpha < 1$. Since Γ is convex,

$$\mathbf{p} + \alpha\mathbf{h} = (1 - \alpha)\mathbf{p} + \alpha(\mathbf{p} + \mathbf{h}) \in \Gamma$$

Since f is convex,

$$(1 - \alpha)f(\mathbf{p}) + \alpha f(\mathbf{p} + \mathbf{h}) \geq f(\mathbf{p} + \alpha\mathbf{h})$$

and

$$(1 + \alpha)^{-1} [f(\mathbf{p} + \alpha\mathbf{h}) + \alpha f(\mathbf{p} - \mathbf{h})] \geq f(\mathbf{p})$$

These rearrange to give

$$\alpha[f(\mathbf{p}) - f(\mathbf{p} - \mathbf{h}) \leq f(\mathbf{p} + \alpha\mathbf{h}) - f(\mathbf{p}) \leq \alpha[f(\mathbf{p} + \mathbf{h}) - f(\mathbf{p})] \tag{3.1}$$

From the left-hand inequality of (3.1), $2f(\mathbf{p} + \alpha\mathbf{h}) \geq -f(\mathbf{p} - \mathbf{h}) + 3f(\mathbf{p})$; thus, since f is bounded above on $\mathbf{p} + N$, f is also bounded below on $\mathbf{p} + N$, say $|f(\mathbf{x})| \leq r$ when $\mathbf{x} \leq \mathbf{p} + N$. Then the inequalities (3.1) show that $|f(\mathbf{p} + \alpha\mathbf{h}) - f(\mathbf{p})| \leq 2\alpha c$, which proves that f is continuous at $\mathbf{p}$. □

Remarks

If $\Gamma \subset \mathbf{R}^n$, the ball N may be replaced by an n-dimensional cube, and then convexity of f shows that f is bounded above on $\mathbf{p} + N$ by the maximum of f at each of the finite number of vertices of the cube. So a convex function in finite dimensions is continuous at each interior point of its domain. It need not be continuous at boundary points; the function f: $[0, 1] \to \mathbf{R}$ given by $f(x) = 0$

when $0 < x < 1$, $f(0) = f(1) = 1$ is convex (each chord lies above the graph), but is discontinuous at the boundary points.

3.2.5 Differentiable convex functions

Let X and Y be normed vector spaces; let $S \subset Y$ be a convex cone; let $f: X \to Y$ be Fréchet differentiable. Then $\mathbf{f}$ is S-convex if and only if

$$(\forall \mathbf{x}, \mathbf{z} \in X)\ \mathbf{f}(\mathbf{x}) - \mathbf{f}(\mathbf{z}) - \mathbf{f}'(\mathbf{z})(\mathbf{x} - \mathbf{z}) \in S \tag{3.2}$$

If $X = \mathbf{R}^n$ and $\mathbf{f}$ is twice Fréchet differentiable, then $\mathbf{f}$ is S-convex if and only if

$$(\forall \mathbf{w}, \mathbf{z} \in X)\ \mathbf{w}^T\mathbf{f}''(\mathbf{z})\mathbf{w} \in S \tag{3.3}$$

Remarks

Equation (3.2) means that the graph of f lies above each tangent – see Figure 3.8(a). Suppose now that int $S \neq \emptyset$; then $\mathbf{f}$ is called *strictly S-convex* if $\mathbf{f}$ is (int S)-convex; thus, if strict inequality $>$ replaces $\geq$. This happens exactly when (3.2) holds with int S replacing S.

Consider $X = \mathbf{R}^n$. If $Y = \mathbf{R}$ and $S = \mathbf{R}_+$, then *twice differentiable* means that

$$\mathbf{f}(\mathbf{z} + \mathbf{w}) = \mathbf{f}(\mathbf{z}) + \mathbf{f}'(\mathbf{z})\mathbf{w} + \tfrac{1}{2}\mathbf{w}^T\mathbf{f}''(\mathbf{z})\mathbf{w} + o(\|\mathbf{w}\|^2)$$

where $\mathbf{f}''(\mathbf{z})$ is the matrix of second partial derivatives $\partial^2\mathbf{f}/\partial\mathbf{x}_i\,\partial\mathbf{x}_j$ at $\mathbf{x} = \mathbf{z}$, and property (3.3) means that $\mathbf{w}^T\mathbf{f}''(\mathbf{z})\mathbf{w} \geq 0$ for every vector $\mathbf{w}$; thus the matrix $\mathbf{f}''(\mathbf{z})$ is *positive semidefinite*. This happens exactly when all eigenvalues of the matrix are nonnegative.

If $Y = \mathbf{R}^m$, so $\mathbf{f} = (f_1, \ldots, f_m)$, then $\mathbf{w}^T\mathbf{f}''(\mathbf{z})\mathbf{w}$ means the vector with components $\mathbf{w}^T\mathbf{f}_k''(\mathbf{z})\mathbf{w}$ $(k = 1, 2, \ldots, m)$. In general, $\mathbf{q}^T\mathbf{f}''(\mathbf{z})\mathbf{w}$ is the gradient of $\mathbf{q}^T\mathbf{f}'(\mathbf{z})\mathbf{w}$, and is linear both in $\mathbf{q} \in X'$ and in $\mathbf{w} \in X$.

Proof

Assume S is closed. Let $\mathbf{z}, \mathbf{x} \in X$ and $0 < \alpha < 1$. If $\mathbf{f}$ is S-convex, then the convexity definition rearranges to

$$\mathbf{f}(\mathbf{x}) - \mathbf{f}(\mathbf{z}) - \alpha^{-1}[\mathbf{f}(\mathbf{z} + \alpha(\mathbf{x} - \mathbf{z})] \in S$$

Letting $\alpha \downarrow 0$ proves (3.2). If S is not closed, let $0 < \beta < \alpha < 1$, $\gamma = \beta/\alpha$ and $\mathbf{y} = \mathbf{x} - \mathbf{z}$; then

$$\beta[\beta^{-1}[\mathbf{f}(\mathbf{p} + \beta\mathbf{y}) - \mathbf{f}(\mathbf{p})] + \alpha^{-1}[\mathbf{f}(\mathbf{p} + \alpha\mathbf{y}) - \mathbf{f}(\mathbf{p})]] = -\mathbf{f}(\gamma(\mathbf{p} + \alpha\mathbf{y}) + (1 - \gamma)\mathbf{p})$$
$$+ \gamma\mathbf{f}(\mathbf{p} + \alpha\mathbf{y}) + (1 - \gamma)\mathbf{f}(\mathbf{p}) \in S$$

As $\beta \downarrow 0$,

$$-\mathbf{f}'(\mathbf{p})\mathbf{y} + \alpha^{-1}[\mathbf{f}(\mathbf{p} + a\mathbf{y}) - \mathbf{f}(\mathbf{p})] \in S$$

Then

$$\mathbf{f}(\mathbf{x}) - \mathbf{f}(\mathbf{z}) \in S + \alpha^{-1}[\mathbf{f}(\mathbf{p} + a\mathbf{y}) - \mathbf{f}(\mathbf{p})] \in S + S + \mathbf{f}'(\mathbf{p})\mathbf{y}$$

Conversely, let (3.2) hold, let $\mathbf{u}, \mathbf{v} \in \Gamma$, let $0 < \alpha < 1$ and let $\xi = \alpha\mathbf{u} + (1 - \alpha)\mathbf{v}$.

Then

$$\alpha\mathbf{f}(\mathbf{u}) + (1-\alpha)\mathbf{f}(\mathbf{v}) - \mathbf{f}(\xi) = \alpha[\mathbf{f}(\mathbf{u}) - \mathbf{f}(\xi)] + (1-\alpha)[\mathbf{f}(\mathbf{v}) - \mathbf{f}(\xi)]$$
$$\in S + \alpha\mathbf{f}'(\xi)(\mathbf{u}-\xi) + (1-\alpha)\mathbf{f}'(\xi)(\mathbf{v}-\xi) = S + 0$$

proving (3.2).

Assume **f** is twice differentiable. If **f** is S-convex, then

$$\mathbf{f}(\mathbf{x}) - \mathbf{f}(\mathbf{z}) \in S + \mathbf{f}'(\mathbf{z})(\mathbf{x}-\mathbf{z})$$

and

$$\mathbf{f}(\mathbf{z}) - \mathbf{f}(\mathbf{x}) \in S + \mathbf{f}'(\mathbf{x})(\mathbf{z}-\mathbf{x})$$

Adding,

$$0 \in S + S - [\mathbf{f}'(\mathbf{x}) - \mathbf{f}'(\mathbf{z})](\mathbf{x}-\mathbf{z}) = S - (\mathbf{x}-\mathbf{z})^{\mathrm{T}}\mathbf{f}''(\mathbf{z})(\mathbf{x}-\mathbf{z}) + o(\|\mathbf{x}-\mathbf{z}\|^2)$$

from which (3.3) follows. Conversely, assume (3.3); let $0 < \alpha < 1$,

$$\xi(\alpha) = \mathbf{u} + \alpha(\mathbf{v}-\mathbf{u}), \quad \varphi(\alpha) = \mathbf{f}(\xi(\alpha)) - \mathbf{f}'(\mathbf{v})(\xi(\alpha) - \mathbf{u})$$

Then, for some $\delta \in (0, 1)$ and some $\mathbf{t} \in [\mathbf{u}, \mathbf{v}]$, using the mean value theorem,

$$\begin{aligned}\mathbf{f}(\mathbf{u}) - \mathbf{f}(\mathbf{v}) - \mathbf{f}'(\mathbf{v})(\mathbf{u}-\mathbf{v}) &= \varphi(1) - \varphi(0)\\ &= \varphi'(\delta)(1-0)\\ &= [\mathbf{f}'(\xi(\delta)) - \mathbf{f}'(\mathbf{z})](\mathbf{v}-\mathbf{u})\\ &= \int_0^\delta (\mathbf{u}-\mathbf{u})^{\mathrm{T}}\mathbf{f}''(\xi(\sigma))(\mathbf{u}-\mathbf{v})\,\mathrm{d}\sigma\\ &\in S\end{aligned}$$

from (3.3); so (3.2) is proved. □

Exercise

Show that the function $f: \mathbf{R} \to \mathbf{R}$, given by $f(\mathbf{x}) = x^{2n}$, is convex when $n = 1, 2, \ldots$.

Example

The quadratic function $f: \mathbf{R}^n \to \mathbf{R}$, given by $f(\mathbf{x}) = \mathbf{x}^{\mathrm{T}}A\mathbf{x} + \mathbf{b}^{\mathrm{T}}\mathbf{x}$, where A is a real symmetric $n \times n$ matrix and **b** is a constant vector, is convex exactly when all eigenvalues of A are nonnegative. (That assumption ensures that A is positive semidefinite: $\mathbf{v}^{\mathrm{T}}A\mathbf{v} \geq 0$ for each vector **v**).

Example

Define the quadratic vector function $\mathbf{F}: \mathbf{R}^n \to \mathbf{R}^k$ by

$$\mathbf{F}(\mathbf{x}) = C\mathbf{x} + (1/2)\mathbf{x}^{\mathrm{T}}A\,\mathbf{x},$$

where C is an $m \times n$ matrix and the notation $A_{.}$ is shorthand, meaning that the component $\mathbf{F}_i(\mathbf{x})$ has quadratic term $(1/2)\mathbf{x}^{\mathrm{T}} A_i \mathbf{x}$. Then $\mathbf{F}$ is $\mathbf{R}_+^k$-convex exactly when each of the matrices of $A_{.}$ is positive semidefinite.

3.3 CONVEX FUNCTIONS AND SUBDIFFERENTIALS

Let $\Gamma \subset X$ be convex; let $f : \Gamma \to \mathbf{R}$ be convex, and bounded above; f need not be differentiable. Then the epigraph of f: $\text{epi} f = \{(\mathbf{x}, y) : \mathbf{x} \in \Gamma, y \geq f(x)\}$ has interior points. Let $\mathbf{p} \in \Gamma$. Since f is convex, epi f is a convex set; and $(\mathbf{p}, f(\mathbf{p})) \notin$ int epi f. By the separation theorem, there exists (at least one) vector $\mathbf{g} \in X'$, such that $\{\mathbf{p} + \mathbf{g}(\mathbf{x} - \mathbf{p}) : \mathbf{x} \in \Gamma\}$ does not meet int epi f. Hence

$$(\forall \mathbf{x} \in \Gamma)\; f(\mathbf{x}) \geq f(\mathbf{p}) + \mathbf{g}(\mathbf{x} - \mathbf{p}) \tag{3.4}$$

The set of such vectors $\mathbf{g}$ is called the *subdifferential* of f at $\mathbf{p}$, and is denoted by $\partial f(\mathbf{p})$. From its definition, $\partial f(\mathbf{p})$ is a convex set. It is also a closed set; if (3.4) is satisfied for a sequence of vectors $\mathbf{g}$, then it also holds for any weak$*$limit (section 1.5.5) of the sequence.

Exercise

If $\Gamma \subset \mathbf{R}^n$ is convex, $f : \Gamma \to \mathbf{R}$ is convex, $0 \in \Gamma$ and $0 \in \partial f(0)$, then $\mathbf{0}$ is a global minimum of f on Γ. (Consider also the case of $f; \Gamma \to \bar{\mathbf{R}} := \mathbf{R} \cup \{\infty\}$, thus allowing f to take the value $+\infty$, with $f(\mathbf{0})$ finite.)

If f happens to be differentiable at $\mathbf{p}$, then $\partial f(\mathbf{p})$ consists of the single vector $\mathbf{f}'(\mathbf{p})$. Figure 3.9 illustrates the construction when $X = \mathbf{R}$. (Note that ∂ meaning *subdifferential* must not be confused with ∂ meaning partial derivative or ∂ meaning boundary.)

It follows that the convex function f is the envelope of a set of affine functions. Thus

$$f(\mathbf{x}) = \max\,\{h_{\mathbf{p},\mathbf{g}}(\mathbf{x}) = f(\mathbf{p}) + \mathbf{g}(\mathbf{x} - \mathbf{p}) : \mathbf{p} \in \Gamma, \mathbf{g} \in \partial f(\mathbf{p})\}$$

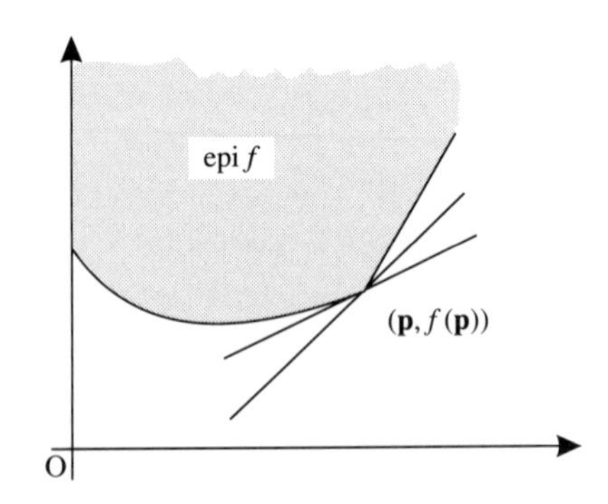

Figure 3.9 Construction of a subdifferential.

Given $f : \Gamma \to \mathbf{R}$ convex and finite, the proof of Theorem 3.2.5 shows that $\beta^{-1}[f(\mathbf{p} + \beta\mathbf{y}) - f(\mathbf{p})]$ is a decreasing function of $\beta > 0$. Hence there exists

$$f'(\mathbf{p}; \mathbf{y}) = \lim_{\beta \downarrow 0} \beta^{-1}[f(\mathbf{p} + \beta\mathbf{y}) - f(\mathbf{p})]$$

the *directional derivative* of f at $\mathbf{p}$ in direction $\mathbf{y}$. From the definition of subdifferential,

$$\mathbf{g} \in \partial f(\mathbf{p}) \Leftrightarrow (\forall \mathbf{y}, \forall \beta > 0)\ \beta^{-1}[f(\mathbf{p} + \beta\mathbf{y}) - f(\mathbf{p})] \geq \mathbf{gy} \Leftrightarrow f'(\mathbf{p}; \mathbf{y}) \geq \mathbf{gy}$$

Hence

$$f'(\mathbf{p}; \mathbf{y}) = \sup_{\mathbf{g} \in \partial f(\mathbf{p})} \mathbf{gy} \tag{3.5}$$

Exercise
If f is a convex function, show that its directional derivative is a convex function of direction (that is, $f'(\mathbf{p}; .)$ is a convex function. *Hint*: the maximum of a set of convex functions is a convex function.)

Exercise
Define the function $f: \mathbf{R}^n \to \mathbf{R}$ by $f(\mathbf{x}) = (\mathbf{x}^{\mathrm{T}} M \mathbf{x})^{1/2}$, where M is a positive semidefinite (symmetric $n \times n$) matrix. Use the *generalized Schwarz inequality* (section 3.2.2) to show that

$$\mathbf{u} \in \partial f(\mathbf{0}) \Leftrightarrow [\mathbf{u} = \mathbf{w}^{\mathrm{T}} M, \mathbf{w}^{\mathrm{T}} M \mathbf{w} \leq 1]$$

Hint: consider first the case when M is *positive definite*, thus when

$$\mathbf{w}^{\mathrm{T}} M \mathbf{w} > 0 \quad \text{whenever} \quad \mathbf{0} \neq \mathbf{w}$$

The positive semidefinite case can be done by diagonalizing M, giving a diagonal matrix with diagonal elements $\lambda_1, \ldots, \lambda_r, 0, \ldots, 0$, with each $\lambda_i > 0$.

Exercise
If $\varphi : \mathbf{R} \to \mathbf{R}$ is a finite convex function, and $a < b < c$, then $b = \lambda a + (1 - \lambda)c$ for some $l \in (0, 1)$. Show from definition of convex that

$$(c - b)^{-1}[\varphi(c) - \varphi(b)] \geq (b - a)^{-1}[\varphi(b) - \varphi(a)]$$

Deduce that, at each x, φ has both a left derivative $\lim_{t \downarrow 0} t^{-1}[\varphi(x - t) - \varphi(x)]$ and a right derivative $\lim_{t \downarrow 0} t^{-1}[\varphi(x + t) - \varphi(x)]$. (Since a bounded monotonic function can have only countably many jumps, φ is differentiable except on a set that is at most countable.)

Exercise
Show that this function φ is minimized at 0 if and only if $0 \in \partial\varphi(0)$. *Hint*: The minimum implies that $\varphi'(0; y) \geq 0$ for each y. If $0 \notin \partial\varphi(0)$, the separation theorem finds d_0 with $(\forall z \in \partial\varphi(0))\ zd_0 < 0$; by (3.5), $\varphi'(0 : d_0) < 0$.)

A related concept is the *support function* $s(.|C)$ of a convex set $C \subset \mathbf{R}^n$, defined by

$$s(\mathbf{x}|C) = \sup_{\mathbf{v} \in C} \mathbf{v}^{\mathrm{T}} \mathbf{x}$$

When C is closed and bounded, the supremum is reached at some point of C, so that max can replace sup in the definition. For example, if C is the unit ball with centre 0, then the supremum is reached when $\mathbf{v} = \hat{\mathbf{x}}$, the unit vector in the direction of $\mathbf{x}$; hence the support function $s(\mathbf{x}|C) = \|\mathbf{x}\|$.

Example
When f is a real convex function, bounded near $\mathbf{p}$, the directional derivative

$$f'(\mathbf{p}; \mathbf{y}) = s(\mathbf{y}|\partial f(\mathbf{p}))$$

Exercise
Let $E = \{\mathbf{x} = (x_1, x_2, \ldots, x_n) \in \mathbf{R}^n_+ : \sum x_i = 1\}$. Show that the support function

$$s(\mathbf{y}|E) = \max_{\mathbf{x} \in E} \mathbf{y}^{\mathrm{T}}\mathbf{x} = \max y_i$$

3.4 ALTERNATIVE THEOREMS

An *alternative theorem* states that two systems are so related that exactly one of them has a solution. These theorems are used to show the existence of Lagrange multipliers for optimization problems. All these alternative theorems are consequences of the separation theorem for convex sets (section 3.1.2).

3.4.1 Positive lemma

If S is a convex cone with interior, $\mathbf{0} \neq \mathbf{p} \in S^*$, and $\mathbf{s} \in \operatorname{int} S$, then $\mathbf{p}(\mathbf{s}) > 0$.

Proof
Since $\mathbf{p} \in S^*$ and $\mathbf{s} \in S$, $\mathbf{p}(\mathbf{s}) \geq 0$. Since $\mathbf{s} \in \operatorname{int} S$, $\mathbf{s} + N \subset S$ for some ball N with centre 0. Suppose, if possible, that $\mathbf{p}(\mathbf{s}) = 0$; then

$$\mathbf{R}_+ \supset \mathbf{p}(\mathbf{s} + N) = \mathbf{p}(N)$$

Since $\mathbf{p} \neq \mathbf{0}$, $\mathbf{p}(\mathbf{n}) \neq 0$ for some $\mathbf{n} \in N$. Either $\mathbf{p}(\mathbf{n}) < 0$ or $\mathbf{p}(\mathbf{n}) > 0$, so that $\mathbf{p}(-\mathbf{n}) < 0$; in either case $\mathbf{p}(N) \subset \mathbf{R}_+$ is contradicted. Hence $0 \neq \mathbf{p}(\mathbf{s}) \geq 0$. □

3.4.2 Basic alternative theorem

Let X and Y be real normed vector spaces; let $S \subset Y$ be a convex cone with non-empty interior; let $\Gamma \subset X$ be convex; let the mapping $f : \Gamma \to Y$ be S-convex. Then exactly one of the two following systems has a solution:

(I) $(\exists \mathbf{x} \in \Gamma)\, -f(\mathbf{x}) \in \operatorname{int} S$

(II) $(\exists \mathbf{p} \in S^*)\, \mathbf{p} \neq 0, (\mathbf{p} \circ f)(\Gamma) \subset \mathbf{R}_+$

Proof
If both (I) and (II) hold, then both $(\mathbf{p} \circ f)(\mathbf{x}) < 0$ from (I), $\mathbf{0} \neq \mathbf{p} \in S^*$ and the positive lemma, and $(\mathbf{p} \circ f)(\mathbf{x}) \geq 0$ from (II), a contradiction.

Suppose that (I) has no solution. Let $K = f(\Gamma) + \text{int } S$. If $\mathbf{k} \in K$, then $\mathbf{k} = \mathbf{s} + f(\mathbf{x})$ for some $\mathbf{s} \in \text{int } S$ and $\mathbf{x} \in \Gamma$, and $\mathbf{s} + N \subset S$ for some ball N, so $\mathbf{k} + N \subset K$; thus K is an open set. If, for $i = 1, 2$, $\mathbf{k}_i = f(x_i) + \mathbf{s}_i \in K$ and $0 < \tau < 1$, $\mathbf{x} = \tau\mathbf{x}_2 + (1 - \tau)\mathbf{x}_1$, then

$$\begin{aligned}(1-\tau)\mathbf{k}_1 + \tau\mathbf{k}_2 &= (1-\tau)(\mathbf{k}_1 - f(\mathbf{x}_1)) + \tau(\mathbf{k}_2 - f(\mathbf{x}_2)) + ((1-\tau)f(\mathbf{x}_1) \\ &\quad + \tau f(\mathbf{x}_2) - f(\mathbf{x})) + f(\mathbf{x}) \\ &\in \text{int } S + \text{int } S + S + f(\mathbf{x}) \\ &= \mathbf{s} + f(\mathbf{x}) \quad \text{for some } \mathbf{s} \in \text{int } S \\ &\in K\end{aligned}$$

Thus K is a convex open set. Since (I) has no solution, $\mathbf{0} \notin K$. By the separation theorem, there exists nonzero $\mathbf{p} \in Y'$ such that $\mathbf{p}(K) \subset \mathbf{R}_+$. Fix $\mathbf{x}_0 \in \Gamma$. If $\mathbf{s} \in \text{int } S$, then $\mathbf{s} + N \subset \text{int } S$ for some ball N, and for θ large enough, $\theta^{-1}(f(\mathbf{x}_0)) \in N$, so $\mathbf{s} - \theta^{-1}f(\mathbf{x}_0) \in \text{int } S$; since S is a cone, $\theta\mathbf{s} - f(\mathbf{x}_0) \in \text{int } S$, so $\theta\mathbf{s} \in K$. Then $\mathbf{p}(\mathbf{s}) = \theta^{-1}\mathbf{p}(\theta\mathbf{s}) \geq 0$. Since $\mathbf{p}$ is continuous, $\mathbf{p}(S) \subset \mathbf{R}_+$; thus $\mathbf{p} \in S^*$. For any $\mathbf{x} \in \Gamma$ and any $\mathbf{e} \in \text{int } S$, $\mathbf{k} = f(\mathbf{x}) + \varepsilon\mathbf{e} \in K$ for each $\varepsilon > 0$, so

$$(\mathbf{p} \circ f)(\mathbf{x}) = \mathbf{p}(\mathbf{k}) - \varepsilon\mathbf{p}(\mathbf{e}) \geq 0 - \varepsilon\mathbf{p}(\mathbf{e}) \to 0 \quad \text{as } \varepsilon \downarrow 0$$

Thus (II) holds.

This proof has shown that NOT (I) $\Rightarrow$ (II). The contrapositive of this statement gives NOT (II) $\Rightarrow$ NOT NOT (I) = (I). □

3.4.3 Remarks

This theorem does not assume that f is continuous.

For a convex constraint $-\mathbf{g}(\mathbf{x}) \in S$ (thus S is a convex cone and $\mathbf{g}$ is an S-convex function) and a convex domain Γ for $\mathbf{x}$, *Slater's constraint qualification* is the statement that $-\mathbf{g}(\hat{\mathbf{x}}) \in \text{int } S$ for some $\hat{\mathbf{x}} \in \Gamma$. (Of course, this requires int S nonempty.) From the basic alternative theorem (substituting $\mathbf{g}$ for $\mathbf{f}$), Slater's constraint qualification is equivalent to *Karlin's constraint qualification*, that there exists no nonzero $\mathbf{p} \in S^*$ for which $(\mathbf{p} \circ f)(\Gamma) \subset \mathbf{R}_+$. These hypotheses are used in the next section.

3.4.4 Necessary condition for convex minimization

Consider the convex minimization problem:

$$(P_0)\text{: Minimize } f(\mathbf{x}) \quad \text{subject to} \quad \mathbf{x} \in \Gamma, -\mathbf{g}(\mathbf{x}) \in S$$

Here S is a closed convex cone with interior, Γ is a convex set, f is a convex function on Γ and $\mathbf{g}$ is an S-convex function on Γ. Then the *feasible set*

$$\Delta = \{\mathbf{x} \in \Gamma : -\mathbf{g}(\mathbf{x}) \in S\}$$

is convex, since if $0 < \alpha < 1$ and $\mathbf{v}, \mathbf{w} \in \Delta$, then $\mathbf{y} = \alpha\mathbf{v} + (1-\alpha)\mathbf{w} \in \Gamma$ since Γ is convex, and, since $\mathbf{g}$ is S-convex,

$$-\mathbf{g}(\mathbf{y}) = [-\mathbf{g}(\mathbf{y}) + \alpha\mathbf{g}(\mathbf{v}) + (1-\alpha)\mathbf{g}(\mathbf{w})] + \alpha[-\mathbf{g}(\mathbf{v})] + (1-\alpha)[-\mathbf{g}(\mathbf{w})]$$
$$\in S + S + S \subset S$$

Thus $\mathbf{y} \in \Delta$. Let (P_0) reach a minimum at $\mathbf{x} = \mathbf{x}_0$. Then there is no solution $\mathbf{x}$ to the system

$$-[f(\mathbf{x}) - f(\mathbf{x}_0)] \in \text{int } \mathbf{R}_+, \quad -\mathbf{g}(\mathbf{x}) \in S, \mathbf{x} \in \Gamma$$

Therefore there is no solution $\mathbf{x} \in \Gamma$ to the more restricted system:

$$-\begin{bmatrix} f(\mathbf{x}) - f(\mathbf{x}_0) \\ \mathbf{g}(\mathbf{x}) \end{bmatrix} \in \text{int } (\mathbf{R}_+ \times S)$$

From the basic alternative theorem, whose convex hypotheses are satisfied here, there exists nonzero $\mathbf{p} \in (\mathbf{R}_+ \times S^*)$; thus $\mathbf{p} = (\tau, \mathbf{v})$ with $\tau \in \mathbf{R}_+^* = \mathbf{R}_+$ and $\mathbf{v} \in S^*$, such that

$$(\forall \mathbf{x} \in \Gamma)\ [\tau \quad \mathbf{v}] \begin{bmatrix} f(\mathbf{x}) - f(\mathbf{x}_0) \\ \mathbf{g}(\mathbf{x}) \end{bmatrix} = \tau[f(\mathbf{x}) - f(\mathbf{x}_0)] + \mathbf{v}\mathbf{g}(\mathbf{x}) \geq 0$$

Here τ and $\mathbf{v}$ are not both zero. If $\tau = 0$, then $(\exists \mathbf{v} \in S^*, \mathbf{v} \neq 0)\ \mathbf{v}\mathbf{g}(\Gamma) \subset \mathbf{R}_+$. This does not happen if Karlin's (or equivalently Slater's) constraint qualification is also assumed; and then (dividing by τ) $\tau = 1$ can be assumed.

Now replace the objective function $f : \Gamma \to \mathbf{R}$ by a vector objective function $\mathbf{f} : \Gamma \to Z$, where Z is a normed vector space, and let $Q \subset Z$ be a closed convex cone with interior. The point $\mathbf{x}_0 \in \Gamma$ is called a *weak minimum* of $\mathbf{f}(\mathbf{x})$, subject to constraints $\mathbf{x} \in \Gamma, -\mathbf{g}(\mathbf{x}) \in S$, with respect to the *order cone* Q if there is *no* point $\mathbf{x}$ satisfying the constraints for which

$$\mathbf{f}(\mathbf{x}) - \mathbf{f}(\mathbf{x}_0) \in -\text{int } Q$$

If $W = Z \backslash (-\text{int } Q)$, then $\mathbf{f}(\mathbf{x}) - \mathbf{f}(\mathbf{x}_0) \in W$ must hold for each $\mathbf{x}$ satisfying the constraints. The discussion of the previous paragraph then applies, now replacing f by $\mathbf{f}$ and $\mathbf{R}_+$ by Q. Hence a *weak minimum* $\mathbf{x}_0$ implies that

$$(\exists \boldsymbol{\tau} \in Q^*, \mathbf{v} \in S^*, (\boldsymbol{\tau}, \mathbf{v}) \neq (\mathbf{0}, \mathbf{0}))\ (\forall \mathbf{x} \in \Gamma)\ \boldsymbol{\tau}[\mathbf{f}(\mathbf{x}) - \mathbf{f}(\mathbf{x}_0)] + \mathbf{v}\mathbf{g}(\mathbf{x}) \geq 0$$

Figure 3.10 shows the case $Z = \mathbf{R}^2$, $Q = \mathbf{R}_+^2$. Observe that W is a closed set, but not convex. Also shown is the set $U = Y \backslash (-(S \backslash \{0\})$, which arises later in discussion of *Pareto minimum*; note that $U = \{0\} \cup \text{int } W$.

However, and additional assumption, called a constraint qualification (sections 3.5.1 and 3.7) is needed to ensure that $\tau \neq 0$ (or $\boldsymbol{\tau} \neq \mathbf{0}$).

3.4.5 Ideas for Farkas's theorem

The basic alternative theorem needs the assumption int $S \neq \varnothing$. To see what happens when S has no interior points, consider first an $m \times n$ matrix A, a system

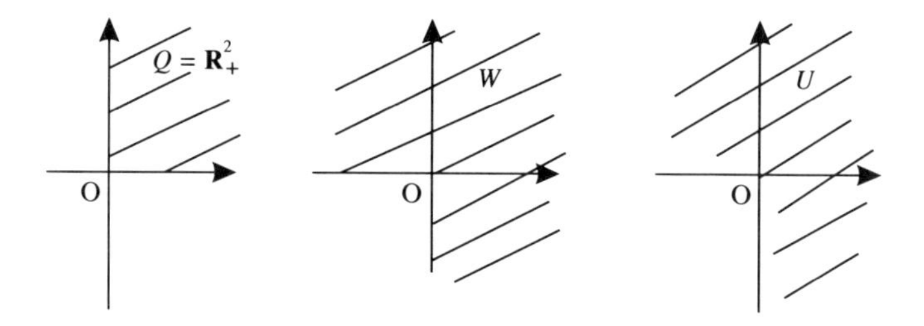

Figure 3.10 Pareto and weak minimum.

of linear inequalities $A\mathbf{x} \geq \mathbf{0}$ ($\mathbf{x} \in \mathbf{R}^n$, $\mathbf{0} \in \mathbf{R}^m$) and a vector $\mathbf{c} \in (\mathbf{R}^n)'$ (thus $\mathbf{c}$ is a row vector with n components). The inequality $\geq$ applies to each component of the vector equation $A\mathbf{x} \geq \mathbf{0}$. Now suppose that A and $\mathbf{c}$ are so related that $\mathbf{cx} \geq 0$ whenever $A\mathbf{x} \geq \mathbf{0}$; this may be written as

$$A\mathbf{x} \geq \mathbf{0} \Rightarrow \mathbf{cx} \geq 0 \tag{3.6}$$

Farkas's theorem states that (3.6) happens exactly when $\mathbf{c} = \boldsymbol{\lambda} A$, for some vector $\boldsymbol{\lambda} \geq \mathbf{0}$ (thus, each component of $\boldsymbol{\lambda}$ is nonnegative). Now (3.6) states that the system

$$A\mathbf{x} \in \mathbf{R}^n_+, \quad -\mathbf{cx} \in \text{int}\,\mathbf{R}_+ \tag{3.7}$$

has no solution $\mathbf{x} \in \mathbf{R}^n$.

The following result generalizes Farkas's theorem to general vector spaces and orderings involving convex cones. Note that here cl means weak$*$closure.

3.4.6 Farkas's theorem

Let X and Y be normed vector spaces, $A: X \to Y$ a continuous linear mapping, $T \subset Y$ a closed convex cone and $\mathbf{c} \in X'$. Then

$$[A\mathbf{x} \in T \Rightarrow \mathbf{cx} \geq 0] \Leftrightarrow [\mathbf{c} \in \text{cl}(A^{\text{T}}(T^*))]$$

Proof

Using the definition of the adjoint linear mapping A^{T},

$$A\mathbf{x} \in T \Leftrightarrow (\forall \mathbf{t}^* \in T^*)\, \mathbf{t}^* A\mathbf{x} \geq 0 \Leftrightarrow (\forall \mathbf{t}^* \in T^*)\, (A^{\text{T}}\mathbf{t}^*)(\mathbf{x}) \geq 0 \Leftrightarrow \mathbf{x} \in (A^{\text{T}}(T^*))^*$$

and similarly

$$\mathbf{cx} \geq 0 \Leftrightarrow \mathbf{x} \in (\mathbf{c}^{\text{T}}(\mathbf{R}^*_+))^*$$

Denote $E = A^{\text{T}}(T^*)$ and $F = \mathbf{c}^{\text{T}}(\mathbf{R}^*_+) = \{r\mathbf{c}^{\text{T}}: r \in \mathbf{R}_+\}$. Then

$$[A\mathbf{x} \in T \Rightarrow \mathbf{cx} \geq 0] \Leftrightarrow [E^* \subset F^*] \Leftrightarrow [F \subset \text{cl}\, E]$$

For if there exists $\mathbf{q} \in F \backslash \text{cl}\, E$, then there exists $\boldsymbol{\lambda} \in (\text{cl}\, E)^* = E^*$ such that $\mathbf{q}\boldsymbol{\lambda} < 0$, contradicting $E^* \subset F^*$; hence $\mathbf{c} \in \text{cl}\, E$. □

Remarks

For the existence of this $\boldsymbol{\lambda}$, see the separation property for a closed convex cone, discussed in section 3.1.3. For the matrix case (3.6) above, where A is an $m \times n$ matrix and T is the convex cone $\mathbf{R}_+^m$ in $\mathbf{R}^m$, the cone $A^T(T^*)$ is closed because $T = \mathbf{R}_+^m$, and then $\mathbf{c} \in \mathrm{cl}(A^T(T^*))$ reduces to $\mathbf{c} = \mathbf{t}^* A$ (thus $\mathbf{c}^T = A^T \mathbf{t}^*$) for some $\mathbf{t}^* \in T^*$. Here $\mathbf{c}$ and $\mathbf{t}^*$ are row vectors.

Exercise

Let P be a polyhedral cone, defined in terms of a matrix A by

$$\mathbf{x} \in P \Leftrightarrow A\mathbf{x} \geq \mathbf{0}$$

Use Farkas's theorem to show that

$$\mathbf{u} \in P^* \Leftrightarrow \mathbf{u} = \boldsymbol{\lambda} A, \boldsymbol{\lambda} \geq \mathbf{0}$$

(If $A\mathbf{x} \geq 0$ is written as $A\mathbf{x} \in \mathbf{R}_+^k$, for suitable k, then

$$[A^{-1}(\mathbf{R}_+^k)]^* = A^T[(\mathbf{R}_+^k)^*]$$

Example

Let B be a matrix, K a polyhedral cone and $\mathbf{z}$ a vector, related by

$$[\mathbf{x} = B\mathbf{v}, \mathbf{v} \in K] \Rightarrow \mathbf{z}^T\mathbf{x} \geq 0$$

Then, for each $\mathbf{v} \in K$, $0 \leq \mathbf{z}^T B\mathbf{v}$, so that $\mathbf{z}^T B \in K^*$. Hence

$$\{B(K)\}^* = B^{-1}(K^*)$$

Exercise

With A and P as above, let $\mathbf{w} \in P^{**} \equiv (P^*)^*$. Show that

$$(\forall \boldsymbol{\lambda} \geq 0)\ \boldsymbol{\lambda} A\mathbf{w} \geq \mathbf{0}$$

and deduce that $\mathbf{w}$ satisfies $A\mathbf{w} \geq \mathbf{0}$, so that $\mathbf{w} \in P$. Complete the proof that $P^{**} = P$.

What additional requirement must be satisfied if $\mathbf{R}_+^k$, is replaced by a general closed convex cone?

3.4.7 Motzkin alternative theorem

Let X, Y, V be normed vector spaces, $A : X \to V$ and $B : X \to Y$ continuous linear mappings, and $S \subset Y$ and $T \subset V$ convex cones with int $S \neq \varnothing$ and $A^T(T^*)$ weak$*$closed. Then either

$$\text{(I}'\text{)} \qquad (\exists \mathbf{x} \in X) - A\mathbf{x} \in T, \quad -B\mathbf{x} \in \mathrm{int}\, S$$

or

$$\text{(II}'\text{)} \qquad (\exists \mathbf{p} \in S^*, \mathbf{p} \neq 0)\ (\exists \mathbf{q} \in T^*)\ A^T\mathbf{q} + B^T\mathbf{p} = 0$$

Remark
Consider in particular $X = \mathbf{R}^n$, $Y = \mathbf{R}^m$, $V = \mathbf{R}^r$, $S = \mathbf{R}^k_+$, $T = \{0\} \subset \mathbf{R}^r$. Then (I′) and (II′) become respectively, in matrix language:

$$(\exists \mathbf{x})\ A\mathbf{x} = \mathbf{0},\ B\mathbf{x} < \mathbf{0} \quad \text{and} \quad (\exists \mathbf{p}, \mathbf{q})\ A^{\mathrm{T}}\mathbf{q} + B^{\mathrm{T}}\mathbf{p} = \mathbf{0},\ \mathbf{0} \neq \mathbf{p} \geq \mathbf{0}$$

Proof
Denote $\Gamma = \{\mathbf{x} \in X : -A\mathbf{x} \in T\}$. Then Γ is a convex set, and

NOT (I′) $\Leftrightarrow (\nexists \mathbf{x} \in \Gamma)\ -B\mathbf{x} \in \operatorname{int} S$

$\Leftrightarrow (\exists \mathbf{p} \in S^*)\ \mathbf{p} \neq \mathbf{0},\ \mathbf{p}B(\Gamma) \subset \mathbf{R}_+$ (by section 3.4.2 (basic alternative theorem))

$\Leftrightarrow (\exists \mathbf{p} \in S^*)\ \mathbf{p} \neq \mathbf{0},\ [(-A)\mathbf{x} \in T \Rightarrow (\mathbf{p}B)\mathbf{x} \geq 0]$

$\Leftrightarrow (\exists \mathbf{p} \in S^*)(\exists \mathbf{p} \in T^*)\ \mathbf{p} \neq \mathbf{0},\ \mathbf{p}B = \mathbf{q}(-A)$ (by section 3.4.6 (Farkas's theorem))

$\Leftrightarrow$ (II′)

Note that Farkas's theorem gives the result stated, since $A^{\mathrm{T}}(T^*)$ is assumed weak$*$closed. □

Exercise
Show that *either* the linear system $A\mathbf{x} = \mathbf{c}$ has a solution $\mathbf{x}$, or the system $A^{\mathrm{T}}\mathbf{v} = \mathbf{0}$ has a solution $\mathbf{v}$ with $\mathbf{v}^{\mathrm{T}}\mathbf{c} = 1$, but not both. (This follows from Motzkin's theorem, by replacing $(\exists \mathbf{x})\ A\mathbf{x} = \mathbf{c}$ by the equivalent system

$$(\exists \mathbf{y}, t)\ A\mathbf{y} - \mathbf{c}t = 0, \quad t > 0$$

obtained by the substitution $\mathbf{x} = y/t$. This equivalent expression can be put into the form of (I′) by writing it as

$$\left(\exists \begin{bmatrix} y \\ t \end{bmatrix}\right) [A \quad -\mathbf{c}] \begin{bmatrix} \mathbf{y} \\ \mathbf{t} \end{bmatrix} = 0, \quad [0 \quad 1] \begin{bmatrix} y \\ t \end{bmatrix} \in \operatorname{int} \mathbf{R}_+$$

The closed-cone hypothesis of Motzkin's theorem must then be given its meaning in the present case. Note that another equivalent expression of the form (II′) can also be found.)

Exercise
Deduce Farkas's theorem from Motzkin's theorem. (Put the expression $A\mathbf{x} \in T \Rightarrow \mathbf{cx} \geq 0$ into an equivalent Motzkin form as

$$(\nexists \mathbf{x})\ -A\mathbf{x} \in -T, \quad \mathbf{cx} \in -\operatorname{int} \mathbf{R}_+)$$

3.4.8 Extended Farkas's theorem
Let X, Y, V be normed vector spaces, $A : X \to V$ and $M : X \to Y$ continuous linear mappings, $T \subset V$ and $Q \subset Y$ convex cones with $\operatorname{int} Q \neq \emptyset$ and $\mathbf{b} \in -T$,

$\mathbf{s} \in -Q$. Let the cone $[A, \mathbf{b}]^{\mathrm{T}}(T^*)$ be weak∗closed. Let $W = Y \backslash (-\mathrm{int}\, Q)$. Then

$$[A\mathbf{x} + \mathbf{b} \in -T \Rightarrow M\mathbf{x} + \mathbf{s} \in W] \Leftrightarrow [(\exists \boldsymbol{\tau} \in Q^*, \boldsymbol{\lambda} \in T^*)\, \boldsymbol{\tau} \neq 0,$$
$$\boldsymbol{\tau} M + \boldsymbol{\lambda} A = 0, \boldsymbol{\lambda}\mathbf{b} = 0, \boldsymbol{\tau}\mathbf{s} = 0] \quad (3.8)$$

Proof

By the substitution $\mathbf{x} = \mathbf{y}/t$, $t > 0$,

$$[A\mathbf{x} + \mathbf{b} \in -T \Rightarrow M\mathbf{x} + \mathbf{s} \in W]$$

$$\Leftrightarrow [(\nexists \mathbf{y}, t)\, A\mathbf{y} + \mathbf{b}t \in T, M\mathbf{y} + \mathbf{s}t \in -\mathrm{int}\, Q, t \notin \mathrm{int}\, \mathbf{R}_+]$$

$$\Leftrightarrow \left[(\nexists \mathbf{y}, t) - [A \;\; \mathbf{b}] \begin{bmatrix} \mathbf{y} \\ t \end{bmatrix} \in -T, \quad -\begin{bmatrix} M & \mathbf{s} \\ 0 & 1 \end{bmatrix} \begin{bmatrix} \mathbf{y} \\ t \end{bmatrix} \in \mathrm{int} \begin{bmatrix} Q \\ \mathbf{R}_+ \end{bmatrix} \right]$$

$$\Leftrightarrow [(\exists \boldsymbol{\tau} \in Q, r \in \mathbf{R}_+, \lambda \in T^*)\, \boldsymbol{\lambda}^{\mathrm{T}}[A \;\; \mathbf{b}] + [\boldsymbol{\tau}^{\mathrm{T}} \;\; -r] \begin{bmatrix} M & \mathbf{s} \\ 0 & 1 \end{bmatrix} = [\mathbf{0} \;\; 0], [\boldsymbol{\tau} \;\; r] \neq [0 \;\; 0]]$$

by Motzkin's theorem (section 3.4.7), applicable since $[A, \mathbf{b}]^{\mathrm{T}}(T^*)$ is weak∗ closed.

Expanding the matrix expressions gives $\boldsymbol{\lambda}^{\mathrm{T}} A + \boldsymbol{\tau}^{\mathrm{T}} M = \mathbf{0}$, $\boldsymbol{\lambda}^{\mathrm{T}}\mathbf{b} + \boldsymbol{\tau}^{\mathrm{T}}\mathbf{s} = r$, where $r \geq 0$, and the assumptions on $\mathbf{b}$ and $\mathbf{s}$ ensure that $\boldsymbol{\lambda}^{\mathrm{T}}\mathbf{b} \leq 0$, $\boldsymbol{\tau}^{\mathrm{T}}\mathbf{s} \leq 0$. Hence $r = \boldsymbol{\lambda}^{\mathrm{T}}\mathbf{b} = \boldsymbol{\tau}^{\mathrm{T}}\mathbf{s} = 0$. Since $r = 0$, $\boldsymbol{\tau} \neq 0$, and the result follows. □

In particular, if $Q = \mathbf{R}_+ \subset \mathbf{R}$, then $W = \mathbf{R}_+$ also; and $0 \neq \tau \in \mathbf{R}_+$ may be replaced by $\tau = 1$, by dividing $\boldsymbol{\lambda}$ by $\tau > 0$. Hence, subject to a 'closed-cone assumption' and $\mathbf{b} \in -T$,

$$[A\mathbf{x} + \mathbf{b} \in -T \Rightarrow \mathbf{c}\mathbf{x} + \mathbf{s} \geq 0] \Leftrightarrow [(\exists \boldsymbol{\lambda} \in T^*)\, \mathbf{c} + \boldsymbol{\lambda} A = \mathbf{0}, \boldsymbol{\lambda}\mathbf{b} = 0] \quad (3.9)$$

3.5 LINEARIZATION AND LAGRANGIAN CONDITIONS

Consider the constrained minimization problem:

$$\text{Minimize } f(\mathbf{x}) \quad \text{subject to} \quad -\mathbf{h}(\mathbf{x}) \in T \quad (3.10)$$

where X and Z are normed vector spaces, $f: X \to R$ and $h: X \to Z$ are (Fréchet) differentiable functions and $T \subset Z$ is a closed convex cone. Suppose that $\mathbf{x} = \mathbf{x}_0$ is a minimum point for (P_1). Consider a linear approximation to the constraint $-\mathbf{h}(\mathbf{x}) \in T$ near the point $\mathbf{x}_0$; this takes the form

$$\mathbf{h}(\mathbf{x}_0) + \mathbf{h}'(\mathbf{x}_0)\mathbf{v} \in -T$$

where $\mathbf{v} = \mathbf{x} - \mathbf{x}_0$.

3.5.1 Counter-example

It is intuitive, but not correct, that this linear approximation to the constraint inequality also gives a good approximation to the feasible set near $\mathbf{x}_0$. Consider the example of Ben-Israel, Ben-Tal and Zlobec (1979):

$$\text{Minimize}_{x \in \mathbf{R}} f(x) = -x \quad \text{subject to} \quad h(x) = (x_+)^2 \leq 0 \tag{3.11}$$

where $x_+ = x$ if $x \geq 0$, $x_+ = 0$ if $x < 0$. The feasible set is $(\infty, 0]$, and the minimum occurs when $x = 0$. Now h is differentiable, with $h(0) = 0$, $h'(0) = 0$, so a linear approximation to the constraint near $x = 0$ is $0 + 0x \leq 0$. So this linear approximation would enlarge the feasible set from $(-\infty, 0]$ to $(-\infty, \infty)$. Some restrictive assumption, called a *constraint qualification*, is required in order to restrict such awkward cases.

3.5.2 Definition of locally solvable

The constraint $-\mathbf{h}(\mathbf{x}) \in T$ is called *locally solvable* at $\mathbf{x}_0$ if, for each direction $\mathbf{v}$ satisfying the linearized constraint $\mathbf{h}(\mathbf{x}_0) + \mathbf{h}'(\mathbf{x}_0)\mathbf{v} \in -T$, there exists a solution $\mathbf{x} = \mathbf{x}_0 + \alpha\mathbf{v} + \zeta(\alpha)$ to $-\mathbf{h}(\mathbf{x}) \in T$, valid whenever α is sufficiently small positive, where $\zeta(\alpha) = o(\alpha)$, meaning that $\zeta(\alpha)/\alpha \to 0$ as $\alpha \to 0$.

Some sufficient conditions for locally solvable are given in section 3.7.

3.5.3 Karush–Kuhn–Tucker (KKT) conditions

Consider the problem (3.10), assuming that a minimum is reached at $\mathbf{x} = \mathbf{x}_0$, and assuming that the constraint is locally solvable there. Assume also that the cone $[\mathbf{h}'(\mathbf{x}_0)\ \mathbf{h}(\mathbf{x}_0)]^{\mathrm{T}}(T^*)$ is weak$*$closed. (The matrix cited adjoins $\mathbf{h}(\mathbf{x}_0)$ as an extra column to the matrix $\mathbf{h}'(\mathbf{x}_0)$.) Consider the linearization of the problem (3.10) at $\mathbf{x}_0$, which is:

$$\text{Minimize } f'(\mathbf{x}_0)\mathbf{v} \quad \text{subject to} \quad \mathbf{h}(\mathbf{x}_0) + \mathbf{h}'(\mathbf{x}_0)\mathbf{v} \in -T \tag{3.12}$$

Suppose that the direction $\mathbf{v}$ satisfies the constraint of (3.12). By the assumed local solvability, there is $\mathbf{x} = \mathbf{x}_0 + \alpha\mathbf{v} + \zeta(\alpha)$, such that $\mathbf{h}(\mathbf{x}) \in -T$ for sufficiently small $\alpha > 0$, and $\zeta(\alpha) = o_1(\alpha)$. (Several different $o(\alpha)$ terms will be distinguished by suffixes.) Since $\mathbf{x}_0$ minimizes (3.10),

$$\begin{aligned} 0 &\leq \alpha^{-1}[f(\mathbf{x}_0 + \alpha\mathbf{v} + \zeta(\alpha)) - f(\mathbf{x}_0)] \\ &= \alpha^{-1}[f'(\mathbf{x}_0)\alpha\mathbf{v} + o_2(\alpha)] \\ &= f'(\mathbf{x}_0)\mathbf{v} + o_2(\alpha)/\alpha \\ &\to f'(\mathbf{x}_0)\mathbf{v} \quad \text{as } \alpha \downarrow 0 \end{aligned}$$

Hence $f'(\mathbf{x}_0)\mathbf{v} \geq 0$. Therefore

$$[\mathbf{h}(\mathbf{x}_0) + \mathbf{h}'(\mathbf{x}_0)\mathbf{v} \in -T] \Rightarrow f'(\mathbf{x}_0)\mathbf{v} \geq 0$$

By the extended Farkas theorem (section 3.4.8) (equation (3.9)), this happens when a Lagrange multiplier $\boldsymbol{\lambda} \in T^*$ exists, satisfying the *Karush–Kuhn–Tucker (KKT) necessary conditions*:

$$f'(\mathbf{x}_0) + \boldsymbol{\lambda}\mathbf{h}'(\mathbf{x}_0) = 0, \quad \boldsymbol{\lambda} \in T^*, \quad \boldsymbol{\lambda}\mathbf{h}(\mathbf{x}_0) = 0 \tag{3.13}$$

together with feasibility: $-\mathbf{h}(\mathbf{x}_0) \in T$. Note that (3.13) says, in part, that the Lagrangian $f(\mathbf{x}) + \boldsymbol{\lambda}\mathbf{h}(\mathbf{x})$ has zero gradient at $\mathbf{x} = \mathbf{x}_0$.

Remark

If *Maximize* replaces *Minimize* in the problem (3.10), it is equivalent to minimize $-f(x)$. So the Lagrangian becomes $-f(x) + \boldsymbol{\lambda}\mathbf{h}(\mathbf{x})$. Since the Karush–Kuhn–Tucker conditions require the gradient of the Lagrangian to be zero, there results $-f'(\mathbf{x}_0) + \boldsymbol{\lambda}\mathbf{h}'(\mathbf{x}_0) = \mathbf{0}$. Hence

$$f'(\mathbf{x}_0) - \boldsymbol{\lambda}\mathbf{h}'(\mathbf{x}_0) = \mathbf{0}, \quad \boldsymbol{\lambda} \in T^*, \quad \boldsymbol{\lambda}\mathbf{h}(\mathbf{x}_0) = 0$$

Thus $-\boldsymbol{\lambda}$ has replaced $\boldsymbol{\lambda}$.

A *critical point* for (3.10) may be a maximum, a minimum or a saddle point. Subject to a constraint qualification, a critical point $\mathbf{x}_0$ satisfies

$$f'(\mathbf{x}_0) \quad + \boldsymbol{\lambda}\mathbf{h}'(\mathbf{x}_0) = \mathbf{0}, \quad \boldsymbol{\lambda}\mathbf{h}(\mathbf{x}_0) = 0, \quad -\mathbf{h}(\mathbf{x}_0) \in T$$

(not necessarily $\boldsymbol{\lambda} \in T^*$), and so may be called a *Karush–Kuhn–Tucker point.*

Exercise

Consider the problem in two dimensions:

$$\text{Minimize } f(x, y) = 2x^3 - 3y^2 \quad \text{subject to} \quad x^2 + 2y^2 \leq 2$$

Show that a critical point $(\bar{x}, \bar{y})$ satisfies

$$6x^2 + 2\lambda x = 0, \quad -6y + \lambda(2y) = 0, \quad \lambda(x^2 + 2y^2 - 2) = 0, \quad x^2 + 2y^2 - 2 \leq 0$$

Show that the set of all points satisfying these conditions comprises

$$(0, \pm 1), \quad (-1, \pm 2^{-1/2}), \quad (0, 0), \quad (\pm 2^{1/2}, 0)$$

Remark

If $f(.)$ is replaced by a vector function $\mathbf{f}(.)$, and a weak minimum (section 3.4.4) with respect to an order cone Q is considered, then a related theorem holds. A weak minimum inmplies that

$$[\mathbf{h}(\mathbf{x}_0) + \mathbf{h}'(\mathbf{x}_0)\mathbf{v} \in -T] \Rightarrow f'(\mathbf{x}_0)\mathbf{v} \notin -\text{int } Q$$

So there is no solution $\mathbf{v}$ to

$$\mathbf{h}(\mathbf{x}_0) + \mathbf{h}'(\mathbf{x}_0)\mathbf{v} \in -T, \quad f'(\mathbf{x}_0)\mathbf{v} \in -\text{int } Q$$

The Motzkin alternative theorem (section 3.4.7) can be applied to show that the (3.13) conditions hold in the form:

$$\boldsymbol{\tau} f'(\mathbf{x}_0) + \boldsymbol{\lambda}\mathbf{h}'(\mathbf{x}_0) = \mathbf{0}, \quad \boldsymbol{\lambda} \in T^*, \quad \boldsymbol{\lambda}\mathbf{h}(\mathbf{x}_0) = \mathbf{0}$$

for some $\boldsymbol{\lambda} \in T^*$ and some nonzero $\boldsymbol{\tau} \in Q^*$.

Exercise
Prove this, substituting $\mathbf{v} = \mathbf{w}/\beta$, $\beta > 0$, and applying the Motzkin alternative theorem.

3.5.4 Fritz John conditions

A weakened version of (3.13) is sometimes required. Adjoint an additional constraint $-\mathbf{g}(\mathbf{x}) \in S$ to (3.10), where $\mathbf{g}$ is differentiable, and the convex cone S has nonempty interior in the normed vector space Y. The problem then becomes

$$\text{Minimize } f(\mathbf{x}) \quad \text{subject to} \quad -\mathbf{h}(\mathbf{x}) \in T, \quad -\mathbf{g}(\mathbf{x}) \in S \tag{3.14}$$

Assume a minimum of (3.14) at $\mathbf{x} = \mathbf{x}_0$, and suppose that the direction $\mathbf{v} = \mathbf{x} - \mathbf{x}_0$ satisfies $\mathbf{h}(\mathbf{x}_0) + \mathbf{h}'(\mathbf{x}_0)\mathbf{v} \in -T$ and $\mathbf{g}(\mathbf{x}_0) + \mathbf{g}'(\mathbf{x}_0)\mathbf{v} \in -\text{int } S$. Assume local solvability of the constraint $-\mathbf{h}(\mathbf{x}) \in T$, but *not* for the whole constraint system; and assume that the cone $[\mathbf{h}'(\mathbf{x}_0)\,\mathbf{h}(\mathbf{x}_0)]^{\mathrm{T}}(T^*)$ is weak$*$closed. As in section 3.5.3, there is a solution $\mathbf{x} = \mathbf{x}_0 + \alpha\mathbf{v} + o_1(\alpha)$ to $-\mathbf{h}(\mathbf{x}) \in T$. Then

$$\begin{aligned} -\mathbf{g}(\mathbf{x}_0 + \alpha\mathbf{v} + o_1(\alpha)) &= -\mathbf{g}(\mathbf{x}_0) - \alpha\mathbf{g}'(\mathbf{x}_0)\mathbf{v} - o_3(\alpha) \\ &= -(1-\alpha)\mathbf{g}(\mathbf{x}_0) - \alpha[\mathbf{g}(\mathbf{x}_0) + \mathbf{g}'(\mathbf{x}_0)\mathbf{v}] - o_3(\alpha) \end{aligned}$$

Now $o_3(\alpha)/\alpha \to 0$ as $\alpha \downarrow 0$, and $-[\mathbf{g}(\mathbf{x}_0) + \mathbf{g}'(\mathbf{x}_0)\mathbf{v}] \in \text{int } S$, so

$$-[\mathbf{g}(\mathbf{x}_0) + \mathbf{g}'(\mathbf{x}_0)\mathbf{v}] - o_3(\alpha)/\alpha \in \text{int } S$$

when α is small enough. Also $-(1-\alpha)\mathbf{g}(\mathbf{x}_0) \in S$ when $\alpha < 1$. Hence $\mathbf{x}_0 + \alpha\mathbf{v} + o_1(\alpha)$ satisfies also the constraint $-\mathbf{g}(\mathbf{x}) \in S$. Since $\mathbf{x}_0$ is a minimum point for (3.14), $f'(\mathbf{x}_0)\mathbf{v} \geq 0$, as in section 3.6.3. So there is no solution $\mathbf{v}$ to

$$\mathbf{h}(\mathbf{x}_0) + \mathbf{h}'(\mathbf{x}_0)\mathbf{v} \in -T, \quad \mathbf{g}(\mathbf{x}_0) + \mathbf{g}'(\mathbf{x}_0)\mathbf{v} \in -\text{int } S, \quad -f'(\mathbf{x}_0)\mathbf{v} > 0$$

Thus

$$\mathbf{h}(\mathbf{x}_0) + \mathbf{h}'(\mathbf{x}_0)\mathbf{v} \in -T \Rightarrow \begin{bmatrix} \mathbf{g}(\mathbf{x}_0) \\ \mathbf{0} \end{bmatrix} + \begin{bmatrix} \mathbf{g}'(\mathbf{x}_0) \\ \mathbf{f}'(\mathbf{x}_0) \end{bmatrix} \mathbf{v} \notin -\text{int} \begin{bmatrix} S \\ -\mathbf{R}_+ \end{bmatrix}$$

An application of extended Farkas (section 3.4.8) shows that there exist Lagrange multipliers $\tau \in \mathbf{R}_+$, $\boldsymbol{\lambda} \in S^*$, $\boldsymbol{\mu} \in T^*$, with τ and $\boldsymbol{\lambda}$ not both zero, satisfying the *Fritz John conditions*:

$$\begin{gathered} \tau f'(\mathbf{x}_0) + \boldsymbol{\lambda}\mathbf{g}'(\mathbf{x}_0) + \boldsymbol{\mu}\mathbf{h}'(\mathbf{x}_0) = 0, \quad \boldsymbol{\lambda}\mathbf{g}(\mathbf{x}_0) = 0, \quad \boldsymbol{\mu}\mathbf{h}(\mathbf{x}_0) = 0, \\ \tau \in \mathbf{R}_+, \quad \boldsymbol{\lambda} \in S^*, \quad (\tau, \boldsymbol{\lambda}) \neq (0, \mathbf{0}), \quad \boldsymbol{\mu} \in T^* \end{gathered} \tag{3.15}$$

together with the constraints of (3.14) at $\mathbf{x}_0$.

Example
Consider the problem

$$\text{Minimize } -x \quad \text{subject to} \quad (x_+)^2 \leq 0$$

At the minimum point $x = 0$, Fritz John conditions

$$0(-1) + \lambda(0) = 0, \quad \lambda \geq 0, \quad \lambda(0) = 0$$

hold, with any $\lambda \geq 0$. But Karush–Kuhn–Tucker conditions do not hold, since they would require

$$-1 + \lambda(0) = 0$$

to hold for some $\lambda \geq 0$. This constraint is not locally solvable at 0 (section 3.5.1).

3.5.5 Theorem (sufficient optimality conditions)

Let the point $\mathbf{x}_0$ satisfy the constraints of (3.14), and let the Karush–Kuhn–Tucker conditions hold there, with multipliers $\boldsymbol{\lambda}$ and $\boldsymbol{\mu}$. If f is convex, $\mathbf{g}$ is S-convex and $\mathbf{h}$ is T-convex, then $\mathbf{x}_0$ is a minimum point for (3.10).

Proof

Let $-\mathbf{g}(\mathbf{x}) \in S$ and $-\mathbf{h}(\mathbf{x}) \in T$. Since $\mathbf{g}$ is S-convex and $\boldsymbol{\lambda} \in S^*$, $\boldsymbol{\lambda}\mathbf{g}$ is convex. Since $\mathbf{h}$ is T-convex and $\boldsymbol{\mu} \in T^*$, $\boldsymbol{\mu}\mathbf{h}$ is convex. Hence the Lagrangian

$$L(\mathbf{x}) = f(\mathbf{x}) + \boldsymbol{\lambda}\mathbf{g}(\mathbf{x}) + \boldsymbol{\mu}\mathbf{h}(\mathbf{x})$$

is a convex function of $\mathbf{x}$. Also $\boldsymbol{\lambda}(-\mathbf{g}(\mathbf{x})) \geq 0$ and $\boldsymbol{\mu}(-\mathbf{h}(\mathbf{x})) \geq 0$. Hence

$$\begin{aligned} f(\mathbf{x}) - f(\mathbf{x}_0) &= L(\mathbf{x}) - L(\mathbf{x}_0) - \boldsymbol{\lambda}\mathbf{g}(\mathbf{x}) - \boldsymbol{\mu}\mathbf{h}(\mathbf{x}) + \boldsymbol{\lambda}\mathbf{g}(\mathbf{x}_0) + \boldsymbol{\mu}\mathbf{h}(\mathbf{x}_0) \\ &\geq L(\mathbf{x}) - L(\mathbf{x}_0) \quad \text{since } \boldsymbol{\lambda}\mathbf{g}(\mathbf{x}_0) = \boldsymbol{\mu}\mathbf{h}(\mathbf{x}_0) = 0 \\ &\geq L'(\mathbf{x}_0)(\mathbf{x} - \mathbf{x}_0) \quad \text{(by section 3.2)} \\ &= 0 \end{aligned}$$

□

3.5.6 Nonsmooth convex problem

Consider the problem:

$$\text{Minimize } f(\mathbf{x}) \quad \text{subject to} \quad -\mathbf{g}(\mathbf{x}) \in S, \quad \mathbf{x} \in \Gamma \tag{3.16}$$

where $\mathbf{x} \in X = \mathbf{R}^n$, Γ is a bounded closed convex set in X, S is a closed convex cone with interior, f is convex and $\mathbf{g}$ is S-convex. The functions f and $\mathbf{g}$ need not be differentiable at all points. Assume a minimum is reached at $\mathbf{x} = \mathbf{x}_0$.

From section 3.4.4, there are multipliers $\tau \in \mathbf{R}_+$ and $\mathbf{v} \in S^*$, not both zero, with

$$(\forall \mathbf{x} \in \Gamma)\ \tau(f(\mathbf{x}) f(\mathbf{x}_0)) + \mathbf{v}\mathbf{g}(\mathbf{x}) \geq 0; \quad \mathbf{v}\mathbf{g}(\mathbf{x}_0) = 0 \tag{3.17}$$

Assume a Slater constraint qualification in the form that $\zeta \in \Gamma$ exists with $-\mathbf{g}(\zeta) \in \text{int } S$. If $\tau = 0$ then $\mathbf{v} \neq \mathbf{0}$, so $\mathbf{v}\mathbf{g}(\zeta) < 0$, contradicting $\mathbf{v}\mathbf{g}(\zeta) \geq 0$ from (3.17). Assume then that $\tau > 0$, so take $\tau = 1$ without loss of generality.

Let $\varphi(z) := f(\mathbf{x}_0 + \mathbf{z}) - f(\mathbf{x}_0)) + \mathbf{v}\mathbf{g}(\mathbf{x}_0 + \mathbf{z})$; then $\varphi(\mathbf{0}) = 0$, φ is convex and φ is minimized at $\mathbf{0}$ over $\Gamma - \mathbf{x}_0$, and hence also over the convex cone $T := \{\alpha\mathbf{z} : \alpha \geq 0,$

$\mathbf{z} \in \Gamma - \mathbf{x}_0\}$. Suppose, if possible, that $\mathbf{0} \notin T^* \cap \partial\varphi(\mathbf{0})$. By the separation theorem, there is $\mathbf{w} \in K$ with $\mathbf{zw} < 0$ for each $\mathbf{z} \in \partial\varphi(\mathbf{0})$. But, from, the minimum, $0 \leq \varphi'(0 : \mathbf{w}) = \sup_{\mathbf{u} \in \partial\varphi(\mathbf{0})} \mathbf{uw}$; so $0 \geq \mathbf{uw}$ for some $\mathbf{u} \in \partial\varphi(\mathbf{0})$. The contradiction shows that $\mathbf{0} \in T^* \cap \partial\varphi(0)$; thus K^* and $\partial\varphi(0)$ have a common element. Hence

$$\mathbf{0} \in \partial(f + \mathbf{vg})(\mathbf{x}_0) - (\Gamma - \mathbf{x}_0)^*, \quad \mathbf{vg}(\mathbf{x}_0) = 0, \quad \mathbf{v} \in \mathbf{S}^* \tag{3.18}$$

The dual cone $(\Gamma - \mathbf{x}_0)^*$ may be denoted by $N(\Gamma, \mathbf{x}_0)$, the normal cone to Γ at $\mathbf{x}_0$.

Exercise

Show that a point $\mathbf{x}_0$ satisfying the constraints, with (3.18), is a minimum if f is convex, $\mathbf{g}$ is S-convex and Γ is convex. (Let $L(\mathbf{x}) = f(\mathbf{x}) + \boldsymbol{\lambda}\mathbf{g}(\mathbf{x})$; replace $L(\mathbf{x}) - L(\mathbf{x}_0) \geq L'(\mathbf{x}_0)(\mathbf{x} - \mathbf{x}_0)$ in section 3.5.5 by $L(\mathbf{x}) - L(\mathbf{x}_0) \geq (\boldsymbol{\alpha} + \boldsymbol{\beta})(\mathbf{x} - \mathbf{x}_0)$ whenever $\boldsymbol{\alpha} \in \partial f(\mathbf{x}_0)$ and $\boldsymbol{\beta} \in \partial(\boldsymbol{\lambda}\mathbf{g})(\mathbf{x}_0)$ for some $\boldsymbol{\alpha}, \boldsymbol{\beta}, \boldsymbol{\alpha} + \boldsymbol{\beta} \in N(\Gamma, \mathbf{x}_0)$ by (3.18).

3.5.7 KKT with additional constraint

Now assume instead that f and $\mathbf{g}$ are Fréchet differentiable, but not necessarily convex. Assume still that S is a convex cone with interior, and Γ is a closed convex set. Let $K := \{(t\mathbf{x}, t) \in X \times \mathbf{R} : \mathbf{x} \in \Gamma, t \in \mathbf{R}_+\}$. Since Γ is a convex set, K is a convex cone. The minimization problem (3.16) is equivalent to

$$\text{Minimize}_{\mathbf{x},t}\, f(\mathbf{x}) \quad \text{subject to} \quad -\mathbf{g}(\mathbf{x}) \in S, \quad (t\mathbf{x}, t) \in K, \quad t = 1$$

Define a Lagrangian for this problem as

$$L(\mathbf{x}, t) := \tau f(\mathbf{x}) + \boldsymbol{\lambda}\mathbf{g}(\mathbf{x}) - \boldsymbol{\rho} t\mathbf{x} - \theta t - v(t-1)$$

Subject to verification (below) that a certain cone is weak$*$closed, the necessary Fritz John conditions for a minimum at $(\mathbf{x}_0, \bar{t})$ with $\bar{t} = 1$ are

$$\bar{\tau} f'(\mathbf{x}_0) + \bar{\boldsymbol{\lambda}}\mathbf{g}'(\mathbf{x}_0) - \bar{\boldsymbol{\rho}}\bar{t} = \mathbf{0}; \quad -\bar{\boldsymbol{\rho}}\mathbf{x}_0 - \bar{\theta} - \bar{v} = 0$$

$$\bar{\tau} \geq 0, \quad \bar{\boldsymbol{\lambda}} \in S^*, \quad (\bar{\tau}, \bar{\boldsymbol{\lambda}}) \neq (0, \mathbf{0}), \quad (\bar{\boldsymbol{\rho}}, \bar{\theta}) \in K^*, \quad \bar{\boldsymbol{\lambda}}\mathbf{g}(\mathbf{x}_0) = 0, \quad \bar{\boldsymbol{\rho}}\bar{t}\mathbf{x}_0 + \bar{\theta}\bar{t} = 0$$

Then $\bar{\boldsymbol{\rho}}(\Gamma) + \bar{\theta} \subset \mathbf{R}_+$. From $\bar{\boldsymbol{\rho}}\mathbf{x}_0 + \bar{\theta} + \bar{v} = 0$ and $\bar{\boldsymbol{\rho}}(\mathbf{x}_0) + \bar{\theta} = 0$, $\bar{v} = 0$. From $\bar{\boldsymbol{\rho}}(\Gamma) + \bar{\theta} \subset \mathbf{R}_+$ and $\bar{\boldsymbol{\rho}}\mathbf{x}_0 + \bar{\theta} = 0$, $\bar{\boldsymbol{\rho}}(\Gamma - \mathbf{x}_0) \subset \mathbf{R}_+$, hence $\bar{\boldsymbol{\rho}} \in (\Gamma - \mathbf{x}_0)^*$. Hence

$$\bar{\tau} f'(\mathbf{x}_0) + \bar{\boldsymbol{\lambda}}\mathbf{g}'(\mathbf{x}_0) = \bar{\boldsymbol{\rho}} \in (\Gamma - \mathbf{x}_0)^*, \quad \bar{\tau} \geq 0, \quad \bar{\boldsymbol{\lambda}} \in S^*, \quad (\bar{\tau}, \bar{\boldsymbol{\lambda}}) \neq (0, \mathbf{0}), \quad \bar{\boldsymbol{\lambda}}\mathbf{g}(\mathbf{x}_0) = 0 \tag{3.19}$$

Denote $h(\mathbf{x}, t) := (t\mathbf{x}, t, t-1)$; write ∇ for gradient with respect to $(\mathbf{x}, t)$. The closed-cone condition mentioned requires that $[\nabla h|h](K^* \times \{0\}^*)$ is (weak$*$) closed. This requires that the cone

$$C := \{(\boldsymbol{\rho}, \boldsymbol{\rho}\mathbf{x}_0 + \theta + r, \boldsymbol{\rho}\mathbf{x}_0 + \theta) : (\rho, \theta) \in K^*, r \in \mathbf{R}\}$$

is closed. In finite dimensions, suppose there are sequences $\{\boldsymbol{\rho}_j\}$, $\{\theta_j\}$, $\{r_j\}$ such that $\{(\boldsymbol{\rho}_j, \boldsymbol{\rho}_j\mathbf{x}_0 + \theta_j + r_j, \boldsymbol{\rho}_j\mathbf{x}_0 + \theta_j)\}$ converges. Then $\{\boldsymbol{\rho}_j\} \to \tilde{\boldsymbol{\rho}} \in (\Gamma - \mathbf{x}_0)^*$ as in the previous paragraph. Hence $\{\boldsymbol{\rho}_j\mathbf{x}_0\}$ converges, hence $\{\theta_j\}$ converges, hence $\{r_j\}$ converges. Hence C is closed, as required. (In infinite dimensions, sequences must be replaced by *nets* (e.g. Dugundji, 1966).)

A more general treatment is given in section 3.9 of nonsmooth problems with nonconvex functions, satisfying Lipschitz conditions.

Remark

If f and $\mathbf{g}$ are also convex, then they are directionally differentiable (section 3.3). If $\mathbf{x}_0$ is a minimum point, then there is no solution $\mathbf{d}$ to

$$f'(\mathbf{x}_0; \mathbf{d}) < 0, \quad \mathbf{g}(\mathbf{x}_0) + \mathbf{g}'(\mathbf{x}_0; \mathbf{d}) \in -\operatorname{int} S, \quad \mathbf{d} \in \Gamma_{\mathbf{x}_0}$$

where $\Gamma_{\mathbf{x}_0} := \{\alpha(\mathbf{x} - \mathbf{x}_0) : \alpha \geq 0, \mathbf{x} \in \Gamma\}$. Then applying the basic alternative theorem (section 3.4.2) leads to a version of (3.19) using directional derivatives instead of Fréchet derivatives.

3.5.8 Nature of critical points

For the problem (3.10), with objective function $f(\mathbf{x})$ and constraint $-\mathbf{h}(\mathbf{x}) \in T = \mathbf{R}^m_+$, assume that the functions f and $\mathbf{h}$ are twice continuously differentiable, $\mathbf{x} \in \mathbf{R}^n$ with $m < n$, and that the point $\mathbf{x}_0$ satisfies the conditions

$$f'(\mathbf{x}_0) + \boldsymbol{\mu}\mathbf{h}'(\mathbf{x}_0) = 0, \quad \boldsymbol{\mu}\mathbf{h}(\mathbf{x}_0) = 0, \quad -\mathbf{h}(\mathbf{x}_0) \in \mathbf{R}^m_+ \tag{3.20}$$

Thus, the KKT necessary conditions (3.13) are assumed (and feasibility), *except* for the sign constraint $\mu \in \mathbf{R}^m_+$ on the Lagrange multiplier.

In order to discuss what kind of critical point $\mathbf{x}_0$ is, define the *Lagrangian* $L(\mathbf{x}) = f(\mathbf{x}) + \boldsymbol{\mu}\mathbf{h}(\mathbf{x})$, and also the *modified Lagrangian*

$$L^{\#}(\mathbf{x}, \mathbf{s}) = f(\mathbf{x}) + \boldsymbol{\mu}[\mathbf{h}(\mathbf{x}) + \mathbf{s}]$$

in which the constraint is expressed as $\mathbf{h}(\mathbf{x}) + \mathbf{s} = \mathbf{0}$, in terms of a *slack variable* $\mathbf{s} \in \mathbf{R}^m_+$. If $\mathbf{x}$ is a feasible point, and $\mathbf{h}(\mathbf{x}_0) + \mathbf{s}_0 = \mathbf{0}$, then

$$\begin{aligned} f(\mathbf{x}) - f(\mathbf{x}_0) &= L^{\#}(\mathbf{x}, \mathbf{s}) - L^{\#}(\mathbf{x}_0, \mathbf{s}_0) \\ &= L'(\mathbf{x}_0)(\mathbf{x} - \mathbf{x}_0) + \boldsymbol{\mu}(\mathbf{s} - \mathbf{s}_0) + \tfrac{1}{2}(\mathbf{x} - \mathbf{x}_0)^{\mathrm{T}} L''(\mathbf{x}_0)(\mathbf{x} - \mathbf{x}_0) + o(\|\mathbf{x} - \mathbf{x}_0\|^2) \\ &= 0 + \boldsymbol{\mu}(\mathbf{s} - \mathbf{s}_0) + \tfrac{1}{2}(\mathbf{x} - \mathbf{x}_0)^{\mathrm{T}} L''(\mathbf{x}_0)(\mathbf{x} - \mathbf{x}_0) + o(\|\mathbf{x} - \mathbf{x}_0\|^2) \end{aligned} \tag{3.21}$$

Denote $\mathbf{v} = \mathbf{x} - \mathbf{x}_0$.

Assume that the matrix $M = L''(\mathbf{x}_0)$ of the second partial derivatives of L at $\mathbf{x}_0$ is *nondegenerate*, i.e. M has no zero eigenvalues. A suitable rotation of coordinates expresses the quadratic form $\mathbf{v}^{\mathrm{T}} M \mathbf{v}$ in the form $\sum \lambda_i \xi_i^2$, in which the new coordinates ξ_i are linear combinations of the components of $\mathbf{v}$ and the λ_i are the eigenvalues of M. Since all λ_i are assumed nonzero, the higher-order terms, represented above by $o(\|\mathbf{x} - \mathbf{x}_0\|^2)$ are dominated by the second-order terms. This means that they are small compared with the second-order terms when $\|\mathbf{v}\|$ is sufficiently small. Hence they can be neglected in discussing the behaviour of $f(\mathbf{x})$ for points $\mathbf{x}$ close to $\mathbf{x}_0$.

If *inactive constraints* (for which $h_i(\mathbf{x}_0) < 0$) are omitted, because they do not affect the behaviour of the problem near $\mathbf{x}_0$, then $\mathbf{s}_0 = \mathbf{0}$. Assume also that the $m \times n$ matrix $\mathbf{h}'(\mathbf{x}_0)$ has full rank, and is thus of rank m, since $m < n$. Then the constraint is locally solvable at $\mathbf{x}_0$ (section 3.7), so that the linear approximation

$$\mathbf{h}(\mathbf{x}_0 + \mathbf{v}) \approx \mathbf{h}(\mathbf{x}_0) + \mathbf{h}'(\mathbf{x}_0)\mathbf{v} = \mathbf{h}'(\mathbf{x}_0)\mathbf{v}$$

is a good local approximation to the feasible set near $\mathbf{x}_0$. Moreover, the rank condition ensures that

$$\mathbf{h}'(\mathbf{x}_0)\mathbf{v} + \mathbf{s} = 0 \tag{3.22}$$

can be solved for some m components of $\mathbf{v}$ (making a vector $\mathbf{w}$, say) in terms of $\mathbf{s}$ and the remaining n–m components of $\mathbf{v}$. If this approximate solution is substituted into $\frac{1}{2}\mathbf{v}^{\mathrm{T}}M\mathbf{v}$, there results a quadratic form in $\mathbf{w}$, and the error due to approximating $\mathbf{h}(\mathbf{x}_0 + \mathbf{v})$ is $o(\|\mathbf{v}\|^2)$, and so is dominated by quadratic terms. Hence, for $\|\mathbf{v}\|$ small, the behaviour of $f(\mathbf{x})$ subject to the constraint is described by

$$f(\mathbf{x}_0 + \mathbf{v}) - f(\mathbf{x}_0) \approx \tfrac{1}{2}\mathbf{w}^{\mathrm{T}}K\mathbf{w} + \boldsymbol{\mu}\mathbf{s}$$

for some matrix K calculated as above.

Hence the point $\mathbf{x}_0$ can be described as follows:

- If $\boldsymbol{\mu} > \mathbf{0}$, and all eigenvalues of K are ≥ 0, then $\mathbf{x}_0$ is a *minimum*.
- If $\boldsymbol{\mu} < \mathbf{0}$, and all eigenvalues of K are ≤ 0, then $\mathbf{x}_0$ is a *maximum*.

In other cases, $\mathbf{x}_0$ is a saddle point, since $f(\mathbf{x}_0 + \mathbf{v})$ increases as $\mathbf{v}$ moves away from $\mathbf{0}$ in some directions allowed by the constraints, and $f(\mathbf{x}_0 + \mathbf{v})$ decreases as $\mathbf{v}$ moves away from $\mathbf{0}$ in some other directions allowed by the constraints.

These conditions are often called *second-order conditions*, since they involve second derivatives. (More details, for the case of several constraints, may be found in Craven (1981a, pp. 60–6) and Craven (1979). For second-order conditions formulated specifically for a minimum, see also Fiacco and McCormick (1968).)

However, if the *nondegenerate* hypothesis is not satisfied, then terms of higher order than second must be examined. The following example shows how to proceed. Consider the critical points of $f(x, y) = 2y^2 - x^2y$ subject to the constraint $\frac{1}{2}x^2 + y^2 \leq 1$. The Lagrangian is

$$L(x, y; \mu) = 2y^2 - x^2y + \mu(\tfrac{1}{2}x^2 + y^2 - 1)$$

and $L^{\#}$ adds an extra term μs. The conditions (3.21) require that

$$-2xy + \mu x = 0; \quad 4y - x^2 + 2\mu y = 0; \quad \mu(\tfrac{1}{2}x^2 + y^2 - 1) = 0$$

From the first equation, $x = 0$ or $\mu = 2y$. Substituting into the second equation, if $x = 0$ then $y = 0$ or $\mu = -2$. So $(x, y) = (0, 0)$ is one critical point, with the constraint inactive, so $\mu = 0$ there; and another has $\mu = -2$, $y = \frac{1}{2}\mu = -1$. If instead

Table 3.1 Critical points of the function $f(x, y) = 2y^2 - x^2y$

$\bar{x}$	$\bar{y}$	$\bar{\mu}$	$\bar{s}$	$f(\bar{x}, \bar{y})$	
$\pm 4/3$	$1/3$	$2/3$	0	$-10/27$	Min
0	-1	-2	0	2	Max
0	1	-2	0	2	Max
0	0	0	1	0	Saddle

$\mu = 2y$ then, substituting into the second equation, $4y - x^2 + 2(2y)y = 0$. Substituting for x^2 into the third equation (supposing $\mu \neq 0$) gives $3y^2 + 2y - 1 = 0$, whence $y = 1/3, -1$, so that (from the third equation) $x = \pm 4/3, 1$. So a list of all the critical points, $(\bar{x}, \bar{y})$ say, with multiplier $\bar{\mu}$, is as shown in Table 3.1. The tabulated descriptions of the kinds of critical points will result from the following calculations.

The matrix of second derivatives of L and the gradient of the constraint are

$$L''(\bar{x}, \bar{y}) = \begin{bmatrix} -2\bar{y} + \bar{\mu} & -2\bar{x} \\ -2\bar{x} & 4 + 2\bar{\mu} \end{bmatrix}; \; h'(\bar{x}, \bar{y}) = [\bar{x}, 2\bar{y}]$$

At $(\bar{x}, \bar{y}) = (4/3, 1/3)$,

$$L''(\bar{x}, \bar{y}) = \begin{bmatrix} 0 & -8/3 \\ -8/3 & 16/3 \end{bmatrix}$$

is nondegenerate, and $h'(\bar{x}, \bar{y}) = [4/3 \; 2/3]$ has full rank; (3.22) gives $4/3 v_1 + 2/3 v_2 + s = 0$, hence $v_1 = -1/2 v_2 - 3/4 s$; substituting this into (3.21) gives (to sufficient approximation)

$$f(4/3 + v_1, 1/3 + v_2) - f(4/3, 1/3) \approx 4v_2^2 + (2/3 + 2v_2)s \geq 0$$

so the point is a (local) minimum. (Note that the slack variable s must *not* be eliminated, since then the requirement $s \geq 0$ would be lost.)

At $(\bar{x}, \bar{y}) = (0, -1)$, $L''(\bar{x}, \bar{y})$ is the zero matrix, and so is degenerate. Here it is required to set $x = 0 + v_1$ and $y = -1 + v_2$ and expand, including higher-order terms. This gives (with $\mu = -2$) that

$$\begin{aligned} f(0 + v_1, -1 + v_2) - f(0, -1) &= L^{\#}(0 + v_1, -1 + v_2; s) - L^{\#}(0, 1 -1; 0) \\ &= -2(-1 + v_2)^2 - v_1^2(-1 + v_2) - 2(1/2 v_1^2 \\ &\quad + (-1 + v_2)^2 + s - 1) - (2 + 2) \\ &= -v_1^2 v_2 - 2s \end{aligned} \tag{3.23}$$

subject to the constraint (3.22), which gives $1/2 v_1^2 + (-1 + v_2)^2 + s = 1$, so that $2v_2 = s + 1/2 v_1^2 + v_2^2$. Substituting this into (3.23) gives

$$f(0 + v_1, -1 + v_2) - f(0, -1) = -2s(1 - 1/2 v_1^2) - 1/2 v_1^2(1/2 v_1^2 + v_2^2) \leq 0$$

noting that $s \geq 0$, so the point $(0, -1)$ is a (local) maximum. So, similarly, is $(0, 1)$. The point $(0, 0)$ is an (unconstrained) saddle point.

Exercise
Discuss the nature of the critical points for the first exercise in section 3.5.3.

Exercise
Discuss all the critical points of the function

$$f(x, y) = \tfrac{1}{3}x^3 - y^2$$

subject to the constraint

$$\tfrac{1}{2}x^2 + y^2 \leq 1$$

3.6 INVEX FUNCTIONS

Some consequences of convexity extend to wider classes of functions. The *invex* property is useful for dual problems and elsewhere. It is motivated as follows (Hanson, 1980; Craven, 1981).

Let $\mathbf{F} : \mathbf{R}^n \to \mathbf{R}^p$ be a differentiable K-convex function, where K is a convex cone in $\mathbf{R}^p$; let $\mathbf{a} \in \mathbf{R}^n$. Let $\varphi : \mathbf{R}^p \to \mathbf{R}^p$ be a differentiable function, with $\varphi(\mathbf{a}) = \mathbf{a}$; assume that the gradient $\varphi'(\mathbf{a})$ is invertible. Let $\mathbf{\Theta} = \mathbf{F} \circ \varphi$. Let $\mathbf{x}, \mathbf{a} \in \mathbf{R}^n$; let $\mathbf{u} = \mathbf{x} - \mathbf{a}$. Then

$$\begin{aligned}
\mathbf{\Theta}(\mathbf{x}) - \mathbf{\Theta}(\mathbf{a}) &= \mathbf{F}(\varphi(\mathbf{x})) - \mathbf{F}(\varphi(\mathbf{a})) \\
&\in \mathbf{F}'(\varphi(\mathbf{a}))\,[\varphi(\mathbf{x}) - \varphi(\mathbf{a})] + K \qquad \text{(since } \mathbf{F} \text{ is } K\text{-convex)} \\
&= \mathbf{\Theta}'(\mathbf{a}) \circ \varphi'(\mathbf{a})^{-1}\,[\varphi(\mathbf{x}) - \varphi(\mathbf{a})] + K \quad \text{(by the chain rule)} \\
&= \mathbf{\Theta}'(\mathbf{a})\,\kappa(\mathbf{x} - \mathbf{a}, \mathbf{a}) + K \qquad (3.24)
\end{aligned}$$

if $\kappa(\mathbf{u}, \mathbf{a})$ is defined as $\varphi'(\mathbf{a})^{-1}\,[\varphi(\mathbf{a} + \mathbf{u}) - \varphi(\mathbf{a})]$. Observe that (3.24) shows what has become of the convexity of $\mathbf{F}$ after its domain space has been distorted by the function φ. The property thus remaining is called invex (for *invariant convex* – what remains of convex.)

3.6.1 Definition

A differentiable function $\mathbf{\Theta} : \mathbf{R}^n \to \mathbf{R}^p$ will be called *K-invex* at the point $\mathbf{a}$ (where K is a convex cone in $\mathbf{R}^p$) if, for some function $\kappa(.,.) : \mathbf{R}^n \times \mathbf{R}^n \to \mathbf{R}^n$ (called a *scale function*) and all $\mathbf{x}$,

$$\mathbf{\Theta}(\mathbf{x}) - \mathbf{\Theta}(\mathbf{a}) \in \mathbf{\Theta}'(\mathbf{a})\,\kappa(\mathbf{x} - \mathbf{a}, \mathbf{a}) + \mathbf{K}$$

In particular, if $\mathbf{\Theta}$ is a real function, then it is invex ($\equiv \mathbf{R}_+$-invex) at $\mathbf{a}$ if

$$\mathbf{\Theta}(\mathbf{x}) - \mathbf{\Theta}(\mathbf{a}) \geq \mathbf{\Theta}'(\mathbf{a})\,\kappa(\mathbf{x} - \mathbf{a}, \mathbf{a})$$

for some function κ.

3.6.2 Some invex properties

Assume the functions are differentiable. Then the following properties hold.

(a) (*Relation to convex*) *K-convex* implies *K-invex.*
(b) (*Pre-invex*) If $\mathbf{f}$ is *K-pre-invex* at $\mathbf{a}$, thus if

$$\mathbf{f}(\mathbf{a} + \alpha\boldsymbol{\kappa}(\mathbf{x} - \mathbf{a}, \mathbf{a})) - (1 - \alpha)f(\mathbf{a}) + \alpha f(\mathbf{x}) \in -K$$

for some function $\boldsymbol{\kappa}$ and $0, \alpha < 1$, then $\mathbf{f}$ is *K-invex.*

Exercise

Verify this from definition of K-invex.

(c) (*Global minimum*) If $f: X \to \mathbf{R}$ is invex at $\mathbf{a}$, and $\mathbf{a}$ is a stationary point (thus, $f'(\mathbf{a}) = \mathbf{0}$), then $\mathbf{a}$ is a *global minimum* of f (thus, $f(\mathbf{x}) - f(\mathbf{a}) \geq 0$ for all $\mathbf{x}$). (Compare section 3.2.3.)
(d) (*Sufficient Karush–Kuhn–Tucker conditions*) Let $\mathbf{F} = (\mathbf{f}, \mathbf{g})$ be $(Q \times S)$-invex at $\mathbf{a}$, where $\mathbf{f}: X \to Y$, $\mathbf{g}: X \to Z$ are functions, and $Q \subset Y$ and $S \subset Z$ are convex cones, with int $Q \neq \emptyset$. Assume that $\mathbf{f}$ is $(Q \times S)$-invex at the point $\mathbf{a}$, and that

$$\tau\mathbf{f}'(\mathbf{a}) + \lambda\mathbf{g}'(\mathbf{a}) = \mathbf{0}, \quad \mathbf{0} \neq \tau \in Q^*, \lambda \in S^*, \quad \lambda\mathbf{g}(\mathbf{a}) = 0, \quad -\mathbf{g}(\mathbf{a}) \in S \quad (3.25)$$

Define the Lagrangian $L(\mathbf{x}) = \tau\mathbf{f}(\mathbf{x}) - \lambda\mathbf{g}(\mathbf{x})$. Then $L(.)$ is invex at $\mathbf{a}$, and then, for all $\mathbf{x} \in X$,

$$\tau[\mathbf{f}(\mathbf{x}) - \mathbf{f}(\mathbf{a})] = \tau\mathbf{f}(\mathbf{x}) - \tau\mathbf{f}(\mathbf{a}) \geq L(\mathbf{x}) - L\mathbf{a}) \geq L'(\mathbf{a})\boldsymbol{\kappa}(\mathbf{x} - \mathbf{a}, \mathbf{a}) = 0$$

If $\mathbf{f}(\mathbf{x}) - \mathbf{f}(\mathbf{a}) \in -\text{int } Q$, then this, with $\mathbf{0} \neq \tau \in Q^*$, contradicts section 3.4.1. Hence $\mathbf{f}(\mathbf{x}) - \mathbf{f}(\mathbf{a}) \notin -\text{int } Q$. This means that the Karush–Kuhn–Tucker conditions (3.25), with the invex assumption, imply that the point $\mathbf{a}$ is a weak minimum (section 3.2.3) of the vector function $\mathbf{f}(\mathbf{x})$, subject to the constraint $-\mathbf{g}(\mathbf{x}) \in S$.

In brief, Karush–Kuhn–Tucker *necessary* conditions also become *sufficient* conditions if an invex hypothesis is made. If invex holds globally (thus, for all $\mathbf{x}$), then the sufficient conditions lead to a global minimum.
(e) (*When is a vector function invex?*) Assume now twice-differentiable functions. Let $\mathbf{v} = \mathbf{x} - \mathbf{a}$, and assume expansions, up to second-order terms, for $\mathbf{f}: \mathbf{R}^n \to \mathbf{R}^r$ and $\boldsymbol{\kappa}: \mathbf{R}^n \to \mathbf{R}^n$:

$$\mathbf{f}(\mathbf{x}) = \mathbf{f}(\mathbf{a}) + \mathbf{f}'(\mathbf{a})\mathbf{v} + (1/2)\mathbf{v}^{\mathrm{T}}\mathbf{f}''(\mathbf{a})_{.}\mathbf{v} + o(\|\mathbf{v}\|^2)$$

where the notation $\mathbf{v}^{\mathrm{T}}\mathbf{f}''(\mathbf{a})_{.}\mathbf{v}$ means that component $\mathbf{f}_i$ of $\mathbf{f}$ has second-order terms $(1/2)\mathbf{v}^{\mathrm{T}}\mathbf{f}_i''(\mathbf{a})\mathbf{v}$ and a similar notation will be used for the other functions:

$$\boldsymbol{\kappa}(\mathbf{x}) = \boldsymbol{\kappa}'(\mathbf{a})\mathbf{v} + (1/2)\mathbf{v}^{\mathrm{T}}\boldsymbol{\kappa}''(\mathbf{a})_{.}\mathbf{v} + o(\|\mathbf{v}\|^2)$$

Then, substituting,

$$\mathbf{f}(\mathbf{x}) - \mathbf{f}(\mathbf{a}) - \mathbf{f}'(\mathbf{a})\boldsymbol{\kappa}(\mathbf{v}) = [\mathbf{f}'(\mathbf{a}) - \mathbf{f}'(\mathbf{a})\boldsymbol{\kappa}'(\mathbf{a})]\mathbf{v} + (1/2)\mathbf{v}^{\mathrm{T}}\,\mathbf{f}''(\mathbf{a})_{.} - \mathbf{f}'(\mathbf{a})\boldsymbol{\kappa}''(\mathbf{a})_{.}]\mathbf{v} + o(\|\mathbf{v}\|^2) \quad (3.26)$$

in which matrix k of $\mathbf{f}'(\mathbf{a})\boldsymbol{\kappa}''(\mathbf{a})_.$ has elements

$$\sum_l [\mathbf{f}_i'(\mathbf{a})]_l [\boldsymbol{\kappa}_k''(\mathbf{a})]_{ij}$$

Since any element in the nullspace of $\mathbf{f}'(\mathbf{a})$ can be added to $\boldsymbol{\kappa}(\mathbf{v})$ without altering the invex property, it can be assumed that $\boldsymbol{\kappa}'(\mathbf{a})$ is the identity; so the terms in (3.26) that are linear in $\mathbf{v}$ can be dropped.

In relation to the cones $\mathbf{R}_+^r$ and $\mathbf{R}_+^n$, $\mathbf{f}$ is invex when

$$\mathbf{f}(\mathbf{x}) - \mathbf{f}(\mathbf{a}) - \mathbf{f}'(\mathbf{a})\boldsymbol{\kappa}(\mathbf{v}) \in \mathbf{R}_+^n$$

This happens if each of the n matrices of $\mathbf{f}''(\mathbf{a})_. - \mathbf{f}'(\mathbf{a})\boldsymbol{\kappa}''(\mathbf{a})_.$ is *positive definite* for suitable choice of the matrices $\boldsymbol{\kappa}_k''(\mathbf{a})$; and then the invex property holds also at other points near enough to $\mathbf{a}$. Conversely, invex implies that all the n matrices are *positive semidefinite*.

Exercise

Discuss whether the vector function $F(\mathbf{x})/h(\mathbf{x})$ is K-invex in a ball around $\mathbf{a} \in \mathbf{R}^n$, given F differentiable K-convex and h differentiable concave, and $h(.) > 0$. (*Hint*: $F(\mathbf{a}+\mathbf{z}) \geq F(\mathbf{a}) + F'(\mathbf{a})\mathbf{z}$ and $h(\mathbf{a}+\mathbf{z}) \leq h(\mathbf{a}) + h'(\mathbf{a})\mathbf{z}$; with $h(.) > 0$ this gives $1/h(\mathbf{a}+\mathbf{z}) \geq \ldots$.) Also discuss when $G(\mathbf{x}) = C\mathbf{x}/\|\mathbf{x}\|$ is cone-invex at a point $\mathbf{a} \neq \mathbf{0}$, when C is a (suitable) matrix.

Exercise

Show that Karush–Kuhn–Tucker conditions are sufficient for a minimum at $\mathbf{a}$ of $f(\mathbf{x})$ subject to $-\mathbf{g}(\mathbf{x}) \in S$, $\mathbf{x} \in \Gamma$ if $(f, \mathbf{g})$ satisfies an invex property also with $\boldsymbol{\kappa}(\mathbf{x}-\mathbf{a}, \mathbf{a}) \in \Gamma - \mathbf{a}$.

3.7 CONDITIONS FOR LOCAL SOLVABILITY

Consider the constraint $-\mathbf{g}(\mathbf{x}) \in S$, where X is a Banach space, Y is a normed space, $S \subset Y$ is a closed convex cone and $\mathbf{g}: X \to Y$ is a continuously differentiable function. Let $-\mathbf{g}(\mathbf{x}_0) \in S$. *Robinson's stability condition* is the assumption that

$$0 \in \operatorname{int}[\mathbf{g}(\mathbf{x}_0) + \mathbf{g}'(\mathbf{x}_0)(X) + S]$$

Local solvability theorem (Robinson, 1976; Craven, 1978): if Robinson's stability condition holds, then the constraint $-\mathbf{g}(\mathbf{x}) \in S$ is locally solvable at $\mathbf{x}_0$. The proof is omitted because of its length.

A nonsmooth extension of Robinson's result for functions that need not be differentiable at all points but satisfy a Lipschitz condition is stated in section 7.5.2.

The following special cases are worth attention. If $\mathbf{g}'(\mathbf{x}_0)$ maps X *onto* Y, then the Robinson condition holds automatically. In particular, this is the case when $Y = \mathbf{R}^m$ and the gradients $\mathbf{g}_i'(\mathbf{x}_0)$ of the m components of $\mathbf{g}$ are *linearly independent*.

Consider now a continuously differentiable function $\mathbf{h} : \mathbf{R}^n \to \mathbf{R}^m$, and let $-\mathbf{h}(\mathbf{x}_0) \in \mathbf{R}^m_+$. Then $\mathbf{h}$ can be partitioned into two vector components $\mathbf{p}$ and $\mathbf{q}$ such that $-\mathbf{p}(\mathbf{x}_0) > 0$ and $\mathbf{q}(\mathbf{x}_0) = \mathbf{0}$. (Thus $\mathbf{q}(.)$ describes the *active constraints.*) Let the direction $\mathbf{d}$ satisfy

$$-\mathbf{h}(\mathbf{x}_0) - \mathbf{h}'(\mathbf{x}_0)\mathbf{d} \in \mathbf{R}^m_+$$

This can be written as

$$-\mathbf{p}(\mathbf{x}_0) - \mathbf{p}'(\mathbf{x}_0)\mathbf{d} \geq 0 \tag{3.27}$$

and

$$-\mathbf{q}'(\mathbf{x}_0)\mathbf{d} \geq 0 \tag{3.28}$$

Assume that $-\mathbf{q}'(\mathbf{x}_0)$ is an *onto* map; equivalently, assume that *the gradients of the active constraints are linearly independent*. Then there is a local solution

$$\mathbf{x} = \mathbf{x}_0 + \alpha\mathbf{d} + o(\alpha)$$

to $-\mathbf{q}'(\mathbf{x}) \geq \mathbf{0}$. Since $-\mathbf{p}(\mathbf{x}_0) > \mathbf{0}$ and $\mathbf{p}$ is continuous,

$$-\mathbf{p}(\mathbf{x}_0 + \alpha\mathbf{d} + o(\alpha)) \geq 0$$

whenever α is sufficiently small positive. Thus, for such α,

$$-\mathbf{h}(\mathbf{x}_0 + \alpha\mathbf{d} + o(\alpha)) \geq \mathbf{0}$$

establishing local solvability.

3.8 DUALITY AND QUASIDUALITY

The *duality theorem* for linear programming, which relates a given (*primal*) linear program (P) to a *dual linear program* (D):

$$\text{(PL)}\quad \text{Minimize } \mathbf{c}^{\mathrm{T}}\mathbf{x} \quad \text{subject to} \quad A\mathbf{x} = \mathbf{b},\ \mathbf{x} \geq \mathbf{0}$$

$$\text{(DL)}\quad \text{Maximize } \mathbf{b}^{\mathrm{T}}\mathbf{w} \quad \text{subject to} \quad A^{\mathrm{T}}\mathbf{w} \leq \mathbf{c}$$

has various extensions to nonlinear problems.

3.8.1 Duality

A problem (D): Maximize $\varphi(\mathbf{z})$ subject to $\mathbf{z} \in N$ is a *strong dual* to the problem (P): Minimize $f(\mathbf{v})$ subject to $\mathbf{v} \in Q$, if there hold:

(a) *Weak duality*: $(\forall \mathbf{v} \in Q, \forall \mathbf{z} \in N)\quad f(\mathbf{v}) \geq \varphi(\mathbf{z})$; and
(b) *Zero duality gap* (*ZDG*): if (P) reaches a minimum at $\mathbf{v}_0 \in Q$, then (D) reaches a maximum for some $\mathbf{z}_0 \in N$ and $\varphi(\mathbf{z}_0) = f(\mathbf{v}_0)$.

The properties (a) and (b) are well known for (PL) and (DL).

Consider a differentiable problem (3.29), and the problem (3.30) obtained from it:

$$\text{Minimize } f(\mathbf{x}) \quad \text{subject to} \quad -\mathbf{h}(\mathbf{x}) \in T \tag{3.29}$$

$$\text{Maximize } f(\mathbf{u}) + \mathbf{vh}(\mathbf{u}) \quad \text{subject to} \quad \mathbf{v} \in T^*, f'(\mathbf{u}) + \mathbf{vh}'(\mathbf{u}) = 0 \tag{3.30}$$

3.8.2 Theorem (Wolfe duality)

Let (3.29) reach a minimum at $\mathbf{x} = \mathbf{x}_0$; let KKT conditions hold there with multiplier $\boldsymbol{\lambda}$; let f be convex; let $\mathbf{h}$ be T-convex; then (3.30) is a strong dual to (3.29). This holds also if

$$\begin{bmatrix} f \\ \mathbf{h} \end{bmatrix} \text{ is } (\mathbf{R}_+ \times T)\text{-invex at each point}$$

Proof

Let $\mathbf{x}$ be feasible for (3.29) and $(u, \mathbf{v})$ feasible for (3.30). Then, for some function $\omega(\mathbf{x} - \mathbf{u})$ (with $\omega(\mathbf{x} - \mathbf{u}) = \mathbf{x} - \mathbf{u}$ in the convex case), using (3.13) for (P),

$$f(\mathbf{x}) - [f(\mathbf{u}) + \mathbf{v}\mathbf{h}(\mathbf{u})] \geq f'(\mathbf{u})\,\omega(\mathbf{x} - \mathbf{u}) - \mathbf{v}\mathbf{h}(\mathbf{u}) = -\mathbf{v}\mathbf{h}'(\mathbf{u})\omega(\mathbf{x} - \mathbf{u}) - \mathbf{v}\mathbf{h}(\mathbf{u})$$

$$\geq -\mathbf{v}\mathbf{h}(\mathbf{x}) \geq 0 \quad (\text{since } \mathbf{h}(\mathbf{x}) \in -T \text{ and } \mathbf{v} \in T^*)$$

proving (a). And (b) follows since $(\mathbf{u}, \mathbf{v}) = (\mathbf{x}_0, \boldsymbol{\lambda})$ is feasible for (3.30), and the objective functions are equal since $\boldsymbol{\lambda}\mathbf{h}(\mathbf{x}_0) = 0$, from (3.13). □

The problem (3.30) is called the *Wolfe dual* of problem (3.29).

3.8.3 Quasimin

Problem (3.29) has a *quasimin* at a feasible point $\mathbf{x}_0$ if, for some function $\theta(\mathbf{x}) = o(\|\mathbf{x} - \mathbf{x}_0\|)$, $f(\mathbf{x}) - f(\mathbf{x}_0) - \theta(\mathbf{x}) \geq 0$ as $\mathbf{x} \to \mathbf{x}_0$ through feasible points. Clearly, a (local) minimum is a quasimin. (A *quasimax* of f is a quasimin of $-f$.) The KKT necessary conditions (3.13) (section 3.5.3) for a minimum are also necessary for a quasimin; an additional $o(\alpha)$ term is added to the proof, not affecting the result. Conversely, for a differentiable problem, (3.13) implies a quasimin. Let $-\mathbf{h}(\mathbf{x}) \in T$; the Lagrangian is

$$L(\mathbf{x}) = f(\mathbf{x}) + \boldsymbol{\lambda} h(\mathbf{x})$$

where $\boldsymbol{\lambda}$ is the multiplier from (3.13). Since $\boldsymbol{\lambda} \in T^*$ and $-\mathbf{h}(\mathbf{x}) \in T$, $-\boldsymbol{\lambda}\mathbf{h}(\mathbf{x}) \geq 0$; also $\boldsymbol{\lambda}\mathbf{h}(\mathbf{x}_0) = 0$ from (3.13). So

$$f(\mathbf{x}) - f(\mathbf{x}_0) \geq L(\mathbf{x}) - L(\mathbf{x}_0) = L'(\mathbf{x}_0)(\mathbf{x} - \mathbf{x}_0) + o(\|\mathbf{x} - \mathbf{x}_0\|) = 0 + o(\|\mathbf{x} - \mathbf{x}_0\|)$$

proving the quasimin. Note that no convex hypotheses have been used.

For an unconstrained problem, the local minimum points are a subset of the *stationary points* (where the gradient is zero). For a differentiable constrained problem, the analogue of stationary point is quasimin (and zero gradient is replaced by (3.13)); the local minimum points are a subset of the quasimins.

Consider now problem (3.29) with minimum replaced by quasimin, and (3.30) with maximum replaced by quasimax; (3.30) is then called the quasidual of (3.29) (see Craven (1977)) because of Theorem 3.8.4 below. To illustrate it, $f(x) = x - \frac{1}{2}x^2$, subject to $x \geq 0$ in $\mathbf{R}$, has a quasimin at 0 (a constrained minimum with value 0), and a quasimin at 1 (a maximum at an interior point of the feasible set with value $\frac{1}{1}$) (Figure 3.11). The quasidual has objective $u - \frac{1}{2}u^2 - vu$, which reduces to $\frac{1}{2}u^2$ using the constraints $1 - u - v = 0$, $v \geq 0$ (so $u \leq 1$).

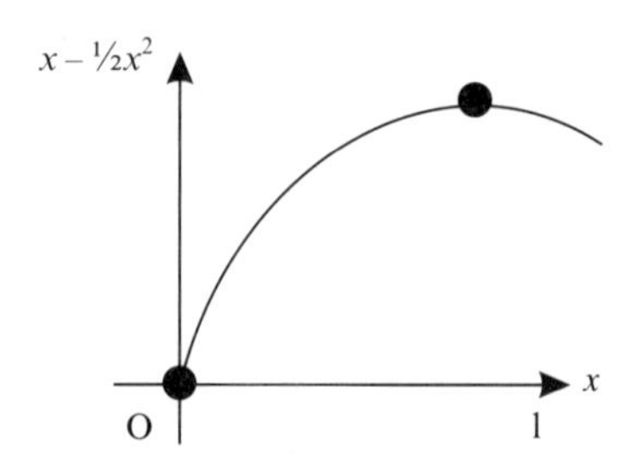

Figure 3.11 Quasimin and quasimax.

Now $\frac{1}{2}u^2$, subject to $u \leq 1$, has a quasimax at 0 (a minimum at an interior point with value 0), and a quasimax at 1 (a constrained maximum with value ½). Each quasimin of (3.29) has a corresponding quasimax of the quasidual, with the same objective value; however, weak duality is lost.

Quasimin and quasidual have application to stability (section 4.7), and to the Pontryagin theory of optimal control (section 7.2.6), where a critical point of proof depends on when a quasimin is actually a minimum.

3.8.4 Theorem (quasiduality)

If (3.29) has a quasimin at $\mathbf{x}_0$ and (3.13) holds there with multiplier $\boldsymbol{\lambda}$, then (3.30) has a quasimax at $(\mathbf{x}_0, \boldsymbol{\lambda})$.

Proof

Let $(\mathbf{u}, \mathbf{v})$ satisfy the constraints of (3.30). Let $L(\mathbf{u}) = f(\mathbf{u}) + \boldsymbol{\lambda}\mathbf{h}(\mathbf{u})$. Then

$$\begin{aligned}[f(\mathbf{u}) + \mathbf{v}\mathbf{h}(\mathbf{u})] - [f(\mathbf{x}_0) + \boldsymbol{\lambda}\mathbf{h}(\mathbf{x}_0)] &= L(\mathbf{u}) - L(\mathbf{x}_0) + (\mathbf{v} - \boldsymbol{\lambda})\mathbf{h}(\mathbf{u}) \\ &= L'(\mathbf{x}_0)\,(\mathbf{u} - \mathbf{x}_0) + o(\|\mathbf{u} - \mathbf{x}_0\|) \\ &\quad + (\mathbf{v} - \boldsymbol{\lambda})\mathbf{h}(\mathbf{u}_0) + (\mathbf{v} - \boldsymbol{\lambda})[\mathbf{h}(\mathbf{u}) - \mathbf{h}(\mathbf{u}_0)] \\ &\leq 0 + o(\|\mathbf{u} - \mathbf{x}_0\|) + (\mathbf{v} - \boldsymbol{\lambda})[\mathbf{h}(\mathbf{u}) - \mathbf{h}(\mathbf{u}_0)] \\ &= o(\|\mathbf{u} - \mathbf{x}_0\| + \|\mathbf{u} - \mathbf{u}_0\|)\end{aligned}$$

using (3.13) for (3.29) and $\mathbf{v}\mathbf{h}(\mathbf{u}_0) \leq 0$. So the quasimax is proved. □

Exercise

Show that a quasimin of a convex problem is a global minimum. (*Hint*: Modify the proof that a local minimum of a convex function over a convex set is a global minimum.)

Exercise

For a convex quadratic program:

$$\text{Minimize } \tfrac{1}{2}\mathbf{x}^{\mathrm{T}}Q\mathbf{x} + \mathbf{c}^{\mathrm{T}}\mathbf{x} \quad \text{subject to} \quad A\mathbf{x} \leq \mathbf{b}$$

write down the Wolfe dual. Subject to the dual constraints, show that an equivalent dual objective is $-\frac{1}{2}\mathbf{u}^T Q\mathbf{u} - \mathbf{v}^T\mathbf{b}$. What can be said if Q is *not* convex (thus, if Q is not positive semidefinite)? When does invex hold?

3.9 NONSMOOTH OPTIMIZATION

Consider the problem:

$$\text{Minimize } f(\mathbf{x}) \quad \text{subject to} \quad -\mathbf{g}(\mathbf{x}) \in S \tag{3.31}$$

where $\mathbf{x} \in \mathbf{R}^n$, $\mathbf{g}(\mathbf{x}) \in \mathbf{R}^m$, S is a closed convex cone in $\mathbf{R}^m$ and the functions f and $\mathbf{g}$ are *not* assumed differentiable at all points. Instead, *Lipschitz conditions* are assumed, namely

$$(\forall \mathbf{x}, \mathbf{y} \in E)\, |f(\mathbf{x}) - f(\mathbf{y}) \leq \kappa_1 \|\mathbf{x} - \mathbf{y}\|, \quad \|\mathbf{g}(\mathbf{x}) - \mathbf{g}(\mathbf{y})\| \leq \kappa_2 \|\mathbf{x} - \mathbf{y}\|$$

Here E is a bounded set containing an open region around a minimum point of (3.31), say at $\bar{\mathbf{x}}$, and the *Lipschitz constants* κ_1 and κ_2 may depend on E but do not depend on $\mathbf{x}$ and $\mathbf{y}$. A famous theorem of Rademacher states that a Lipschitz function on $\mathbf{R}^n$ is differentiable except on a set of zero measure. Consequently, if f (or $\mathbf{g}$) is not differentiable at $\bar{\mathbf{x}}$, there are very many points near $\bar{\mathbf{x}}$, where f (and $\mathbf{g}$) are differentiable.

3.9.1 Smoothed problem

Define *smoothed functions* $f(.:\varepsilon)$ and $\mathbf{g}(.:\varepsilon)$ by the following integrals:

$$f(\mathbf{x}:\varepsilon) := \int f(\mathbf{x} - \mathbf{s})\varepsilon^{-1}\varphi(\varepsilon^{-1}\mathbf{s})\, \mathrm{d}\mathbf{s}$$

$$\mathbf{g}(\mathbf{x}:\varepsilon) := \int \mathbf{g}(\mathbf{x} - \mathbf{s})\varepsilon^{-1}\varphi(\varepsilon^{-1}\mathbf{s})\, \mathrm{d}\mathbf{s}$$

where the integration is over $\mathbf{R}^n$, and the *mollifier function* $\varphi(.)$ has the properties

$$(\forall \mathbf{x})\ \varphi(\mathbf{x}) \geq 0; \quad \int_{E_0} \varphi(\mathbf{x})\, \mathrm{d}\mathbf{x} = 1; \quad (\forall \mathbf{x} \notin E_0)\ \varphi(\mathbf{x}) = 0$$

where E denotes the cube $\{\mathbf{x} \in \mathbf{R}^n : (|x_i|) \leq 1\ (i = 1, 2, \ldots, n)\}$. In particular, $\varphi(.)$ could be the function $2^{-n}\chi_{E_0}(.)$, where the indicator function $\chi_{E_0}(\mathbf{x}) = 1$ for $\mathbf{x} \in E_0$, and 0 for $\mathbf{x} \notin E_0$. Note that

$$\int_{\varepsilon E_0} \varepsilon^{-1}\varphi(\varepsilon^{-1}\mathbf{s})\, \mathrm{d}\mathbf{s} = \int_{E_0} \varphi(\mathbf{s})\, \mathrm{d}\mathbf{s} = 1$$

Then $f(. : \varepsilon)$ and $\mathbf{g}(. : \varepsilon)$ are differentiable, with derivatives

$$f'(\mathbf{x}:\varepsilon) := \int f'(\mathbf{x} - \mathbf{s})\varepsilon^{-1}\varphi(\varepsilon^{-1}\mathbf{s})\, \mathrm{d}\mathbf{s}$$

$$\mathbf{g}'(\mathbf{x}:\varepsilon) := \int \mathbf{g}'(\mathbf{x} - \mathbf{s})\varepsilon^{-1}\varphi(\varepsilon^{-1}\mathbf{s})\, \mathrm{d}\mathbf{s}$$

obtained by differentiating under the integral sign, noting that, by a property of the Lebesgue integral, the value of the integral is not affected by the values of the integrand on the set of zero measure on which the derivatives $f'(\mathbf{x}-\mathbf{s})$ and $\mathbf{g}'(\mathbf{x}-\mathbf{s})$ may not exist.

Denote by $\partial f(\mathbf{x})$ the closed convex hull of limit points of the sequence $\{f'(\mathbf{x}_j)\}$ for sequences $\{\mathbf{x}_j\} \to \mathbf{x}$ such that the derivative exists at each $\mathbf{x}_j$. Similarly, define the *generalized Jacobian* $\partial\mathbf{g}(\mathbf{x})$ as the closed convex hull of limit points of the sequence $\{\mathbf{g}'(\mathbf{x}_j)\}$ for sequences $\{\mathbf{x}_j\} \to \mathbf{x}$ such that the derivative exists at each $\mathbf{x}_j$. Since $f'(\mathbf{x}:\varepsilon)$ is, from the integral, an average of values of $f'(\mathbf{x}-\mathbf{s})$ over values of $\mathbf{s} \in \varepsilon E_0$ where this derivative exists, it follows that all limit points of $f'(\mathbf{x}:\varepsilon)$ as $\varepsilon \downarrow 0$ lie in $\partial f(\mathbf{x})$. Similarly, all limit points of $\mathbf{g}'(\mathbf{x}:\varepsilon)$ as $\varepsilon \downarrow 0$ lie in $\partial\mathbf{g}(\mathbf{x})$.

Assume that (3.31) reaches a local *strict minimum* at $\bar{\mathbf{x}}$ (section 4.6.1). This requires that $f(\mathbf{x}) > f(\bar{\mathbf{x}})$ whenever $\mathbf{x}$ is feasible, $\mathbf{x} \neq \bar{\mathbf{x}}$ and $\|\mathbf{x}-\bar{\mathbf{x}}\|$ is small; if this does not hold, add a term $\varepsilon_0\|\mathbf{x}-\bar{\mathbf{x}}\|^2$ to make it so, with some small positive ε_0. Consider the *smoothed problem*:

$$\text{Minimize } f(\mathbf{x}:\varepsilon) \quad \text{subject to} \quad -\tilde{\mathbf{g}}(\mathbf{x}:\varepsilon) \in S \tag{3.32}$$

where

$$\tilde{\mathbf{g}}(\mathbf{x}:\varepsilon) := \mathbf{g}(\mathbf{x}:\varepsilon) - \mathbf{g}(\bar{\mathbf{x}}:\varepsilon) + \mathbf{g}(\bar{\mathbf{x}})$$

so that $-\tilde{\mathbf{g}}(\bar{\mathbf{x}}:\varepsilon) \in S$; thus (3.32) still has feasible points. The strict minimum implies a stability property, illustrated in section 4.6.1 and proved in section 7.4, namely that whenever ε is sufficiently small, the smoothed problem (3.32) also reaches a local minimum at some point $\bar{\mathbf{x}}(\varepsilon)$, such that

$$\bar{\mathbf{x}}(\varepsilon) \to \bar{\mathbf{x}} \quad \text{as } \varepsilon \to 0$$

Assuming local solvability and a closed-cone condition for any equality constraints, the Fritz John conditions (section 3.5.4) apply to the differentiable problem (3.32) at $\bar{\mathbf{x}}(\varepsilon)$. Hence there are multipliers $\tau(\varepsilon) \geq 0$ and $\boldsymbol{\lambda}(\varepsilon) \in S^*$, not both zero, such that

$$\tau(\varepsilon)f'(\bar{\mathbf{x}}(\varepsilon):\varepsilon) + \boldsymbol{\lambda}(\varepsilon)\mathbf{g}'(\bar{\mathbf{x}}(\varepsilon)) = 0, \quad \boldsymbol{\lambda}(\varepsilon)\mathbf{g}(\bar{\mathbf{x}}(\varepsilon)) = 0$$

The vector $(\tau(\varepsilon), \boldsymbol{\lambda}(\varepsilon)) \in \mathbf{R}^{n+1}$ can be scaled to have unit length.

Consider a sequence $\{\varepsilon_j\} \downarrow 0$, say with $\varepsilon_j = 1/j$ $(j = 1, 2, \ldots)$. Since the unit sphere in $\mathbf{R}^{n+1}$ (the set of all unit vectors in $\mathbf{R}^{n+1}$) is a compact set, some subsequence of the sequence $\{(\tau(\varepsilon_j), \boldsymbol{\lambda}(\varepsilon_j)) : j = 1, 2, \ldots\}$ converges, say to $(\bar{\tau}, \bar{\boldsymbol{\lambda}})$. For some subsequence of this subsequence, the values of $f'(\bar{\mathbf{x}}(\varepsilon_j), \varepsilon_j)$ converge to some element $\boldsymbol{\xi} \in \partial f(\bar{\mathbf{x}})$. For some further subsequence, the values of $\mathbf{g}'(\bar{\mathbf{x}}(\varepsilon_j), \varepsilon_j)$ converge to some (matrix) element $\boldsymbol{\eta} \in \partial\mathbf{g}(\bar{\mathbf{x}})$. Hence $0 = \bar{\tau}\boldsymbol{\xi} + \bar{\boldsymbol{\lambda}}\eta$. Alternatively, there is some subsequence for which the values of

$$\tau(\varepsilon_j)f'(\bar{\mathbf{x}}(\varepsilon_j), \varepsilon_j) + \boldsymbol{\lambda}(\varepsilon_j)\mathbf{g}'(\bar{\mathbf{x}}(\varepsilon_j), \varepsilon_j)$$

converge to some element of $\partial(\bar{\tau}f + \bar{\boldsymbol{\lambda}}\mathbf{g})(\bar{\mathbf{x}})$. If a term $\varepsilon_0\|\mathbf{x}-\bar{\mathbf{x}}\|^2$ is added to the objective its contribution vanishes in the limit.

The appropriate generalized version of Robinson's stability condition (due to Yen (1990)) is given in section 7.5.1; it replaces $\mathbf{g}'(\bar{\mathbf{x}})$ by elements of $\partial \mathbf{g}(\bar{\mathbf{x}})$. Similarly, the closed-cone condition, that $[\mathbf{g}'(\bar{\mathbf{x}}), \mathbf{g}(\bar{\mathbf{x}})]$ is closed, is replaced by the condition that $[M, \mathbf{g}(\bar{\mathbf{x}})]$ is closed for each $M \in \partial \mathbf{g}(\bar{\mathbf{x}})$. These imply the usual Robinson and closed-cone conditions for the smoothed problem.

3.9.2 Theorem (Lagrangian conditions for Lipschitz problem)

Let $f: \mathbf{R}^n \to \mathbf{R}$ and $\mathbf{g}: \mathbf{R}^n \to \mathbf{R}^m$ satisfy Lipschitz conditions; let $S \subset \mathbf{R}^m$ be a closed convex cone; let $f(.)$ reach a local minimum, subject to $-\mathbf{g}(.) \in S$, at $\bar{\mathbf{x}}$; let the generalized Robinson stability condition hold there for the constraint, with a closed-cone condition for any equality constraint. Then, for some Lagrange multiplier $\bar{\boldsymbol{\lambda}} \in S^*$, there hold Karush–Kuhn–Tucker conditions in the form

$$0 \in \partial(f + \bar{\boldsymbol{\lambda}}\mathbf{g})(\bar{\mathbf{x}}); \quad \bar{\boldsymbol{\lambda}}\mathbf{g}(\bar{\mathbf{x}}) = 0 \tag{3.33}$$

Proof

The preceding discussion showed that

$$0 \in \partial(\bar{\tau} f + \bar{\boldsymbol{\lambda}}\,\mathbf{g})(\bar{\mathbf{x}})$$

where $\bar{\tau} \geq 0$ and $\bar{\boldsymbol{\lambda}} \in S^*$ are not both zero. Also $\bar{\boldsymbol{\lambda}}\mathbf{g}(\bar{\mathbf{x}}) = 0$ follows from a similar limiting argument, applied to $\bar{\boldsymbol{\lambda}}(\varepsilon)\mathbf{g}(\bar{\mathbf{x}}(\varepsilon), \varepsilon) = 0$ for the smoothed problem.

If $\bar{\tau} = 0$, then $\bar{\boldsymbol{\lambda}} \neq \mathbf{0}$, $\bar{\boldsymbol{\lambda}}M = 0$ for some $M \in \partial \mathbf{g}(\bar{\mathbf{x}})$, and the generalized Robinson condition, with $\bar{\boldsymbol{\lambda}}\mathbf{g}(\bar{\mathbf{x}}) = 0$, gives $0 \in \text{int}[M(\mathbf{R}^n) + S]$. Multiplying by $\bar{\boldsymbol{\lambda}}$ gives $0 \in \text{int}[0 + \mathbf{R}_+]$, a contradiction. □

3.9.3 Remarks

The Lagrangian condition is also obtained in the form

$$0 \in \partial f(\bar{\mathbf{x}}) + \bar{\boldsymbol{\lambda}}\partial \mathbf{g}(\bar{\mathbf{x}}); \quad \bar{\boldsymbol{\lambda}}\mathbf{g}(\bar{\mathbf{x}}) = 0 \tag{3.34}$$

In fact, (3.33) implies (3.34).

The corresponding Karush–Kuhn–Tucker necessary condition for weak minimum of a vector objective function $\mathbf{f}(\mathbf{x}) \in \mathbf{R}^r$ (compare sections 3.4.4 and 3.5.3) with respect to an order cone $Q \subset \mathbf{R}^r$ (replacing $\mathbf{R}_+ \subset \mathbf{R}$) is

$$0 \in \partial(\bar{\boldsymbol{\tau}} f + \bar{\boldsymbol{\lambda}}\mathbf{g})(\bar{\mathbf{x}}); \quad \bar{\boldsymbol{\lambda}}\mathbf{g}(\bar{\mathbf{x}}) = 0$$

with $0 \neq \bar{\boldsymbol{\tau}} \in Q^*$ and $\bar{\boldsymbol{\lambda}} \in S^*$. The proof is similar.

If $f(.)$ is convex, then $\partial f(\bar{\mathbf{x}})$ becomes the convex subdifferential, as defined in section 3.3.

The set $\partial f(\bar{\mathbf{x}})$ defined in this section agrees with the *generalized gradient* defined by Clarke (1983). Clarke starts from a different point, by defining a generalized directional derivative applicable to a Lipschitz function f as

$$f^{\circ}(\bar{\mathbf{x}}; \mathbf{d}) := \sup_{\mathbf{y} \to \bar{\mathbf{x}}, \alpha \downarrow 0} \alpha^{-1}[f(\mathbf{y} + \alpha\mathbf{d}) - f(\mathbf{y})]$$

and then defining

$$\partial f(\bar{\mathbf{x}}) := \{\boldsymbol{\zeta} : (\forall \mathbf{d})\, f^{\circ}(\bar{\mathbf{x}}; \mathbf{d}) \geq \boldsymbol{\zeta}^{\mathrm{T}}\mathbf{d}\}$$

The *generalized Jacobian* $\partial\mathbf{g}(\bar{\mathbf{x}})$ agrees with Clarke's definition. The present approach is based on Craven (1986).

Exercise
Define (nonconvex) $f(x) = x - x^2$ $(x \geq 0)$, $\frac{1}{2}x$ $(x < 0)$. Calculate $\partial f(0)$.

Sufficiency conditions require the following extension of invex.

3.9.4 Definition
The Lipschitz function $\mathbf{g}: \mathbf{R}^n \to \mathbf{R}^m$ is *generalized K-invex* at the point $\mathbf{x}$, with respect to the convex cone $K \subset \mathbf{R}^m$ if, for some function $\boldsymbol{\kappa}: \mathbf{R}^n \to \mathbf{R}^n$,

$$(\forall M \in \partial\mathbf{g}(\bar{\mathbf{x}}))\ \mathbf{g}(\mathbf{x}) - \mathbf{g}(\bar{\mathbf{x}}) \in M\boldsymbol{\kappa}(\mathbf{x} - \bar{\mathbf{x}}) + \mathbf{S}$$

3.9.5 Theorem (sufficient KKT conditions for Lipschitz problem)
The necessary Karush–Kuhn–Tucker conditions of Theorem 3.9.2 for a Lipschitz problem are also sufficient for a minimum at a feasible point $\bar{\mathbf{x}}$ if the vector function $(f, \mathbf{g})$ is generalized invex at $\bar{\mathbf{x}}$ with respect to the cone $\mathbf{R}_+ \times S$.

Proof
Let $-\mathbf{g}(\mathbf{x}) \in S$. In terms of the Lagrangian $L(.) := f(.) + \bar{\boldsymbol{\lambda}}\mathbf{g}(.)$,

$$f(\mathbf{x}) - f(\bar{\mathbf{x}}) = L(\mathbf{x}) - L(\bar{\mathbf{x}}) - \bar{\boldsymbol{\lambda}}\mathbf{g}(\mathbf{x}) \geq L(\mathbf{x}) - L(\bar{\mathbf{x}}) \geq M\boldsymbol{\kappa}(\mathbf{x} - \bar{\mathbf{x}})$$

for each $M \in \partial L(\bar{\mathbf{x}})$. But $0 \in \partial L(\bar{\mathbf{x}})$ by the necessary Karush–Kuhn–Tucker conditions. So, taking $M = 0, f(\mathbf{x}) - f(\bar{\mathbf{x}}) \geq 0$. □

Exercise
Extend Theorem 3.9.5 to a Lipschitz problem with a vector objective $\mathbf{f}(.)$. *Hint*: Show that $\bar{\boldsymbol{\tau}}[f(\mathbf{x}) - f(\bar{\mathbf{x}})] \geq 0$ whenever $\mathbf{x}$ is feasible for the problem. If $\bar{\mathbf{x}}$ is not a weak minimum, then for some sequence of vectors $\mathbf{x} \to \bar{\mathbf{x}}, f(\mathbf{x}) - f(\bar{\mathbf{x}}) \in -Q$.

Remark
See section 7.9 for a related discussion of an optimal control problem under Lipschitz assumptions.

Example
Consider the following problem involving a norm which is nondifferentiable at 0 in an infinite-dimensional space X, either $C(I)$ or $L^2(I)$ with $I = [0, 1]$ and Fréchet differentiable functions $f: X \to \mathbf{R}$ and $\mathbf{g}: X \to \mathbf{R}^m$.

$$\text{Minimize}_{\mathbf{x} \in X}\ f(\mathbf{x}) + \varepsilon\|\mathbf{x}\| \quad \text{subject to} \quad -\mathbf{g}(\mathbf{x}) \in S$$

Suppose, by shift of origin, that the minimum occurs at $\mathbf{x} = \mathbf{0}$. If $\mathbf{x}$ is restricted to a subspace S_n of finite dimension n, then Theorem 3.9.2 can be applied, showing that

$$\tau_n f'(\mathbf{0}) + \boldsymbol{\lambda}_n\mathbf{g}'(\mathbf{0}) + \varepsilon\zeta_n + \boldsymbol{\mu}_n = \mathbf{0}, \quad \boldsymbol{\lambda}_n \in S^*, \quad \boldsymbol{\mu}_n \in S_n^*, \quad \boldsymbol{\lambda}_n\mathbf{g}(\mathbf{0}) = 0$$

for some $\zeta_n \in \partial\|\mathbf{0}\|$; thus $\|\zeta_n\| \leq 1$ and $\tau_n = 1$. Components of $f'(\mathbf{0})$ and $\mathbf{g}'(\mathbf{0})$ applying to dimensions greater than n are absorbed into $\boldsymbol{\mu}_n$. It is assumed here that S_n is *complemented*, thus that $X = S_n + S_n^+$ for some subspace S_n^+ with $S_n \cap S_n^+ = \{0\}$. This is automatic for $L^2(I)$, and suitable subspaces for $C(I)$ can be constructed, as described in section 7.6, by truncating a Fourier series.

Now fix a direction $\mathbf{d} \in S_k$. For $n > k$, $\boldsymbol{\mu}_n\mathbf{d} = 0$. By Alaoglu's theorem in functional analysis (section 1.5.6), the closed unit ball in X^* is weak$*$compact; hence there is some ζ with $\|\zeta\| \leq 1$, and some subset of values of n, such that $\zeta_n\mathbf{d} \to \zeta\mathbf{d}$ for each $\mathbf{d}$. Scaling τ_n so that $\|(\tau_n, \boldsymbol{\lambda}_n)\| = 1$, a subsequence converges, say to $(\tau, \boldsymbol{\lambda})$. Hence

$$\tau f'(\mathbf{0}) + \boldsymbol{\lambda}\mathbf{g}'(\mathbf{0}) + \varepsilon\zeta = 0, \quad \boldsymbol{\lambda}\mathbf{g}(\mathbf{0}) = 0, \quad \boldsymbol{\lambda} \in S^*, \quad \|\zeta\| \leq 1$$

where $\tau = 1$ holds if a constraint qualification is assumed, giving Karush–Kuhn–Tucker conditions.

If $\mathbf{g} : X \to \mathbf{R}^m$ is no longer differentiable, but satisfies a Lipschitz condition, then the above argument is readily modified to obtain

$$\tau f'(\mathbf{0}) + \boldsymbol{\lambda}\boldsymbol{\sigma} + \varepsilon\zeta = 0, \quad \boldsymbol{\lambda}\mathbf{g}(\mathbf{0}) = 0, \quad \boldsymbol{\lambda} \in S^*, \quad |\zeta\| \leq 1, \quad \boldsymbol{\sigma} \in \partial\mathbf{g}(\mathbf{0})$$

The multiplier $\tau \neq 0$ if a constraint qualification holds (in particular, the generalized Robinson condition; see section 7.5.1).

3.10 REFERENCES

Ben-Israel, A., Ben-Tal, A. and Zlobec, S. (1979) Optimality conditions in convex programming, in Prékopa, A. (ed.), *Survey of Mathematical Programming*, North-Holland, Amsterdam.

Ben-Israel, A. (1969) Linear equations and inequalities in finite dimensional, real or complex, vector spaces: a unified theory, *J. Math. Anal. Appl.*, **27** 367–89.

Clarke, F. (1983) *Optimization and Nonsmooth Analysis*, Wiley, New York.

Craven, B. D. (1977) Lagrangean conditions and quasiduality, *Bull. Austral. Math. Soc.*, **16** 325–39.

Craven, B. D. (1978) *Mathematical Programming and Control Theory*, Chapman & Hall, London.

Craven, B. D. (1979) On constrained maxima and minima, *Gazette of the Australian Mathematical Society*, **6**(2) 46–50.

Craven, B. D. (1981a) *Functions of Several Variables*, Chapman & Hall, London.

Craven, B. D. (1981b) Duality for generalized convex fractional programs, in Schaible, S. and Ziemba, W. T. (eds) *Generalized Concavity in Optimization and Economics*, Academic Press, New York, pp. 473–90.

Craven, B. D. (1986) Nondifferentiable optimization by smooth approximations, *Optimization*, **17** 3–17.

Dugundji, J. (1966) *Topology*, Allyn & Bacon, Boston.

Fiacco, A. V. and McCormick, G. P. (1968) *Nonlinear Programming: Sequential Unconstrained Minimization Techniques*, Wiley, New York.

Hanson, M. A. (1980) On the sufficiency of the Kuhn–Tucker conditions, *Journal of Mathematical Analysis and Applications*, **80** 545–50.

Robinson, S. M. (1976) Stability theory for systems of inequalities, Part II, *SIAM Journal of Numerical Analysis*, **13** 497–513.
Schaefer, H. H. (1966) *Topological Vector Spaces*, Macmillan, New York. (The separation theorem is proved in section II.9.)
Yen, N. D. (1990) *Stability of the Solution Set of Perturbed Nonsmooth Inequality Systems*, International Centre for Theoretical Physics, International Atomic Energy Agency, UNESCO, Trieste, Italy.

4

Optimality conditions for control problems

4.1 FORMULATION

Consider the discrete-time and the continuous-time optimal control models, as formulated in section 2.1. Assume that the functions describing these models, thus $\psi^{(t)}(.,.)$, $\varphi^{(t)}(.,.)$ for discrete-time, and $\Phi(.)$, $f(.,.,.)$, $m(.,.,.)$ for continuous-time, are differentiable.

The discrete-time problem:

$$\min \; J(u) \equiv F(x, u) = \sum_{t=0}^{N} \psi^{(t)}(x_t, u_t)$$

$$\text{subject to} \quad \Delta x_t \equiv x_{t+1} - x_t = \varphi^{(t)}(x_t, u_t) \;\; (t = 0, 1, 2, \ldots, N-1); \quad x_0 = c;$$

$$\alpha \leq u_t \leq \beta \;\; (t = 0, 1, 2, \ldots, N); \quad \zeta(x_{N+1}) = 0$$

can be expressed as a mathematical programming problem, with finitely many variables, namely those comprised in the vectors $\mathbf{x}$ and $\mathbf{u}$. Put these variables together into a single vector $\mathbf{z}$, and consider a problem of the form

$$\min F(\mathbf{z}) \quad \text{subject to} \quad -\mathbf{g}(\mathbf{z}) \in S$$

For this problem, the Lagrangian function is $F(\mathbf{z}) + \boldsymbol{\theta}\mathbf{g}(\mathbf{z})$, where the (row) vector $\boldsymbol{\theta}$ is a Lagrange multiplier. Necessary conditions for a minimum will be derived from the Lagrangian. (This assumes, of course, that some minimum is reached.)

4.1.1 Discrete-time formulation

For the discrete-time optimal control problem, the Lagrangian is then

$$L(\mathbf{x}, \mathbf{u}; \boldsymbol{\lambda}, \mathbf{p}, \mathbf{q}, \sigma) = \sum_{t=0}^{N} \{\psi^{(t)}(x_t, u_t) + p_t(u_t - \beta) - q_t(u_t - \alpha)$$

$$- \lambda_t(\Delta x_t - \varphi^{(t)}(x_t, u_t)\} + \sigma\zeta(x_{N+1}))$$

Here $p_t \geq 0$ is the Lagrange multiplier for the constraint $u_t \leq \beta$, $q_t \geq 0$ is the Lagrange multiplier for the constraint $u_t \geq \alpha$ ($\Leftrightarrow -(u_t - \alpha) \leq 0$), λ_t (without restriction of sign) is the Lagrange multiplier for the equality constraint $\Delta x_t = \varphi^{(t)}(x_t, u_t)$ and σ is the Lagrange multiplier for the terminal constraint $\zeta(x_{N+1}) = 0$. The minus sign before λ_t is for convenience, to make later formulae tidy. Note that $\mathbf{x}, \mathbf{u}, \boldsymbol{\lambda}, \mathbf{p}, \mathbf{q}$ are the vectors respectively of $x^{(t)}$, $u^{(t)}$, $\lambda^{(t)}$, $p^{(t)}$ and $q^{(t)}$. The vectors of Lagrange multipliers are considered here as row vectors. (If u_t is itself a column vector, then α and β are also vectors.) It is arbitrary whether the initial condition and the terminal constraint are included in the Lagrangian or are excluded and treated separately as boundary conditions. To show what happens, the first has been included and the second excluded. Note that α and β can be given functions of time t, instead of constants.

4.1.2 Continuous-time formulation

Consider an optimal control problem:

$$\text{Minimize } F(x, u) = \int_0^T f(x(t), u(t), t)\,\mathrm{d}t + \Phi(x(T))$$

over the fixed time interval $[0, T]$, subject to the differential equation with initial conditions:

$$\mathrm{d}x(t)/\mathrm{d}t = m(x(t), u(t), t) \quad (0 \leq t \leq T)$$

$$x(0) = x_0$$

and to the constraints

$$g(u(t), t) \in S(t) \quad \text{and} \quad n(x(t), t) \in V(t) \quad (0 \leq t \leq T)$$

Here $x(t) \in \mathbf{R}^n$, $u(t) \in \mathbf{R}^k$; $f(., ., .)$, $m(., ., .)$, $g(., .)$ and $n(., .)$ are continuously differentiable functions, and $S(t) \subset \mathbf{R}^r$ and $V(t) \subset \mathbf{R}^h$ are closed convex cones, which may vary with t.

A *control constraint* $g(u(t), t) \in S(t)$ will be written also in an equivalent form $g^*(u(t) \in \Gamma(t))$ for suitable sets $\Gamma(t) \subset \mathbf{R}^k$, for example, a constraint $(\forall t \in [0, T])$ $|u(t)| \leq 1$ can be expressed by

$$g^*(u(t)) := 1 - |u(t)|^2 \in \mathbf{R}_+$$

If $u(t) \in \mathbf{R}^k$ with $k > 1$, then $|u(t)|$ means here an appropriate norm of $u(t)$ in $\mathbf{R}^k$; this could be the Euclidean norm $|u(t)| := (\sum u_i(t)^2)^{1/2}$, or it could instead be chosen as $|u(t)| := \max_i |u_i(t)|$; the latter corresponds to a control constraint

$$(\forall_i) \quad -1 \leq u_i(t) \leq 1$$

A similar remark about norms applies when $x(t) \in \mathbf{R}^n$.

The *state constraint* $n(x(t), t) \in V(t)$ is not always present. It arises, for example, in inventory problems, often as a bound, $x(t) \leq \beta$ or $x(t) \geq \alpha$, for each $t \in I \equiv [0, T]$.

Assume temporarily that the term $\Phi(x(T))$ is absent from the objective function $F(x, u)$.

In order to analyse the control problem, it is necessary to specify appropriate vector spaces of functions. These must be large enough to contain the optimum, yet small enough to be manageable. It turns out (section 4.3) that an optimal function $u(.)$ commonly has discontinuities, so the space $C[0, T]$ of continuous functions on $[0, T]$ would not be large enough. Therefore, consider now $u(.) \in U$, the space of *piecewise continuous* functions (Figure 4.1(a)), namely functions that are continuous except for a finite number of simple jumps, from $I = [0, T]$ into $\mathbf{R}^k$, with the uniform norm

$$\|u(.)\|_\infty \equiv \|u\|_\infty := \sup_{0 \le t \le T} |u(t)|$$

where $|u(t)|$ means the norm of $u(t)$ in $\mathbf{R}^k$. The norm $|u(t)|$ of a function value $u(t) \in$ in $\mathbf{R}^k$ must be distinguished from the norm $\|u\|$ in the function space U of the function u (also denoted by $u(.)$). It is convenient to visualize the function $u(.)$ as its whole graph, as distinct from a point on that graph.

Denote by W the space of piecewise continuous functions from I into $\mathbf{R}^n$, with the uniform norm. Denote by X the space of functions $y(.) : I \to \mathbf{R}^n$, such that each $y \in X$ is the indefinite integral of some function in W, with $y(0) = 0$. Such functions $y \in X$ are called *piecewise smooth* functions (Figure 4.1(b)). Then

$$y(t) = \int_0^t w(s)\,\mathrm{d}s$$

may be expressed as $w = \mathrm{D}x$, where $\mathrm{D} = \mathrm{d}/\mathrm{d}t$ except at discontinuities of w. The linear map D is now made continuous by giving X the *graph norm*

$$\|y\| := \|y\|_\infty + \|\mathrm{D}y\|_\infty$$

(Note that $\|y - z\|$ is small exactly when both $\|y - z\|_\infty$ and $\|\mathrm{D}y - \mathrm{D}z\|_\infty$ are small. Thus, the graph norm corresponds to uniform approximation of both the function y and its derivative $\mathrm{D}y$.) Other choices of norms can be

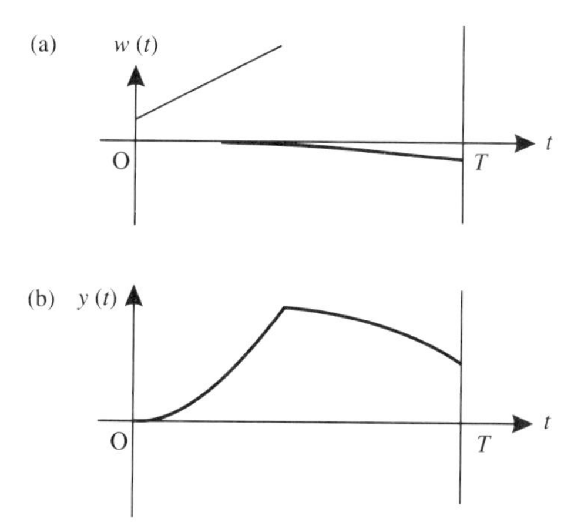

Figure 4.1 (a) Piecewise continuous function; (b) piecewise smooth function.

appropriate – see section 4.2.8. Using the graph norm means that a local minimum point $(\bar{x}, \bar{u})$ minimizes the objective function with respect to points satisfying the constraints and close to $(\bar{x}, \bar{u})$ in the sense that u is near $\bar{u}$, x is near $\bar{x}$, and also $\mathrm{D}x$ is near $\mathrm{D}\bar{x}$, each in terms of its appropriate norm.

The differential equation for $x(t)$, with initial condition, is now written as the integral equation

$$x(t) = x_0 + \int_0^t m(x(s), u(s), s)\,\mathrm{d}s$$

This will be expressed as $\mathrm{D}x = M(x, u)$ where the map $M : X \times U \to W$ is defined as $M(x, u)(t) = m(x(t), u(t), t)$. This map M may be visualized as mapping the graphs of $x(.)$ and $u(.)$ to the graph of $m(x(.), u(.), .)$. The differential equation has now been slightly extended to allow a finite number of exceptional points where x is not differentiable.

To simplify the presentation, a shift of origin will be assumed, to replace the initial condition $\mathbf{x}_0$ by $\mathbf{0}$. As mentioned in section 1.3, this makes the space of state functions considered into a vector space, since now the sum of two such functions satisfies the zero initial condition. At the end of the calculation, the shift of origin can be reversed to express the results in terms of the original variables. There is then no need actually to calculate the shift of origin, and it will henceforth be ignored for simplicity.

The control and state constraints can also be put into abstract terms as follows. Let Q (resp. P) denote the space of piecewise continuous functions from I into $\mathbf{R}^r$ (resp. into $\mathbf{R}^s$). Then $(\forall t)\ g(u(t), t) \in S(t)$ is equivalent to $G(u) \in K_u$, when the map $G : U \to Q$ is defined by $G(u)(t) := g(u(t), t)$, and $K_u \subset Q$ is the convex cone

$$K_u = \{q \in Q : (\forall t \in I)\ q(t) \in S(t)\}$$

Similarly, $(\forall t)\ n(x(t), t) \in V(t)$ is equivalent to $N(x) \in K_x$, where $N(x)(t) := n(x(t), t)$ and $K_x = \{p \in P : (\forall t \in I)\ p(t) \in V(t)\}$.

The given minimization problem is now expressed as:

$$\text{Minimize}_{x \in x_0 + X, u \in U}\ F(x, u) \quad \text{subject to} \quad \mathrm{D}x = M(x, u), \quad G(u) \in K_u, \ N(x) \in K_x \tag{4.1}$$

This looks exactly the same as a minimization problem with finite-dimensional vectors, and will be treated similarly.

The term $\Phi(x(T))$ can now be put back into $F(x, u)$ by replacing the integrand $f(x(t), u(t), t)$ by $f(x(t), u(t), t) + \Phi(x(t))\delta(t - T)$, where $\delta(.)$ denotes Dirac's delta function. As discussed in section 1.5.8, it can be assumed that $\delta(t - T)$ vanishes when $t \geq T$.

The Lagrangian for problem (4.1) is then

$$F(x, u) + \bar{\lambda}[-\mathrm{D}x + M(x, u)] - \bar{\mu}G(u) - \bar{\nu}N(x)$$

where $\bar{\lambda}$, $\bar{\mu}$, and $\bar{\nu}$ are Lagrange multipliers.

4.1.3 Fréchet derivatives

Since derivatives will be required to describe a minimum, the functions F, M, G and N must be shown to be differentiable in some suitable sense.

Since $m(.,.,.)$ is continuously differentiable

$$[M(x+z)-M(x)](t) = m(x(t)+z(t), u(t), t) - m(x(t), u(t), t)$$
$$= m_x(x(t), u(t), t)z(t) + \zeta(t)$$

if $x, x+z \in x_0 + X$, m_x denotes a partial derivative and the remainder term is small in the sense that

$$|\zeta(t)| \leq \varepsilon|z(t)| \leq \varepsilon\|z\|_\infty \quad \text{whenever } \|z\|_\infty < \delta(\varepsilon)$$

for some function $\delta(.)$ of positive ε; thus

$$\|\zeta\| \leq \varepsilon\|z\|_\infty \quad \text{whenever } \|z\|_\infty < \delta(\varepsilon)$$

Consequently, $M(x, u)$ is partially Fréchet differentiable with respect to x, with derivative $M_x(x, u)$ given by

$$[M_x(x, u)z](t) = m_x(x(t), u(t), t)z(t)$$

Consider now, for $x \in X$ and $z \in X$,

$$\left|F(x+z, u) - F(x, u) - \int_0^T f_x(x(t), u(t), t)\, z(t)\, \mathrm{d}t\right|$$
$$= \left|\int_0^T [f(x(t)+z(t), u(t), t) - f(x(t), u(t), t) - f_x(x(t), u(t), t)]\, z(t)\, \mathrm{d}t\right|$$
$$\leq \int_0^T |f(x(t)+z(t), u(t), t) - f(x(t), u(t), t) - f_x(x(t), u(t), t)||z(t)|\, \mathrm{d}t$$
$$\leq \int_0^T (\varepsilon/T)\, \|z\|_\infty\, \mathrm{d}t$$
$$= \varepsilon\|z\|_\infty$$

whenever $\|z\|_\infty$ is sufficiently small, since $f(.,.,.)$ is differentiable. Thus $F(x, u)$ is Fréchet differentiable with respect to x, with Fréchet derivative given by $F_x(x, u)z = \int_0^T f_x(x(t), u(t), t)z(t)\, \mathrm{d}t$. Other derivatives may be similarly discussed.

4.2 LAGRANGE MULTIPLIERS

Consider the optimization problem:

$$\min_{\mathbf{w} \in W} F(\mathbf{w}) \quad \text{subject to} \quad -\mathbf{C}(\mathbf{w}) \in K \tag{4.2}$$

where the variable $\mathbf{w}$ runs over a normed vector space W, the functions $F: W \to \mathbf{R}$ and $\mathbf{C}: W \to Z$ are differentiable, Z is a normed vector space and K is a closed convex cone. For the control problem (4.1), $\mathbf{w}$ has components x

and u. Assume that (4.2) reaches a (local) minimum at the point $\bar{w}$. The Lagrangian for (4.2) is $L(\mathbf{w};\boldsymbol{\lambda}) = F(\mathbf{w}) + \boldsymbol{\lambda}\mathbf{C}(\mathbf{w})$. The following theorem was proved in section 3.5.3, using differentiability to linearize the problem (4.2), thus to approximate it by a linear problem near the point $\bar{\mathbf{w}}$.

The following results are cited from Chapter 3.

4.2.1 Karush–Kuhn–Tucker theorem

In (4.2), let the functions be differentiable; let (4.2) reach a minimum when $\mathbf{w} = \bar{\mathbf{w}}$; let the constraint be locally solvable at $\bar{\mathbf{w}}$; then there exists a Lagrange multiplier $\bar{\boldsymbol{\lambda}} \in Z'$ such that the *Karush–Kuhn–Tucker* conditions hold:

$$\mathbf{F}'(\bar{\mathbf{w}}) + \bar{\boldsymbol{\lambda}}\mathbf{C}'(\bar{\mathbf{w}}) = \mathbf{0}; \quad \bar{\boldsymbol{\lambda}}\mathbf{C}(\bar{\mathbf{w}}) = 0, \quad \bar{\boldsymbol{\lambda}} \in \mathbf{K}^* \tag{4.3}$$

4.2.2 Remarks

$\mathbf{F}'(\bar{\mathbf{w}}) + \bar{\boldsymbol{\lambda}}\mathbf{C}'(\bar{\mathbf{w}})$ is the gradient $L_{\mathbf{w}}(\bar{\mathbf{w}}, \bar{\boldsymbol{\lambda}})$. (The gradient $L_{\boldsymbol{\lambda}}$ of L with respect to $\boldsymbol{\lambda}$ does not enter here.) If the *local solvability* assumption is somewhat weakened (section 3.5.4), (4.3) may be replaced by *Fritz John conditions*, in which $\mathbf{F}'(\bar{\mathbf{w}})$ is replaced by $\bar{\tau}\mathbf{F}'(\bar{\mathbf{w}})$, with some multiplier $\bar{\tau} \geq 0$. However, it will appear that the possibility $\bar{\tau} = 0$ seldom occurs in optimal control problems.

4.2.3 Converse Karush–Kuhn–Tucker theorem

If (4.3) holds at a point $(\bar{\mathbf{w}})$ which is feasible for (4.2) (thus $-\mathbf{C}(\bar{\mathbf{w}}) \in K$), and if F and $\mathbf{C}$ are convex at $(\bar{\mathbf{w}})$, then $(\bar{\mathbf{w}})$ is a minimum point for (4.2).

Proof (sections 3.5.5 and 3.6.2(d))

Let $-\mathbf{C}(\mathbf{w}) \in K$. By the assumed convexity, $L(., \bar{\boldsymbol{\lambda}})$ is convex. Then

$$\begin{aligned} F(\mathbf{w}) - F(\bar{\mathbf{w}}) &= L(\mathbf{w}, \bar{\boldsymbol{\lambda}}) - L(\bar{\mathbf{w}}, \bar{\boldsymbol{\lambda}}) && (\text{since } \bar{\boldsymbol{\lambda}}\,\mathbf{C}(\bar{\mathbf{w}}) = 0) \\ &\geq L_{\mathbf{w}}(\bar{\mathbf{w}}, \bar{\boldsymbol{\lambda}})(\mathbf{w} - \bar{\mathbf{w}}) && (\text{by the convexity}) \\ &= 0 && (\text{by } (4.3)) \end{aligned}$$

□

4.2.4 Remarks

The convex hypotheses can be replaced by the weaker hypothesis that $(F, \mathbf{C})$ is $(\mathbf{R}_+ \times K)$-invex at $\bar{\mathbf{w}}$ (section 3.6.2). This fact will be used later to obtain sufficient conditions for a minimum of (4.1).

4.2.5 Conditions 4.3 for discrete optimal control

For the Lagrangian function $L(.)$ (from section 4.2.1):

$$\sum_{t=0}^{N} \{\psi^{(t)}(x_t, u_t) + p_t(u_t - \beta) - q_t(u_t - \alpha) - \lambda_t(\Delta x_t - \varphi^{(t)}(x_t, u_t)\} + \sigma\zeta(x_{N+1})$$

KKT (4.3) divides into parts (i) involving L_x and (ii) involving L_u. In differentiating $L(.)$ with respect to x_t, note that x_t occurs in exactly two places in the summation, namely in the terms:

$$\lambda_{t-1}\Delta x_{t-1} + \lambda_t \Delta x_t = \lambda_{t-1}(x_t - x_{t-1}) + \lambda_t(x_{t+1} - x_t)$$

Thus differentiating this expression with respect to x_t gives the two terms $\lambda_t - \lambda_{t-1} = \Delta\lambda_{t-1}$. Thus $L_x(.) = 0$ includes the terms $\Delta\bar{\lambda}_{t-1} = \bar{\lambda}_t - \bar{\lambda}_{t-1}$; because of the discrete time t, $\Delta\bar{\lambda}_{t-1}$ occurs, rather than $\Delta\bar{\lambda}_t$. Now the Karush–Kuhn–Tucker necessary conditions required both $L_x(.) = 0$ and $L_u(.) = 0$. Thus, at the minimum point $(\bar{\mathbf{x}}, \bar{\mathbf{u}})$ with optimal multipliers $\bar{\boldsymbol{\lambda}}$, $\mathbf{p}$, $\mathbf{q}$, $L_x = 0$ gives

$$\psi_x^{(t)}(\bar{x}_t, \bar{u}_t) - \bar{\lambda}_{t-1} + \bar{\lambda}_t + \bar{\lambda}_t(\varphi_x^{(t)}(\bar{x}_t, \bar{u}_t)) = 0 \quad (t = 1, 2, \ldots, N) \tag{4.4}$$

$L_u = 0$ gives

$$\psi_u^{(t)}(\bar{x}_t, \bar{u}_t) + \bar{\lambda}_t\, \varphi_u^{(t)}(\bar{x}_t, \bar{u}_t) + p_t - q_t = 0, \quad p_t(u_t - \beta) = 0, \quad q_t(u_t + \alpha) = 0$$

$$(t = 0, 1, 2, \ldots, N) \tag{4.5}$$

and

$$\sigma\zeta_x(\bar{x}_{N+1}) - \bar{\lambda}_N = 0 \tag{4.6}$$

comes from differentiating L with respect to x_{N+1}. Note that (4.4) does not apply when $t = 0$ since $\bar{x}_0$ is a fixed initial condition, and (4.6) relates to the terminal constraint on x_{N+1}. If u_t is vector valued, the *transversality conditions* (= *complementary slackness* conditions in (4.3)) $p_t(u_t - \beta) = 0$, $q_t(u_t + \alpha) = 0$, apply to each component separately. The vector $\lambda = (\lambda_0, \lambda_1, \ldots, \lambda_N)$ is called the costate variable (or function).

Define now the *discrete-time Hamiltonian*

$$h^{(t)}(x_t, u_t, \lambda_t) = \psi^{(t)}(x_t, u_t) + \lambda_t \varphi^{(t)}(x_t, u_t)$$

Then (4.4) can be written as the *adjoint difference equation*:

$$-\Delta\bar{\lambda}_{t-1} = h_x^{(t)}(\bar{x}_t, \bar{u}_t, \bar{\lambda}_t) \quad (t = 1, 2, \ldots, N) \tag{4.7}$$

with (4.6) giving a boundary condition on $\bar{\lambda}_{N+1}$.

Equation (4.5) coincides with the Karush–Kuhn–Tucker necessary conditions for a minimum at $u_t = \bar{u}_t$ of the associated problem (for $t = 0, 1, 2, \ldots$):

$$\text{Minimize}_{u_t}\, h^{(t)}(\bar{x}_t, u_t, \bar{\lambda}_t) \quad \text{subject to} \quad \alpha \le u_t \le \beta$$

in which only u_t varies. But these necessary conditions are not always sufficient; they do not always imply a minimum for the associated problem.

4.2.6 Theorem

For the discrete-time optimal control problem, assume for each t that $\psi^{(t)}(x_t, .)$ is convex, and $\varphi^{(t)}(x_t, .)$ is linear. Then necessary conditions for a local minimum at $(x_t, u_t) = (\bar{x}_t, \bar{u}_t)$ $(t = 0, 1, 2, \ldots)$ are that the costate λ_t $(t = 0, 1, 2, \ldots)$ satisfies the adjoint difference equation (4.7) with boundary condition (4.6), and that the associated problem is minimized when $u_t = \bar{u}_t$ $(t = 0, 1, 2, \ldots)$.

Proof

It was shown above that a minimum of the control problem implies both the adjoint equation and necessary KKT conditions for the associated problem. If these necessary conditions for the associated problem are also sufficient, then this implies a minimum for the associated problem. By hypothesis, $\varphi^{(t)}(x_t, u_t)$ is linear in u_t, hence $\lambda_t \varphi^{(t)}(x_t, u_t)$ is convex in u_t, whatever the sign of λ_t; also $\psi^{(t)}(x_t, u_t)$ is convex in u_t. Hence the Hamiltonian

$$h^{(t)}(\bar{x}_t, u_t, \bar{\lambda}_t) = \sum_{t=0}^{N} \{\psi^{(t)}(x_t, u_t) + \lambda_t \varphi^{(t)}(x_t, u_t)\} + \sigma\zeta(x_{N+1})$$

is convex in u_t. The associated problem has this as objective and linear constraints, and so is a convex problem. Therefore, by Theorems 3.5.5 or 4.2.3, necessary KKT conditions at a point satisfying the constraints are also sufficient. □

4.2.7 KKT conditions (4.3) for the continuous-time control problem (4.1)

Since $\mathbf{w}$ consists of $\mathbf{x}$ and $\mathbf{u}$, (4.3) divides into parts (i) involving L_x and (ii) involving L_u. These give respectively, at the point $(\bar{\mathbf{x}}, \bar{\mathbf{u}})$, with $\bar{\boldsymbol{\lambda}} \in W'$:

$$F_x + \bar{\boldsymbol{\lambda}}(-D + M_x) - \bar{\boldsymbol{\nu}} N_x = 0, \quad \bar{\boldsymbol{\nu}} \in K_{X^*}, \quad \bar{\boldsymbol{\nu}} N(\bar{x}) = 0 \tag{4.8}$$

$$F_u + \bar{\boldsymbol{\lambda}}(-M_u) - \bar{\boldsymbol{\mu}} G_u = 0, \quad \bar{\boldsymbol{\mu}} \in K_{U^*}, \quad \bar{\boldsymbol{\mu}} G(\bar{u}) = 0 \tag{4.9}$$

Observe that F and M occur here in the combination $F + \bar{\boldsymbol{\lambda}} M$. Fritz John conditions (section 3.5.4), with F replaced by τF, hold here if the differential equation $Dx = M(x, u)$ is locally solvable (section 3.5.2), which happens when the linear operator $-D + M_{\mathbf{x}}$ is assumed to be *onto* (section 3.7), and when the cones K_U and K_X in (4.1) have interiors. If there is no state constraint, it is shown below from the adjoint differential equation that $\tau = 1$. (If there is a state constraint, then $\tau = 1$ if the whole constraint system for (4.1) is locally solvable.)

The vector $\bar{\boldsymbol{\lambda}}$ is an element of the dual space of the function space W. In order to obtain a workable expression for $\bar{\boldsymbol{\lambda}}$, assume (as in section 1.5.9) that $\bar{\boldsymbol{\lambda}}$ has a *representation* by a function $\lambda(.)$. Thus assume that

$$(\forall \mathbf{w} \in W)\ \bar{\boldsymbol{\lambda}}\mathbf{w} = \int_I \boldsymbol{\lambda}(t)\mathbf{w}(t)\,\mathrm{d}t$$

Then, applying the expression $\bar{\boldsymbol{\lambda}} D$ applied to a vector $\mathbf{z}$, and integrating by parts,

$$-\bar{\boldsymbol{\lambda}} D\mathbf{z} = -\int_0^T \boldsymbol{\lambda}(t)\,[D\mathbf{z}(t)]\,\mathrm{d}t = \int_0^T [D\boldsymbol{\lambda}(t)]\mathbf{z}(t)\,\mathrm{d}t + [\boldsymbol{\lambda}(t)\mathbf{z}(t)]_0^{\mathrm{T}}$$

Since $\mathbf{x}(0) = \mathbf{0}$ can be assumed, by shift of origin (see comment in section 4.1.1), $\boldsymbol{\lambda}(0)\mathbf{z}(0) = 0$. If $\mathbf{x}(T)$ is *free*, thus not restricted by any boundary condition, the condition $\boldsymbol{\lambda}(T) = 0$ will be imposed. Thus the integrated parts vanish.

Define the function

$$H(\mathbf{x}, \mathbf{u}; \boldsymbol{\lambda}) = F(\mathbf{x}, \mathbf{u}) + \bar{\boldsymbol{\lambda}} M(\mathbf{x}, \mathbf{u}) - \bar{\boldsymbol{\nu}} N(\mathbf{x})$$

Note that H differs from the Lagrangian L by omitting the term $-\bar{\lambda}D\mathbf{x}$, and all terms in $G(.)$. Then H is the integral of the Hamiltonian

$$h(\mathbf{x}(t), \mathbf{u}(t), t; \boldsymbol{\lambda}(t)) = f(\mathbf{x}(t), \mathbf{u}(t), t) + \boldsymbol{\lambda}(t)m(\mathbf{x}(t), \mathbf{u}(t), t) - \boldsymbol{\nu}(t)n(\mathbf{x}(t), t)$$

where $\boldsymbol{\lambda}(.)$ and $\boldsymbol{\nu}(.)$ are functions (or sometimes generalized functions – see section 1.5.9), representing the Lagrange multipliers $\bar{\boldsymbol{\lambda}}$ and $\bar{\boldsymbol{\nu}}$.

Now, applying the equation $F_x + \bar{\boldsymbol{\lambda}}(-D + M_x) - \bar{\boldsymbol{\nu}}N_x = 0$ to a vector $\mathbf{z}$, and substituting the expressions for Fréchet derivatives from section 4.1.3, gives

$$\int_0^T \{f_x(\bar{\mathbf{x}}(t), \bar{\mathbf{u}}(t), t) + D\boldsymbol{\lambda}(t) + \boldsymbol{\lambda}(t)m(\bar{\mathbf{x}}(t), \bar{\mathbf{u}}(t), t) - \boldsymbol{\nu}(t)\mathbf{n}(\bar{\mathbf{x}}(t), t)\}\, \mathbf{z}(t)\, \mathrm{d}t = 0$$

for every continuous function $\mathbf{z}(.)$ satisfying the boundary condition $\mathbf{z}(0) = 0$. It follows that the expression in $\{\ldots\}$ must vanish. Hence the *adjoint differential equation* (with boundary condition) is obtained as:

$$-\dot{\boldsymbol{\lambda}}(t) = f_{\mathbf{x}}(\bar{\mathbf{x}}(t), \bar{\mathbf{u}}(t), t) + \boldsymbol{\lambda}(t)m_{\mathbf{x}}(\bar{\mathbf{x}}(t), \bar{\mathbf{u}}(t), t) - \boldsymbol{\nu}(t)\mathbf{n}_{\mathbf{x}}(\bar{\mathbf{x}}(t), t), \quad \boldsymbol{\lambda}(T) = \mathbf{0},$$

$$\boldsymbol{\nu}(t) \in V(t), \quad \boldsymbol{\nu}(t)n(x(t), t) = 0 \quad (0 \le t \le T) \qquad (4.10)$$

(The last expression is obtained by a similar calculation from $\bar{\boldsymbol{\nu}}N(\bar{x}) = 0$.)

If K_Y is a cone of nonnegative functions, then $\bar{\boldsymbol{\nu}} \in K_Y^*$ requires that $\boldsymbol{\nu}(t) \ge 0$ $(0 \le t \le T)$. Note that the first line of (4.10) can be summarized as

$$-\dot{\boldsymbol{\lambda}}(t) = h_x(\mathbf{x}(t), \mathbf{u}(t), t; \boldsymbol{\lambda}(t), \boldsymbol{\nu}(t))$$

A similar treatment of (4.9) leads to differential equations

$$f_u(\mathbf{x}(t), \mathbf{u}(t), t) + \boldsymbol{\lambda}(t)m_u(\mathbf{x}(t)), u(t), t) + \boldsymbol{\mu}(t)g_u(u(t), t) = 0,$$

$$\boldsymbol{\mu}(t)g(\mathbf{u}(t), t) = 0, \quad \boldsymbol{\mu}(t) \in S(t)$$

Now these differential equations, with (4.10), are first-order in $\boldsymbol{\lambda}(t)$. Assuming that the optimal $\mathbf{x}(.)$ and $\mathbf{u}(.)$ are respectively piecewise smooth and piecewise continuous, it follows that the generalized function $\boldsymbol{\lambda}(.)$ is, in fact, an ordinary function, and likewise for $\boldsymbol{\nu}(.)$ and $\boldsymbol{\mu}(.)$. So the assumed representation of $\bar{\boldsymbol{\lambda}}$ and $\bar{\boldsymbol{\nu}}$ is validated. Under some different boundary conditions (section 4.5), $\boldsymbol{\lambda}(.)$ may also include a delta function or its first derivative.

If the term in $\mathbf{n}$ is absent, and f is replaced by τf, then (4.10) has the form

$$\dot{\boldsymbol{\lambda}} + \boldsymbol{\lambda}m_{\mathbf{x}} = \tau r, \quad \boldsymbol{\lambda}(T) = 0$$

where r does not depend on $\boldsymbol{\lambda}$. If $\tau = 0$, then $\boldsymbol{\lambda} = \mathbf{0}$ follows; but τ and $\boldsymbol{\lambda}$ cannot both be zero, from the Fritz John conditions. Hence $\tau \ne 0$ (see also section 4.4.1).

One technicality should be mentioned. In deriving (4.10), $\{\ldots\}$ can fail to vanish on a set E_0 to values of t, where E_0 has zero measure (section 1.5.10). Consequently, (4.10) is more precisely stated as an integral equation:

$$\boldsymbol{\lambda}(t) = \int_t^T [f_x(\mathbf{x}(\mathrm{s}), \mathbf{u}(\mathrm{s}), \mathrm{s}) + m(\mathbf{x}(\mathrm{s}), \mathbf{u}(\mathrm{s}), \mathrm{s}) - \boldsymbol{\nu}(\mathrm{s})\mathbf{n}(\mathbf{x}(\mathrm{s}), \mathrm{s})]\, \mathrm{ds}$$

This is analogous to the given differential equation for $\dot{\mathbf{x}}(t)$, whose precise statement is also as an integral equation.

4.2.8 Hadamard derivatives

A function whose domain is itself a vector space of functions – for example a typical objective function for a control problem, specified by an integral – is often not Fréchet differentiable, even for apparently smooth functions, except for the case of uniform norms, discussed above. Consider for example the function $f(x) := \int_I x(t)^3\,dt$, where $I = [0, 1]$ and $x(.) \in L^2(I)$; thus $\int_I x(t)^2\,dt$ is finite. If $x(t) = t^{-3/8}$, then $x(.) \in L^2(I)$, but $f(x)$ is infinite. So $x(.)$ must be further restricted, say to bounded functions (as they would be in most applications). Consider then, with $x(.)$ bounded,

$$f(x+v) - f(x) = \int_I 3x(t)^2 v(t)\,dt + \int_I 3x(t)v(t)^2\,dt + \int_I v(t)^3\,dt$$

The first term on the right is linear in v, giving $f'(x)v$; the second term on the right is bounded by $c_1\|v\|_2^2$, where $c_1\|v\|_2$ is the L^2-norm. But the last term on the right is not generally $o(\|v\|_2)$. Consider, for example, the sequence

$$v_n(t) := n^{3/8}(0 \le t \le n^{-1}), \quad 0(n^{-1} < t \le 1)$$

Then each $v_n(.)$ is bounded, $\|v_n\|_2 = n^{-1/8} \to 0$ as $n \to \infty$, but

$$|f(v_n)|/\|v\|_2 = n^{1/8}/n^{-1/8} \to \infty$$

So f is not Fréchet differentiable on the subspace (call it $L_b^2(I)$) of those functions $L^2(I)$, each of which is bounded. Also f is not Lipschitz, since there is no constant κ for which $(\forall v)\ |f(x+v) - f(v)| \le \kappa\|v\|_2$. However, f is linearly Gâteaux differentiable at x since, for each fixed $v(.) \in L_b^2(I)$,

$$(\forall \alpha > 0)\ f(x+\alpha v) - f(x) = \alpha \int_I 3x(t)^2 v(t)\,dt + \alpha^2 \int_I 3x(t)v(t)^2\,dt + \alpha^3 \int_I v(t)^3\,dt$$

This example suggests that the space of functions may have to be restricted, and also that a less stringent notion of derivative may be needed. The following theorem indicates an approach to less stringent derivatives.

Theorem

Let X and Y be normed spaces; let the function $\mathbf{f}: X \to Y$ have a linear Gâteaux derivative at the point $\mathbf{a}$, and let f satisfy the Lipschitz condition

$$(\exists \kappa, 0 < \kappa < \infty) \quad (\forall \mathbf{v}, \mathbf{w} \in X)\ \|\mathbf{f}(\mathbf{v}) - \mathbf{f}(\mathbf{w})\| \le \kappa\|\mathbf{v} - \mathbf{w}\|$$

Then, for each continuous function $\boldsymbol{\omega}: [0, 1] \to X$ having $\boldsymbol{\omega}(0) = \mathbf{a}$, and having initial slope $\mathbf{s} = \boldsymbol{\omega}'(0+)$,

$$\mathbf{f}(\boldsymbol{\omega}(\alpha)) - f(\mathbf{a}) - \alpha\mathbf{f}'(\mathbf{a})\boldsymbol{\omega}'(0) = o(\alpha) \quad \text{as } \alpha \downarrow 0$$

Remark

A function $\mathbf{f}$ with this property is called *Hadamard differentiable* at $\mathbf{a}$. This property is intermediate between linear Gâteaux differentiable, where the limit defining the derivative converges for each direction separately, and the Fréchet derivative, where this convergence is uniform over directions $\mathbf{d}$ satisfying $\|\mathbf{d}\| \leq 1$. In fact, Hadamard differentiable corresponds to uniform convergence over compact sets of directions; the proof is omitted, since this fact is not used here. Note also that

$$\omega'(0+) = \lim_{\alpha \downarrow 0} \alpha^{-1}[\boldsymbol{\omega}(\alpha) - \boldsymbol{\omega}(0)]$$

the *one-sided derivative* of $\boldsymbol{\omega}$ at 0. The Hadamard derivative is the weakest kind of derivative for which the usual chain rule always holds.

Proof

$$\begin{aligned}
\|\mathbf{f}(\boldsymbol{\omega}(\alpha)) - \mathbf{f}(\mathbf{a}) - \alpha \mathbf{f}'(\mathbf{a})\mathbf{s}\| &\leq \|\mathbf{f}(\boldsymbol{\omega}(\alpha)) - \mathbf{a} - \alpha\mathbf{s}\| + \|\mathbf{f}(\mathbf{a} + \alpha\mathbf{s}) - \mathbf{f}(\mathbf{a}) - \alpha\mathbf{f}'(\mathbf{a})\mathbf{s}\| \\
&\leq \kappa\|\boldsymbol{\omega}(\alpha) - \mathbf{a} - \alpha\mathbf{s}\| + \|\mathbf{f}(\mathbf{a} + \alpha\mathbf{s}) - \mathbf{f}(\mathbf{a}) - \alpha\mathbf{f}'(\mathbf{a})\mathbf{s}\| \\
&= \kappa.o(\alpha) + o(\alpha) \\
&= o(\alpha) \quad \text{as } \alpha \downarrow 0
\end{aligned}$$

since $\mathbf{s} = \boldsymbol{\omega}'(0+)$ and $\mathbf{f}'(\mathbf{a})$ is the Gâteaux derivative of $\mathbf{f}$. □

Remark

The linearization used in the proof (section 3.5.3) of the Karush–Kuhn–Tucker conditions is equally valid when the functions are Hadamard differentiable as when they are Fréchet differentiable. Hence the KKT necessary conditions still hold for a minimum of a constrained problem, when the functions in the problem are Gâteaux differentiable and satisfy a Lipschitz condition, and the derivatives in KKT are linear Gâteaux derivatives.

Hence the necessary conditions for (4.1) in section 4.2.7 apply also when the control problem replaces piecewise continuous functions with the uniform norm by a different normed space of functions, provided that Lipschitz conditions and linear Gâteaux differentiability are assumed. One instance is $L^2_b(I)$, considering however only those increments $v(.)$ that are uniformly bounded (thus, consider a set E on v with $(\forall v \in E)\ (\forall t \in I)\ |v(t)| \leq k$).

Exercise

Prove the *generalized Schwarz inequality*, with a positive semidefinite matrix M, from

$$(\forall \alpha \in \mathbf{R}) \quad (\mathbf{x} + \alpha\mathbf{w})^{\mathrm{T}} M(\mathbf{x} + \alpha\mathbf{w}) \geq 0$$

Hint: This quadratic in α has no real roots (except maybe two equal roots).

Remark

If M is a positive semidefinite $m \times m$ matrix, then M can be diagonalized by a rotation of coordinates, described by an orthogonal matrix L. Denote by

$\lambda_1, \lambda_2, \ldots, \lambda_m$ the eigenvalues of M. Then

$$\mathbf{z}^{\mathrm{T}}M\mathbf{z} = \mathbf{y}^{\mathrm{T}}L^{\mathrm{T}}ML\mathbf{y} = \mathbf{y}^{\mathrm{T}}\Lambda\mathbf{y} \quad \text{where } \Lambda \text{ is a diagonal matrix of eigenvalues}$$
$$\leq \theta\mathbf{y}^{\mathrm{T}}\mathbf{y} \quad \text{where } \theta = \max_i \lambda_i$$
$$= \theta\mathbf{z}^{\mathrm{T}}\mathbf{z} \quad \text{since } L \text{ is an orthogonal matrix}$$

Exercise

Define a quadratic function $f: L^2(I, \mathbf{R}^m) \to \mathbf{R}$, where $I = [0, 1]$ and $L^2(I, \mathbf{R}^m)$ denotes the space of functions from I into $\mathbf{R}^m$ whose squares have finite integrals, by

$$f(x) = \int_I \mathbf{x}(t)^{\mathrm{T}} A(t)\mathbf{x}(t)\,\mathrm{d}t$$

where $A(t)$ is a continuous (hence bounded) positive semidefinite matrix function of $t \in I$. Discuss whether f is (i) Lipschitz, (ii) Fréchet differentiable at $\mathbf{x}$, (iii) Hadamard differentiable at $\mathbf{x}$. It is appropriate to restrict $\mathbf{x}$ and $\mathbf{x} + \mathbf{v}$ to a bounded set E; thus $E := \{\mathbf{z} \in L^2(I) : \|\mathbf{z}\|_2 \leq k\}$.

Remark

For $\mathbf{x}, \mathbf{x} + \mathbf{v} \in E$,

$$|(\mathbf{x}(t) + \mathbf{v}(t))^{\mathrm{T}}A(t)\,(\mathbf{x}(t) + \mathbf{v}\mathbf{z}(t)) - \mathbf{x}(t)^{\mathrm{T}}A(t)\mathbf{x}(t)| = |(2\mathbf{x}(t) + \mathbf{v}(t))^{\mathrm{T}}A(t)\mathbf{v}(t)|$$
$$\leq |(2\mathbf{x}(t) + \mathbf{v}t))^{\mathrm{T}}A(t)\,(2\mathbf{x}(t) + \mathbf{v}(t))|^{1/2}.|\mathbf{v}(t)^{\mathrm{T}}A(t)\mathbf{v}(t)|^{1/2}$$

by the generalized Schwarz inequality.

4.3 PONTRYAGIN CONDITIONS

If (4.1) reaches a minimum at $(\bar{x}, \bar{u})$, it is shown in section 4.2.7 that the adjoint differential equation (4.10) is satisfied by some costate function $\lambda(.)$, and also that the conditions (4.9) hold. Define the Hamiltonian $h(.)$ and its integral H, as was done in section 4.2.7. Then, in the symbols of section 4.2.7,

$$H(\mathbf{x}, \mathbf{u}; \bar{\boldsymbol{\lambda}}, \bar{\mathbf{v}}) = F(\mathbf{x}, \mathbf{u}) + \bar{\boldsymbol{\lambda}}M\mathbf{x}, \mathbf{u}; \quad -\bar{\mathbf{v}}N(\bar{\mathbf{x}})$$

The conditions (4.9) are Karush–Kuhn–Tucker *necessary* conditions for a minimum at $\bar{u}(.)$ of the problem

$$\text{Minimize}_{\mathbf{u}}\ H(\bar{\mathbf{x}}, \mathbf{u}, \bar{\boldsymbol{\lambda}}) \quad \text{subject to} \quad \mathbf{u} \in \Gamma \tag{4.11}$$

where $\mathbf{u} \in \Gamma$ represents the constraints on the control $\mathbf{u}$. Note that only $\mathbf{u}$ is being varied in this problem; $\mathbf{x}$ and $\boldsymbol{\lambda}$ stay at their optimal values $\bar{\mathbf{x}}$ and $\bar{\boldsymbol{\lambda}}$.

Under any conditions that make these necessary conditions also *sufficient*, a minimum of (4.1) will imply a minimum of (4.11). Moreover, when the constraints on the control $\mathbf{u}$ are of the form

$$(\forall t)\ \mathbf{u}(t) \in \Gamma(t) \tag{4.12}$$

and are thus a separate constraint on $\mathbf{u}(t)$ for each t in the interval, without any constraints involving two or more different times, the analogy with the discrete problem suggests that the *associated problem*:

$$\text{Minimize } h(\bar{\mathbf{x}}(t), \mathbf{u}(t), t; \bar{\boldsymbol{\lambda}}(t), \bar{\mathbf{v}}(t)) \quad \text{subject to} \quad \mathbf{u}(t) \in \Gamma(t) \qquad (4.13)$$

should reach a minimum when $\mathbf{u}(t) = \bar{\mathbf{u}}(t)$, for each time t in the interval.

The following theorem gives exact conditions for this to happen. The proof is given in Chapter 7. In order to state the theorem, the following stronger definition of differentiability is required. The function F is partially Fréchet differentiable with respect to $\mathbf{x}$, *uniformly in* $\mathbf{u}$ *near* $\bar{\mathbf{u}}$, if

$$F(\mathbf{x}, \mathbf{u}) - F(\bar{\mathbf{x}}, \mathbf{u}) = F_x(\bar{\mathbf{x}}, \bar{\mathbf{u}})(\mathbf{x} - \bar{\mathbf{x}}) + o(\|\mathbf{x} - \bar{\mathbf{x}}\| + \|\mathbf{u} - \bar{\mathbf{u}}\|)$$

Note that this differs from the usual Fréchet differentiability at $(\bar{\mathbf{x}}, \mathbf{u})$ by replacing $F_x(\bar{\mathbf{x}}, \mathbf{u})$ by $F_x(\bar{\mathbf{x}}, \bar{\mathbf{u}})$ in the definition. A similar property will be required for M. In addition, the L^1-norm will be used for the vector space of controls, i.e. $\|\mathbf{u}\|_1 = \int_I |\mathbf{u}(t)|\, dt$, because a critical step in the proof requires this choice of norm. But then differentiation with respect to $\mathbf{u}$ does not work.

A set $E_0 \subset \mathbf{R}$ is a *null set* if E_0 has zero measure. (Note that E_0 need not be empty!)

4.3.1 Theorem (Pontryagin principle)

Let the optimal control problem (4.1) reach a (local) minimum at $(\bar{\mathbf{x}}, \bar{\mathbf{u}})$ with respect to the L_1-norm for $\mathbf{u}$. Assume that the constraints on the control $\mathbf{u}$ have the separated form (4.12); assume that there is no state constraint on $\mathbf{x}$; and assume that F and M are partially Fréchet differentiable with respect to $\mathbf{x}$, uniformly in $\mathbf{u}$ near $\bar{\mathbf{u}}$. Then necessary conditions for the minimum are that:

(a) a costate function $\boldsymbol{\lambda}(.)$ satisfies the adjoint differential equation (4.10),with boundary conditions, and
(b) that the associated problem is minimized at $\bar{\mathbf{u}}(t)$, for all t except a null set.

4.3.2 Remarks

Part (b) is often expressed as *Pontryagin's maximum principle*, taking the negative of the Hamiltonian used here, and also changing the sign of the costate $\lambda(.)$. See Luenberger (1969) for an alternative treatment.

Some tedious calculations, using the first mean-value theorem of the differential calculus, show that the required property of *partial differentiability, uniformly in* $\mathbf{u}$, holds if f and m are twice differentiable (so that h is also), and the second partial derivatives h_{xx} and h_{xu} are bounded in a region defined by $\|\mathbf{x} - \bar{\mathbf{x}}\| < \delta$, $\|\mathbf{u} - \bar{\mathbf{u}}\| < \delta$, for some $\delta > 0$. (Such a bounded region is enough; such bounds will seldom hold over the whole space.)

In the absence of a state constraint, $\mathbf{v}(.)$ is also absent. If there is a state constraint, then the conclusion of the theorem remains valid, under slightly modified hypotheses. Express the state constraint as $N(\mathbf{x}) - \mathbf{s} = \mathbf{0}$ in terms of a slack variable $\mathbf{s} \geq \mathbf{0}$, and adjoin $\mathbf{s}$ to $\mathbf{x}$ as an additional state variable. The hypothesis of *differentiable, uniformly in* $\mathbf{u}$, is then required to hold when the state variable $\mathbf{x}$ is expanded to $(\mathbf{x}, \mathbf{s})$.

Another description of the optimum (also proved in Chapter 7) holds when the functions f, m, Γ in (4.1) do not depend explicitly on t. Thus, $f(\mathbf{x}(t), \mathbf{u}(t), t)$ reduces to $f(\mathbf{x}(t), \mathbf{u}(t))$ only, and similarly for m.

4.3.3 Theorem

If the functions f, m, Γ in (4.1) do not depend explicitly on t, and there is no state constraint, then $h(\bar{\mathbf{x}}(t), \bar{\mathbf{u}}(t); \boldsymbol{\lambda}(t))$ is constant in t.

4.3.4 Summary diagram

Figure 4.2 summarizes the relations between various properties of fixed-time optimal control. (Some are proved later.)

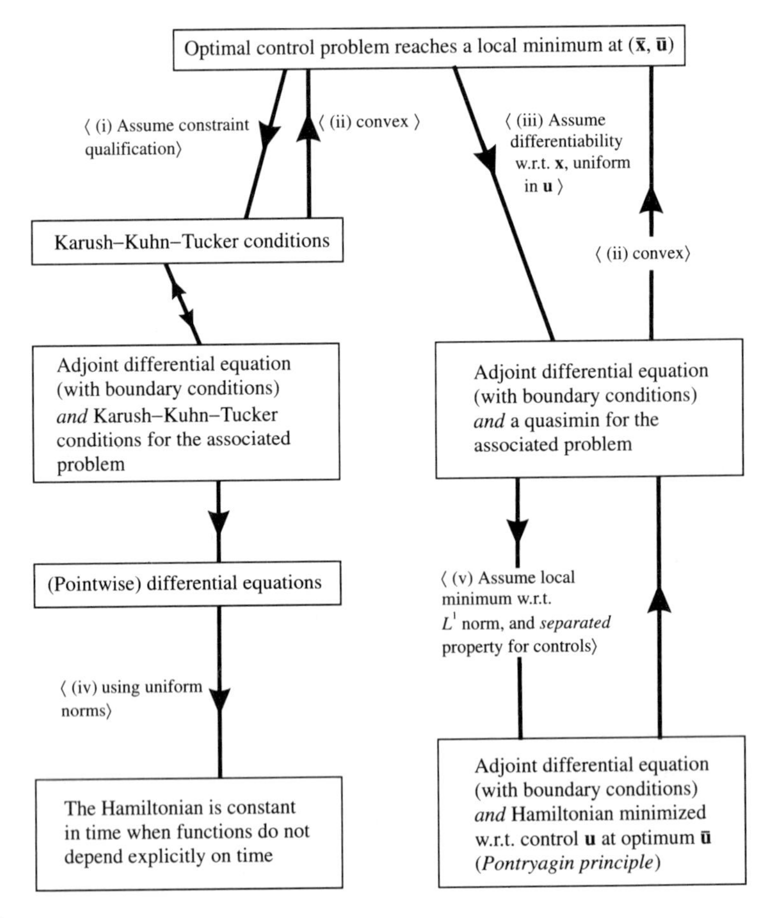

Figure 4.2 Flowsheet for (fixed time) optimal control theorems.

The *assumptions* for this flowsheet are as follows.

(i) The map $[D - M_{\mathbf{x}}, -M_{\mathbf{u}}]$ is surjective, and $\mathbf{u} \in \Gamma$ satisfies a constraint qualification.
(ii) The functions in the problem are convex.
(iii) $H(\mathbf{x}, \mathbf{u}; \boldsymbol{\lambda})$ is continuously differentiable with respect to $\mathbf{x}$ (*not* with respect to $\mathbf{u}$); and $H(\mathbf{x}, \mathbf{u}) - H(\bar{\mathbf{x}}, \mathbf{u}) = H_x(\bar{\mathbf{x}}, \bar{\mathbf{u}})(\mathbf{x} - \bar{\mathbf{x}}) + o(\|\mathbf{x} - \bar{\mathbf{x}}\| + \|\mathbf{u} - \bar{\mathbf{u}}\|)$. Thus this remainder term is small, uniformly in $\mathbf{u}$ near η. This happens if the state space X has norm $\|\mathbf{x}\|_\infty + \|D\mathbf{x}\|_\infty$, and h_{xx} and h_{xu} are bounded.
(iv) X has the norm stated in (iii).
(v) The optimal control problem is minimized at $\mathbf{u} = \bar{\mathbf{u}}$, when the control space has the L^1 norm $\|\mathbf{u}\|_1 = \int |\mathbf{u}(t)|\, dt$; and if $\mathbf{u} \in \Gamma$ and $\mathbf{v} \in \Gamma$, then also $\mathbf{w} \in \Gamma$, if $\mathbf{w}(t) = \mathbf{u}(t)$ for $t \in A$ and $\mathbf{w}(t) = \mathbf{v}(t)$ for $t \in [0, T] \setminus A$, for any set of times $T \subset [0, T]$. This happens if $\mathbf{u} \in \Gamma \Leftrightarrow (\forall t)\ \mathbf{u}(t) \in \Delta$.

4.4 SOME EXAMPLES

In the *associated problem*, consider the case when its objective function is linear in $\mathbf{u}(t)$, and the constraints on $\mathbf{u}(t)$ are also linear – in the simplest case, $\mathbf{u}(t) \in \mathbf{R}$, and the constraint requires $\alpha \le \mathbf{u}(t) \le \beta$, for constant α and β. Denote the coefficient of $\mathbf{u}(t)$ in the associated problem by $c(t)$ (which depends, of course, on the state $\mathbf{x}(t)$ and on the costate $\boldsymbol{\lambda}(t)$.) If $c(t) > 0$, then minimization determines $\mathbf{u}(t) = \alpha$; if $c(t) < 0$, then $\mathbf{u}(t) = \beta$. This describes a phenomenon called *bang-bang control*, where $\mathbf{u}(t)$ jumps between its extreme points, depending on $\boldsymbol{\lambda}(t)$. If $\mathbf{u}(t) \in \mathbf{R}^r$ with $r > 1$, then the analogue of bang-bang control jumps between the vertices of the polyhedron described by the constraints (e.g. section 2.1).

However, there is also the possibility that $c(t) \equiv 0$ for some interval of time t. In this situation, the associated problem does not determine $\mathbf{u}(t)$ in that interval; $\mathbf{u}(t)$ need not be at a vertex, and in that event, that part of the solution is called a *singular arc*. Note then that $c(t) \equiv 0$, say for $\alpha_1 < t < \beta_1$, implies that the derivative $\dot{c}(t) \equiv 0$ for $\alpha_1 < t < \beta_1$, and this often provides further information about the optimal solution.

4.4.1 Example

Consider the optimal control problem:

$$\text{Maximize} \int_0^2 [1 - u(t)]x(t)\, dt \quad \text{subject to} \quad x(0) = a > 0 \quad \text{and}$$
$$\dot{x}(t) = u(t)x(t) - \tfrac{1}{3}x(t), \quad 0 \le u(t) \le 1 \quad (0 \le t \le 2)$$

Appropriate function spaces here are piecewise continuous functions for $u(.)$, and piecewise smooth functions (thus, integrals of piecewise continuous functions) for $x(.)$, with uniform norms as described in sections 1.10 and 4.1.2. The Hamiltonian (introducing a minus sign to change to minimizing) is

$$h(x(t), u(t), \lambda(t)) = -[1 - u(t)]x(t) + \lambda(t)[u(t)x(t) - \tfrac{1}{3}x(t)]$$

The associated problem gives:

$$\text{Minimize}_{0 \le u(t) \le 1} [1 + \lambda(t)]u(t)x(t)$$

(Only terms of h involving $u(t)$ are relevant here.) Assuming (subject to later verification) that $(\forall t)x(t) > 0$, Pontryagin's principle requires that:

$$\text{if } \lambda(t) < -1 \quad \text{then } u(t) = +1$$

$$\text{if } \lambda(t) > -1 \quad \text{then } u(t) = 0$$

(If $\lambda(t)$ remains constant at -1 for some interval of t, then $u(t)$ for that interval must be found by other means.)

The adjoint differential equation (from $-\dot{\lambda}(t) = h_x(.)$) is:

$$-\dot{\lambda}(t) = -1 + u(t) + \lambda(t)[u(t) - \tfrac{1}{3}], \quad \lambda(2) = 0$$

Hence

$$\dot{\lambda}(t) + [u(t) - \tfrac{1}{3}]\lambda(t) = 1 - u(t)$$

For an interval of t on which $u(t) \equiv 0$, the solution is $\lambda(t) = -3 + c_1 e^{\frac{1}{3}t}$; for an interval on which $u(t) \equiv 1$, the solution is $\lambda(t) = c_2 e^{-\frac{2}{3}t}$; c_1 and c_2 are constants of integration. Since $\lambda(2) = 0$, $\lambda(t) > -1$ on some interval $(c, 2)$; hence $u(t) \equiv 0$ on this interval. Since $0 = -3 + c_1 e^{\frac{2}{3}t}$,

$$\lambda(t) = 3(-1 + e^{\frac{1}{3}(t-2)})$$

A *switching time* $t = c$ is determined by $\lambda(c) = -1$; hence $t = 2 + 3\ln(\tfrac{2}{3}) \approx 0.7836$. Consider an interval (c', c) on which $u(t) \equiv 1$. For this interval, noting that $\lambda(c) = -1$,

$$\lambda(t) = -e^{2/3(c-t)} < -1 \quad \text{for } t < c$$

So, for this problem, c' can be as low as 0; there is only one switching time, $t = c$. So the optimum is given by $u(t) \equiv 1$ for $0 < t < c$, $u(t) \equiv 0$ for $c < t < 1$.

Note that, as is usual, the adjoint differential equation is being solved *backwards* in time, commencing with the boundary condition, here $\lambda(2) = 0$. It is commonly the case that numerical methods for integrating ordinary differential equations are *stable* in one direction (thus, errors due to the necessary approximation by finite differences remain small), and *unstable* in the opposite direction (errors tend to grow at an exponential rate). If the given differential equation is stable in the forward direction (which hopefully it is, otherwise the given control problem is not very well defined numerically), then the adjoint differential equation is generally stable when solved in the backwards direction.

Could there be a singular arc, namely an interval of t on which $\lambda(t) \equiv -1$? If so, then the adjoint equation requires

$$0 + (-1)[u(t) - \tfrac{1}{3}] \equiv 1 - u(t)$$

on that interval; but there is no $u(t)$ satisfying this.

Substituting $x = \bar{x} + \alpha$ and $u = \bar{u} + \beta$ into the differential equation for $\dot{x}$, subtracting the differential equation for $\dot{\bar{x}}$, then linearizing by retaining only linear terms in α and β, leads to $\dot{\alpha} = \bar{u}\beta + \bar{x}\alpha - \frac{1}{2}\alpha$. The constraint qualification assumed (section 4.2.7) to obtain Karush–Kuhn–Tucker conditions requires, in part, that the differential equation

$$\dot{\alpha} - [\bar{u}\beta + \bar{x}\alpha - \tfrac{1}{2}\alpha] = \gamma, \quad \alpha(0) = 0$$

shall be solvable, for each given continuous forcing function γ, for some $\bar{u} + \alpha$ satisfying $0 \le (\bar{u} + \alpha)(.) \le 1$. (Thus, the differential operator on the left is *onto*, as required in section 3.7.)

Could only Fritz John conditions hold, in which f has a nonnegative multiplier τ, which may be zero? In the present example, with τ inserted, if $\tau = 0$ then the adjoint differential equation reduces to $\dot{\lambda}(t) + [u(t) - \frac{1}{3}]\lambda(t) = 0$, $\lambda(2) = 0$. Hence

$$\lambda(t) = c_0 e^{-\varphi(t)} \quad \text{where } \varphi(t) = \int_0^T [u(s) - \tfrac{1}{3}]\, ds$$

and $\lambda(2) = 0 \Rightarrow c_0 \Rightarrow 0 \Rightarrow \lambda(.) \equiv 0$. Since τ and $\lambda(.)$ cannot both be zero in the Fritz John conditions, this case does not happen. Solutions for $\lambda(.)$ of exponential type occur commonly in control problems in continuous time, and (as in this example) there results $\tau = 1$.

4.4.2 Example

The problem of maximizing the area enclosed by a differentiable closed plane curve for which the tangent vectors must lie in a given region A can be expressed as the following optimal control problem:

$$\text{Maximize } \tfrac{1}{2} \int_0^T [x_1(t)u_2(t) - x_2(t)u_1(t)]\, dt \quad \text{subject to}$$

$$\dot{x}_1(t) = u_1(t), \quad \dot{x}_2(t) = u_2(t), \quad x_1(T) = x_1(0), \quad x_2(T) = x_2(0), \quad (u_1(t), \quad u_2(t)) \in A$$

The determinantal expression in the integrand represents twice the area of an (approximately) triangular region bounded by a small part of the curve (Figure 4.3) and radius vectors; and so the area enclosed by the whole closed curve is got by integration.

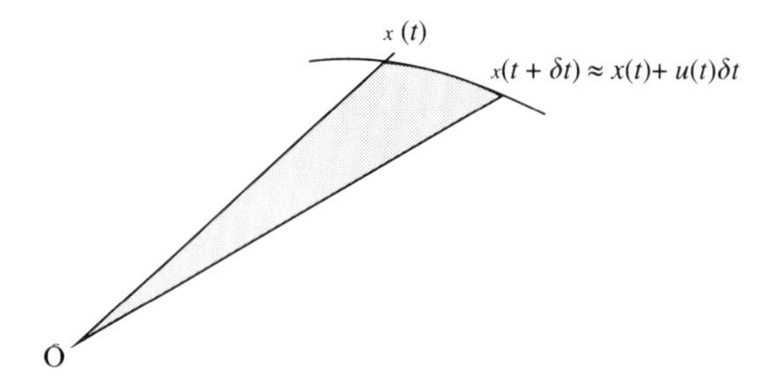

Figure 4.3 Area bounded by a curve.

The Hamiltonian for this problem is:

$$h(.) = \tfrac{1}{2}[-x_1(t)u_2(t) + x_2(t)u_1(t)] + \lambda_1(t)u_1(t) + \lambda_2(t)u_2(t)$$

after conversion to a minimizing problem. The adjoint differential equation is

$$-\dot{\lambda}_1(t) = -\tfrac{1}{2}u_2(t), \quad -\dot{\lambda}_2(t) = \tfrac{1}{2}u_1(t)$$

Since there are terminal conditions, this adjoint equation has no boundary conditions. Substituting from the given differential equations, $2\dot{\lambda}_1(t) = \dot{x}_2(t)$ and $2\dot{\lambda}_2(t) = -\dot{x}_1(t)$. Hence, for some constants α and β, for all t, $2\lambda_1(t) - x_2(t) = \alpha$ and $2\lambda_2(t) + x_1(t) = \beta$. Substituting into the Hamiltonian to eliminate λ_1 and λ_2 gives

$$2h(.) = u_1(t)\,[\alpha + x_2(t)] + u_2(t)\,[\beta - x_1(t)]$$

This is the inner product of the two vectors $u(t) = (u_1(t), u_2(t))$ and $z(t) := (\alpha + x_2(t), \beta - x_1(t))$.

From Pontryagin's principle, $u(t)^{\mathrm{T}}z(t)$ is minimized with respect to $u(t) \in A$ when $u(t) = \eta(t)$. The *support function* of the set A is

$$\mathbf{s}(v|A) := \sup_{w \in A} v^{\mathrm{T}}w$$

So

$$\mathbf{s}(-z(t)|A) = 2h(\bar{x}(t), \bar{u}(t), \bar{\lambda}(t))$$

which is constant in t by Theorem 7.3.3, since, for this problem, the functions specifying the problem have no explicit dependence on t. Therefore $\mathbf{s}(-z(.)|A)$ is constant.

In particular, if A is the closed unit disc, then $\mathbf{s}(v|A) = \|\mathbf{v}\|$. Hence

$$\|(-\alpha - x_2(t), -\beta + x_1(t))\| = \text{constant}$$

Thus $(\alpha + x_2(t))^2 + (\beta - x_1(t))^2 = \text{constant}$. So the optimum curve is a circle.

Exercise

What happens if A is a closed disc whose centre is displaced from the origin? What happens if A is a square region? (These depend on calculating the support function of A; observe that this is an optimization problem in two dimensions only.)

4.5 BOUNDARY CONDITIONS

The endpoint $x(T)$ may be free (unconstrained) or it may be completely specified, by $x(T) = x_{\text{end}}$ say. There are also intermediate cases. If $x(t) \in \mathbf{R}^2$, a terminal condition could require that the endpoint $x(T)$ lies on a given curve, specified by $q(x(T)) = 0$. If instead, $x(t) \in \mathbf{R}^3$, then $q(x(T)) = 0$ requires that $x(T)$ lies on a given surface. Assume first that T is still fixed. The terminal condition can be written as a state constraint:

$$(\forall t \in I) \quad q(x(t)) \in V(t)$$

where the cone $V(t)$ is the whole space when $t < T$ and $V(T) = \{0\}$. This constraint does not restrict $x(t)$ when $t < T$. The adjoint differential equation is then

$$-\dot{\lambda}_2(t) = h_x(\bar{x}(t), \bar{u}(t), t, \lambda(t)) - \beta\delta(t-T)q_x(x(t)), \quad \lambda(T+0) = 0$$

Here $v(t) = 0$ whenever $t < T$, since $V(t)^* = \{0\}$; since $\bar{v}$ is concentrated at $t = T$, a delta function $\delta(t-T)$ is required to represent it. Since $x(T)$ is not fixed, the usual boundary condition is required for $\lambda(.)$; $\lambda(T+0)$ is a conventional symbol for $\lambda(t)$ just to the right of the delta function at T, and $\lambda(T-0)$ means $\lambda(t)$ just to the left of this delta function. Now, integrating each side of the differential equation over the interval $(T-0, T+0)$, there results:

$$\lambda(T-0) - \lambda(T+0) = h_x(\bar{x}(t), \bar{u}(t), t, \lambda(t)).0 - \beta q_x(x(T))$$

since h_x contributes 0 to the integral over an interval that tends to zero. Hence

$$\lambda(T-0) = -\beta q_x(x(T))$$

So it is equivalent to omit the delta function term from the adjoint equation, and to apply the boundary condition $\lambda(T) = -\beta q_x(x(T))$. Here β is a constant, not yet determined.

Consider now the more complicated control problem:

$$\text{Minimize} \quad J = R(x(t^*)) + \int_0^{t^*} f(x(t), u(t), t)\,\mathrm{d}t \quad \text{subject to}$$

$$x(0) = x_0; \quad \dot{x}(t) = m(x(t), u(t), t) \quad (0 \le t \le t^*);$$

$$u(t) \in \Delta(0 \le t \le t^*);$$

$$q(x(t^*)) = 0$$

Note that, in this problem, all of $u(.)$, $x(.)$ and t^* are varied to reach the optimum; t^* denotes the (variable) time to optimality. The objective J includes, as well as the integral, a term depending on the endpoint. The function q, describing the terminal constraint, is real-valued. For brevity, * will be used to label the value of functions evaluated at $t = t^*$.

Now J can be rewritten as

$$J = \int_I [f(x(t), u(t), t)\pi(t^* - t) + R(x(t)\delta(t - t^*)]\,\mathrm{d}t$$

where $I = [0, t^{\#}]$ for some fixed $t^{\#}$ greater than the optimal t^*,

$$\pi(s) = 0 \ \text{ if } s < 0, \quad \pi(s) = 1 \ \text{ if } s \ge 0$$

and $\delta(.)$ is Dirac's delta function. For this resulting fixed-time optimal control problem, the Hamiltonian is:

$$h(x(t), u(t), t, \lambda(t)), t^*) = f(x(t), u(t), t)\pi(t^* - t) + R(x(t)\delta(t - t^*)$$
$$+ \lambda(t)m(x)t), u(t), t) - \beta\delta(t - t^*)q(x(t))$$

Here, as above, the constraint $q(x(t^*))$ has been expressed as

$$(\forall t)\ q(x(t)) \in V(t)$$

where $V(t) = \mathbf{R}$ for $t < t^*$ and $V(t^*) = \{0\}$. So the corresponding costate function $v(t) = 0$ except when $t = t^*$, so that $v(t) = -\beta\delta(t - t^*)$ for some constant β may be appropriately assumed, subject to verification that this solution satisfies all necessary conditions.

The adjoint differential equation is then

$$-\dot{\lambda}(t) = f_x(x(t), u(t), t)\pi(t^* - t) + \lambda(t)m_x(x(t), u(t), t) - \beta\delta(t - t^*)q_x(x(t))$$
$$+ R_x(x(t)\delta(t - t^*)), \quad \lambda(t^{\#}) = 0$$

It suffices to set $t^{\#} = t^* + 0$ in the following derivation of necessary conditions. Integrating the adjoint differential equation from $t = t^* - 0$ to $t = t^* + 0$, only the delta function terms contribute, and there follows:

$$\lambda(t^* - 0) = R_x(x(t^*)) - \beta q_x(x(t^*))$$

which may be briefly written as

$$\lambda^* \equiv \lambda(t^* - 0) = R_x^* - \beta q_x^* \tag{4.14}$$

So far, the optimality of t^* has not been used in the calculation. To do this, t^* may be regarded as a parameter. For optimality, the derivative of J with respect to t^* should be equated to zero. From the formula obtained in section 4.7.7 for the derivative of the objective function of a control problem with respect to a parameter, this derivative equals the integral over I of $\partial h/\partial t^*$, where h is the Hamiltonian. Now (suppressing some obvious arguments),

$$\partial h/\partial t^* = f.\delta(t^* - t) + \beta\delta'(t - t^*).q - R(x(t))].\delta'(t - t^*)$$

This may be integrated with respect to t over I, and also the δ' terms are integrated by parts, using

$$\int R(x(t)).\delta'(t - t^*)\,\mathrm{d}t = -\int [(\partial/\partial t)R(x(t)).\delta(t - t^*)]\,\mathrm{d}t$$

and

$$(\partial/\partial t)R(x(t)) = R_x(x(t))(\mathrm{d}/\mathrm{d}t)x(t) = R_x(x(t))m(x(t), u(t), t)$$

The requirement thus simplifies to

$$0 = f^* + R_x^* m^* - \beta q_x^* m^*$$

Thus the optimal value of β is determined. Substituting it into (4.14) gives the following boundary condition for the adjoint differential equation:

$$\lambda^* = R_x^* - [f^* + R_x^* m^*]q_x^*/[q_x^* m^*] \tag{4.15}$$

The formula (4.15) was first obtained by Teo *et al.* (1987) by a complicated calculation using the chain rule. The present derivation is from Craven (1989).

This formula is relevant to numerical computation of a time-optimal control problem, since the costate function $\lambda(.)$ is needed in order to obtain the gradient of the objective function with respect to the control u, allowing for the dependence of the state x on u (section 4.6.6).

4.6 TIME-OPTIMAL CONTROL

A typical *time-optimal* control problem is to find the minimum time (and corresponding path) to go from a starting point (say the origin) to a *target* set, subject to constraints. While the theory of section 4.5 includes this case, it is worth looking at the problem rather more geometrically.

Figure 4.4 shows a two-dimensional state space, an *optimal path* (= *trajectory*) from the start (the origin) to the target set, and another (non-optimal) path. The optimal path first reaches the target at $t = t^*$; the other path shown reaches the target at $t = t^\wedge > t^*$. Assume that the target is described by a real function $q(.)$, such that $q(\mathbf{x}) > 0$ for $\mathbf{x}$ outside the target set $q(\mathbf{x}) = 0$ for $\mathbf{x}$ on the boundary of the target set, and $q(\mathbf{x}) < 0$ for $\mathbf{x}$ inside the target set. Fix $t^\# > t^*$. Then the objective function becomes, for the state $\mathbf{x}(.)$ sufficiently close to the optimum $\bar{\mathbf{x}}(.)$:

$$f(\mathbf{x}, \mathbf{u}) = \int_0^{t^*} f(\mathbf{x}(t), \mathbf{u}(t), t)\,\mathrm{d}t = \int_0^{t^\#} f(\mathbf{x}(t), \mathbf{u}(t), t)\,(\pi \circ q)(\mathbf{x}(t))\,\mathrm{d}t$$

in which $f(.) \equiv 1$ for a purely time-optimal problem (minimizing exactly the time to the target), and $\pi(s) = 1$ for $s \geq 0$, $\pi(s) = 0$ for $s < 0$. Assume constraints as:

$$\dot{\mathbf{x}}(t) = m(\mathbf{x}(t), \mathbf{u}(t), t), \quad \mathbf{x}(0) = \mathbf{x}_0, \quad n(\mathbf{x}(t), t) \in V(t)$$

For $\mathbf{x} = \bar{\mathbf{x}}$ and $t < t^*$, the Hamiltonian equals

$$f(\mathbf{x}(t), \mathbf{u}(t), t) + \boldsymbol{\lambda}(t)m(\mathbf{x}(t), \mathbf{u}(t), t) - \mathbf{v}(t)n(\mathbf{x}(t), t)$$

since $(\pi \circ q)(\mathbf{x}(t)) = 1$; hence the adjoint differential equation is

$$-\dot{\boldsymbol{\lambda}}(t) = f_{\mathbf{x}}(\mathbf{x}(t), \mathbf{u}(t), t) + \boldsymbol{\lambda}(t)m_{\mathbf{x}}(\mathbf{x}(t), \mathbf{u}(t), t) - \mathbf{v}(t)n_{\mathbf{x}}(\mathbf{x}(t), t)$$

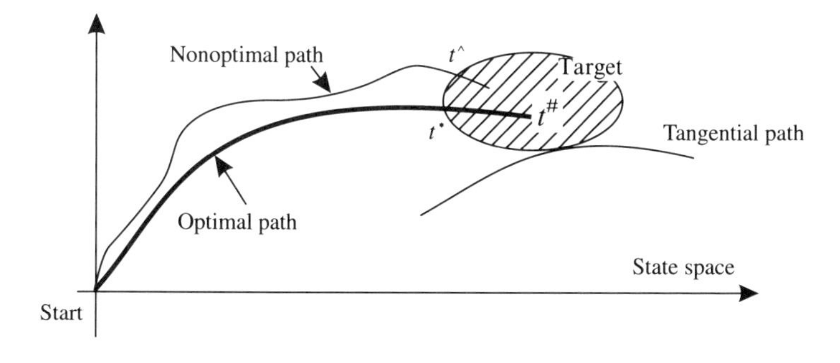

Figure 4.4 Paths in a two-dimensional state space.

for $t < t^*$, noting that $\boldsymbol{v}(t)n(\bar{\mathbf{x}}(t), t) = 0$ and $\boldsymbol{v}(t) \in V(t)^*$. For a purely time-optimal problem without other state constraints, $f_{\mathbf{x}}(.) \equiv 0$, and the adjoint equation becomes $-\dot{\boldsymbol{\lambda}}(t) = \boldsymbol{\lambda}(t)m_{\mathbf{x}}(\mathbf{x}(t), \mathbf{u}(t), t)$. This is enough to solve some problems (e.g. section 5.4.3); for others, the boundary condition for the adjoint equation (section 4.5) is required.

If the space of paths $\mathbf{x}(.)$ is given the norm $\|\mathbf{x}\| = \|\mathbf{x}\|_\infty + \|D\mathbf{x}\|_\infty$, and assuming that $(q \circ \bar{\mathbf{x}})'(t^*) \neq 0$, then t^* is an isolated zero of $q \circ \bar{\mathbf{x}}$ (Craven, 1978). The assumption implies that, for sufficiently small $\|\mathbf{x} - \bar{\mathbf{x}}\|$, $q(\mathbf{x}(t)) < 0$ for some interval $(t^*, t^* + \delta)$; this excludes some cases where $\bar{\mathbf{x}}(.)$ is tangential (Figure 4.4) to the target region. Moreover, $(q \circ \mathbf{x})$ has a zero $\hat{t}(\mathbf{x})$, where $\hat{t}(.)$ is a Fréchet differentiable function at $\bar{\mathbf{x}}$, with $\hat{t}(\bar{\mathbf{x}}) = t^*$.

4.7 SENSITIVITY AND STABILITY

Consider now the mathematical programming problem (3.29), modified to include a parameter $\mathbf{q}$ in the functions. This gives the problem:

$$\text{Minimize}_{\mathbf{x}}\, f(\mathbf{x}, \mathbf{q}) \quad \text{subject to} \quad -\mathbf{h}(\mathbf{x}, \mathbf{q}) \in T \tag{4.16}$$

The suffix $\mathbf{x}$ to *Minimize* indicates that the minimization is only with respect to $\mathbf{x}$. Denote by $V(\mathbf{q})$ the minimum value of the objective function $f(\mathbf{x}, \mathbf{q})$ subject to the constraint $-\mathbf{h}(\mathbf{x}, \mathbf{q}) \in T$. The function V may be called the *perturbation function* (or *value function*) for the problem (4.16). It is assumed that $\mathbf{q}$ is in a normed vector space (in particular, $\mathbf{R}$), and that $\mathbf{q} = \mathbf{0}$ is the original problem which is being perturbed. Denote by $\Gamma(\mathbf{q})$ the feasible set of (4.16), thus

$$\Gamma(\mathbf{q}) = \{\mathbf{x} : -\mathbf{h}(\mathbf{x}, \mathbf{q}) \in T\}$$

A local minimum of (4.16) with $\mathbf{q} = \mathbf{0}$ at $\mathbf{x} = \mathbf{a}$ is called a *strict local minimum* if

$$(\forall^* r > 0)\ (\exists \zeta > 0)\ (\forall \mathbf{x} \in \Gamma(\mathbf{0}), \|\mathbf{x} - \mathbf{a}\| = r), \quad f(\mathbf{x}, \mathbf{0}) \geq f(\mathbf{a}, \mathbf{0}) + \zeta$$

Here $\forall^* r > 0$ means *for all sufficiently small positive r*. The definition means that, if $\mathbf{x}$ moves a distance $r > 0$ from $\mathbf{a}$, and $\mathbf{x}$ is a feasible point for (4.16), then the objective function increases by at least an amount $\zeta = \zeta(r)$, where $\zeta(r)$ is strictly positive. Thus, there is no curve $\mathbf{x} = \boldsymbol{\omega}(\alpha)$, $\alpha \geq 0$ in the feasible region, starting at $\boldsymbol{\omega}(\mathbf{0}) = \mathbf{a}$, with f constant along the curve (thus $(f(\boldsymbol{\omega}(\alpha), \mathbf{0}) = f(\mathbf{a}, \mathbf{0}))$. If such a curve existed, there would then be some small perturbation of f, which would make $f(\boldsymbol{\omega}(\alpha))$ strictly decrease as α increases from 0. That would make the minimum point jump away from $\mathbf{a}$; the perturbed problem (4.16) would not have a local minimum point $\bar{\mathbf{x}}(\mathbf{q})$ close to $\mathbf{a}$, when $\|\mathbf{q}\|$ is small but nonzero.

It is nearly true that, if the constrained problem (4.16) with $\mathbf{q} = \mathbf{0}$ has a strict local minimum, then the problem can be perturbed by a small amount, and there is a perturbed minimum at $\bar{\mathbf{x}}(\mathbf{q})$, where the function $\bar{\mathbf{x}}(.)$ is continuous. A few restrictions are required.

4.7.1 Theorem (perturbation of strict local minimum)
For problem (4.16), assume that:

(i) The unperturbed problem (4.16) with $\mathbf{q} = \mathbf{0}$ reaches a strict local minimum at $\mathbf{x} = \mathbf{a}$.
(ii) For each $\mathbf{q}$, $\mathbf{h}(\mathbf{a}, \mathbf{q}) = \mathbf{h}(\mathbf{a}, \mathbf{0})$.
(iii) The functions $f(\mathbf{x}, \mathbf{q})$ and $\mathbf{h}(\mathbf{x}, \mathbf{q})$ are uniformly continuous on bounded sets.
(iv) When $\mathbf{q} \neq \mathbf{0}$, $f(., \mathbf{q})$ reaches a minimum on each closed bounded set. Then, whenever $\|\mathbf{q}\|$ is sufficiently small, the perturbed problem (4.16) reaches a local minimum at a point $\bar{\mathbf{x}}(\mathbf{q})$, where

$$\bar{\mathbf{x}}(\mathbf{q}) \to \mathbf{a} \quad \text{as} \quad \mathbf{q} \to \mathbf{0}$$

Proof
See Craven (1994) and section 7.4.

Example
Figure 4.5(a) shows a strict local minimum (thick curve), and a small perturbation of it (thin curve), showing that the minimum has moved only a short distance away. Figure 4.5(b) (thick curve) shows a nonstrict local minimum, where the function stays constant in some direction. Here, some small perturbations, e.g. the thin curve, no longer have a minimum near that of the original function.

4.7.2 Remarks
A linear program minimizes a linear function over a polyhedral region. If the minimum is unique, then it is at a vertex (rather than a whole edge or face of the polyhedron consisting of minimum points), and it is a strict minimum. The

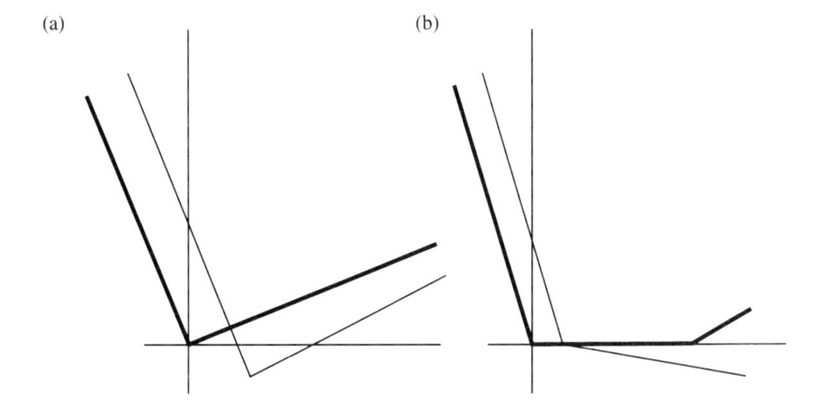

Figure 4.5 (a) Strict and (b) nonstrict minima.

usual sensitivity analysis for linear programming shows then that a small perturbation of the problem only moves the minimum point by a small amount.

A small perturbation of a constraint can remove any feasible points. For example, the constraint $x^2 + q \le 0$ $(x, q \in \mathbf{R})$ is satisfied by $x = 0$ when $q \le 0$, but has no solution x when $q > 0$. The constraint $-\mathbf{h}(\mathbf{x}, \mathbf{q}) \in T$ can be modified to $-\tilde{\mathbf{h}}(\mathbf{x}, \mathbf{q}) \in T$, where $\tilde{\mathbf{h}}(\mathbf{x}, \mathbf{q}) = \mathbf{h}(\mathbf{x}, \mathbf{q}) - \mathbf{h}(\mathbf{a}, \mathbf{q}) + \mathbf{h}(\mathbf{a}, \mathbf{0})$, and then $\tilde{\mathbf{h}}(\mathbf{a}, \mathbf{q}) = \mathbf{h}(\mathbf{a}, \mathbf{0})$. This changes $\mathbf{h}$ by an amount $o(\|\mathbf{q}\|)$; assumption (ii) assumes that this change has been made.

In finite dimensions, continuity implies uniform continuity on bounded sets. But, for control problems, the dimensions are infinite, and (iii) only holds under a further restriction (section 7.6). If these are assumed, then Theorem 4.7.1 leads to a convergence result for certain algorithms for computing control problems. There also the perturbed problems are finite-dimensional computational approximations to the given problems, so that (iv) follows automatically. Note that Theorem 4.7.1 does not need the functions to be differentiable.

Under stronger assumptions, the approximation of $\bar{\mathbf{x}}(\mathbf{q})$ to $\mathbf{a}$ can be expressed more precisely by a Lipschitz condition.

4.7.3 Theorem (Lipschitz perturbations)

For the problem (4.16), assume the hypotheses of Theorem 4.7.1, and in addition:

(v) The functions $f(.,.)$ and $\mathbf{h}(.,.)$ satisfy Lipschitz conditions.

(vi) When $\mathbf{q} = \mathbf{0}$, the constraint satisfies the *Robinson stability condition* (section 3.7).

(vii) The objective function f satisfies the *growth condition*:

$$f(\mathbf{x}, \mathbf{0}) - \mathbf{f}(\mathbf{a}, \mathbf{0}) \ge \kappa\|\mathbf{x} - \mathbf{a}\|^{\sigma}$$

in feasible directions, where κ is constant and $\sigma \ge 1$.

Then, when $\|\mathbf{q}\|$ is sufficiently small, the perturbed problem (4.16) reaches a local minimum at $\mathbf{x} = \bar{\mathbf{x}}(\mathbf{q})$, where

$$\|\bar{\mathbf{x}}(\mathbf{q}) - \mathbf{a}\| \le c\|\mathbf{q}\|^{1/\sigma}$$

where c is a constant.

Proof

See section 7.5.3. □

The following theorem measures the dependence of the objective function value on the parameter $\mathbf{q}$. The Lagrangian for (4.16) is

$$L(\mathbf{u}, \mathbf{v}; \mathbf{q}) := f(\mathbf{u}, \mathbf{q}) + \mathbf{v}\mathbf{h}(\mathbf{u}, \mathbf{q})$$

Let $F(\mathbf{q}) := f(\bar{\mathbf{x}}(\mathbf{q}), \mathbf{q})$.

4.7.4 Theorem (sensitivity of objective)

For problem (4.16), assume the functions f and $\mathbf{h}$ are continuously differentiable, and that:

(i) (4.16) reaches a minimum (or a quasimin) at $\mathbf{x} = \bar{\mathbf{x}}(\mathbf{q})$, satisfying

$$\|\bar{\mathbf{x}}(\mathbf{q}) - \bar{\mathbf{x}}(\mathbf{0})\| \leq c\|\mathbf{q}\|$$

for sufficiently small $\|\mathbf{q}\|$, where c is constant.

(ii) The Karush–Kuhn–Tucker conditions hold:

$$L_{\mathbf{u}}(\bar{\mathbf{x}}(\mathbf{q}), \bar{\mathbf{v}}(\mathbf{q}); \mathbf{q}) = \mathbf{0}; \quad \bar{\mathbf{v}}(\mathbf{q}) \in T^*; \quad \bar{\mathbf{v}}(\mathbf{q})\mathbf{h}(\bar{\mathbf{x}}(\mathbf{q}), \mathbf{q}) = 0$$

with Lagrange multiplier $\bar{\mathbf{v}}(\mathbf{q})$ such that $\bar{\mathbf{v}}(.)$ is continuous at $\mathbf{0}$. Then the perturbation function $V(.)$ is differentiable and satisfies

$$V'(\mathbf{0}) = L_{\mathbf{q}}(\bar{\mathbf{x}}(\mathbf{0}), \bar{\mathbf{v}}(\mathbf{0}); \mathbf{0})$$

Proof

For sufficiently small $\|\mathbf{q}\|$ and $\|\mathbf{q}'\|$,

$$\begin{aligned} V(\mathbf{q}) - V(\mathbf{q}') &= L(\bar{\mathbf{x}}(\mathbf{q}), \bar{\mathbf{v}}(\mathbf{q}); \mathbf{q}) - L(\bar{\mathbf{x}}(\mathbf{q}'), \bar{\mathbf{v}}(\mathbf{q}'); \mathbf{q}') \\ &= L(\bar{\mathbf{x}}(\mathbf{q}), \bar{\mathbf{v}}(\mathbf{q}'); \mathbf{q}) - L(\bar{\mathbf{x}}(\mathbf{q}'), \bar{\mathbf{v}}(\mathbf{q}'); \mathbf{q}') - \bar{\mathbf{v}}(\mathbf{q}')\mathbf{h}(\bar{\mathbf{x}}(\mathbf{q}), \bar{\mathbf{q}}) \\ &\geq L(\bar{\mathbf{x}}(\mathbf{q}), \bar{\mathbf{v}}(\mathbf{q}'); \mathbf{q}) - L(\bar{\mathbf{x}}(\mathbf{q}'), \bar{\mathbf{v}}(\mathbf{q}'); \mathbf{q}') \\ &= L_{\mathbf{u}}(\bar{\mathbf{x}}(\mathbf{q}'), \bar{\mathbf{v}}(\mathbf{q}'); \mathbf{q}')\,(\bar{\mathbf{x}}(\mathbf{q}') - \bar{\mathbf{x}}(\mathbf{q})) + L_{\mathbf{q}}(\bar{\mathbf{x}}(\mathbf{q}'), \bar{\mathbf{v}}(\mathbf{q}'); \mathbf{q}')\,(\mathbf{q} - \mathbf{q}') + r \end{aligned}$$

where the remainder term $r = o(\|\bar{\mathbf{x}}(\mathbf{q}') - \bar{\mathbf{x}}(\mathbf{q})\| + \|\mathbf{q} - \mathbf{q}'\|)$. From the Karush–Kuhn–Tucker conditions, the $L_{\mathbf{u}}$ term vanishes. It follows that

$$V(\mathbf{q}) - V(\mathbf{0}) \geq L_{\mathbf{q}}(\bar{\mathbf{x}}(\mathbf{0}), \bar{\mathbf{v}}(\mathbf{0}); \mathbf{0})\,(\mathbf{q} - \mathbf{0}) + o(\|\mathbf{q}\|)$$

and

$$\begin{aligned} V(\mathbf{0}) - V(\mathbf{q}) &\geq L_{\mathbf{q}}(\bar{\mathbf{x}}(\mathbf{q}), \bar{\mathbf{v}}(\mathbf{q}); \mathbf{q})\,(\mathbf{0} - \mathbf{q}) + o(\|\mathbf{q}\|) \\ &= [L_{\mathbf{q}}(\bar{\mathbf{x}}(\mathbf{0}), \bar{\mathbf{v}}(\mathbf{0}); \mathbf{0}) + o(1)]\,(-\mathbf{q}) + o(\|\mathbf{q}\|) \end{aligned}$$

since $\bar{\mathbf{v}}(.)$ and $L_{\mathbf{q}}(.)$ are assumed continuous. These two inequalities combine to show that

$$V(\mathbf{q}) - V(\mathbf{0}) = L_{\mathbf{q}}(\bar{\mathbf{x}}(\mathbf{0}), \bar{\mathbf{v}}(\mathbf{0}); \mathbf{0})\mathbf{q} + o(\|\mathbf{q}\|)$$

which proves the theorem. □

4.7.5 Remarks

The assumption that the Lagrange multiplier is a continuous function of the parameter $\mathbf{q}$ is *not* fulfilled in all cases. If the problem is a linear program, then the Lagrange multiplier is a dual variable. Some perturbations, such as some changes in the requirements vector of the given linear program, will cause the optimum point of the dual program to jump from one of its vertices to another. In this example, the requirement that $\mathbf{v}(.)$ is continuous is fulfilled exactly when the dual optimum is unique, for then it is a strict maximum. For a nonlinear problem, it suffices if the optimum point of the dual (or quasidual; section 3.8) problem is a continuous function of the parameter.

In order to apply these results to optimal control problems, a technical difficulty must be overcome. If a change in parameter $\mathbf{q}$ changes the optimal control $\mathbf{u}$, then the optimal state $\mathbf{x}$ changes also in consequence of the change in $\mathbf{u}$. So the change in the optimal value of $F(\mathbf{x}, \mathbf{u})$ is partly from the direct change in $\mathbf{u}$ and partly indirectly from the consequent change in $\mathbf{x}$. This can be calculated as follows, using the Hamiltonian.

4.7.6 Gradient of objective in a control problem

The given differential equation for $\dot{\mathbf{x}}(t)$, with initial condition, determines the state $\mathbf{x}(.)$ as a function of the control $\mathbf{u}(.)$; denote this by $\mathbf{x} = \Phi(\mathbf{u})$. Suppose that $(\hat{\mathbf{x}}, \hat{\mathbf{u}})$, where $\hat{\mathbf{x}} = \Phi(\hat{\mathbf{u}})$, is some approximation to the optimum $(\bar{\mathbf{x}}, \bar{\mathbf{u}})$, such as might be obtained in some iteration of an iterative calculation, and that $(\mathbf{x}, \mathbf{u})$, where $\mathbf{x} = \Phi(\mathbf{u})$, is a small modification of $(\hat{\mathbf{x}}, \hat{\mathbf{u}})$. *Assume now the hypotheses of Theorem 4.3.1 (Pontryagin principle)*. From the construction used in the proof of that theorem (section 7.2.3), but with $(\hat{\mathbf{x}}, \hat{\mathbf{u}})$ in place of $(\bar{\mathbf{x}}, \bar{\mathbf{u}})$, the change of F can be expressed as follows in terms of a change in the integrated Hamiltonian H, in which only $\mathbf{u}$ varies. The required expression is

$$F(\hat{\mathbf{x}}, \hat{\mathbf{u}}) - F(\mathbf{x}, \mathbf{u}) = H(\hat{\mathbf{x}}, \hat{\mathbf{u}}, \hat{\rho}) - H(\hat{\mathbf{x}}, \mathbf{u}, \hat{\rho}) + o(\|\hat{\mathbf{u}} - \mathbf{u}\|)$$

provided that $\hat{\rho}(.) = (\hat{\boldsymbol{\lambda}}(.), \hat{\mathbf{v}}(.))$ satisfies the adjoint differential equation with $(\hat{\mathbf{x}}, \hat{\mathbf{u}})$, namely

$$-\dot{\boldsymbol{\lambda}}(t) = f_x(\hat{\mathbf{x}}(t), \hat{\mathbf{u}}(t), t) + \lambda(t)m_x(\hat{\mathbf{x}}(t), \hat{\mathbf{u}}(t), t) - \mathbf{v}(t)n_x(\hat{\mathbf{x}}(t), \hat{t}), \quad \boldsymbol{\lambda}(T) = 0$$

Consequently, if $\mathbf{u} = \hat{u} + \mathbf{w}$ and $J(u) = F(\Phi(\mathbf{u}, u))$, thus expressing the objective function as an explicit function of $\mathbf{u}$, then

$$J(\hat{u} + \mathbf{w}) - J(\hat{\mathbf{u}}) = \int_0^T h_u(\hat{\mathbf{x}}(t), \hat{\mathbf{u}}(t), t; \rho(t))\mathbf{w}(t)\, \mathrm{d}t + o(\|w\|)$$

Thus the gradient of $J(.)$ is given by

$$J'(\hat{\mathbf{u}})\mathbf{w} = \int_0^T h_{\mathbf{u}}(\hat{\mathbf{x}}(t), \hat{\mathbf{u}}(t), t; \rho(t))\mathbf{w}(t)\, \mathrm{d}t$$

so it is calculable, using the Hamiltonian. (This result is used in algorithms in order to compute gradients – section 6.4.)

4.7.7 Control problem with a parameter

Now suppose that the control problem includes also some parameter $\mathbf{q}$; thus the functions in the problem become $f(\mathbf{x}(t), \mathbf{u}(t), t; \mathbf{q})$ and $m(\mathbf{x}(t), \mathbf{u}(t), t; \mathbf{q})$.

In this section, denote the optimal objective function by $\mathbf{J(q)}$, to express its dependence on the parameter $\mathbf{q}$. In the construction used in section 4.7.6, suppose now that $(\mathbf{x}, \mathbf{u})$ corresponds to parameter value $\mathbf{q}$, and $(\hat{\mathbf{x}}, \hat{\mathbf{u}})$ corresponds to parameter value $\hat{\mathbf{q}}$. Assume that $\rho(.) = (\boldsymbol{\lambda}(.), \mathbf{v}(.))$ satisfies the adjoint differential equation with $(\hat{\mathbf{x}}, \hat{\mathbf{u}})$. Hence, showing now the explicit dependence of h on $\mathbf{q}$,

$$\mathbf{J}(\mathbf{q}) - \mathbf{J}(\hat{\mathbf{q}}) = \int_I [h(\hat{\mathbf{x}}(t), \hat{\mathbf{u}}(t), t, \hat{\rho}(t); \mathbf{q})] - h(\mathbf{x}(t), \hat{\mathbf{u}}(t), t, \hat{\rho}(t); \mathbf{q}) + o(|\mathbf{q} - \hat{\mathbf{q}}|)$$

$$= \left[\int_I [h_{\mathbf{q}}(\hat{\mathbf{x}}(t), \hat{\mathbf{u}}(t), \hat{\rho}(t); \mathbf{q})]\right] (\mathbf{q} - \hat{\mathbf{q}})\, \mathrm{d}t + o(|\mathbf{q} - \hat{\mathbf{q}}|)$$

Hence the Fréchet derivative $\mathbf{J}'(\hat{\mathbf{q}})$ is given by

$$\mathbf{J}'(\hat{\mathbf{q}})(\mathbf{q} - \hat{\mathbf{q}}) = \int_I [h_{\mathbf{q}}(\hat{\mathbf{x}}(t), \hat{\mathbf{u}}(t), \hat{\rho}(t); \mathbf{q})](\mathbf{q} - \hat{\mathbf{q}})\, \mathrm{d}t$$

4.7.8 Example

In (4.1), modify the differential equation to

$$\dot{\mathbf{x}}(t) = m(\mathbf{x}(t), \mathbf{u}(t), t) - \mathbf{q}(t) \quad (0 \le t \le T);\ \mathbf{x}(0) = \mathbf{x}_0$$

The optimal objective function value then depends on the *forcing function* $\mathbf{q}(.)$; denote the value by $J(\mathbf{q})$. From section 4.7.7,

$$J(\mathbf{q}) - J(0) = \int_I [-\boldsymbol{\lambda}(t)]\mathbf{q}(t)\, \mathrm{d}t + o(\|\mathbf{q}\|)$$

where $I = [0, T]$, $\boldsymbol{\lambda}(.)$ is the costate function and $\|\mathbf{q}\|$ is the uniform norm of $\mathbf{q}(.)$.

4.7.9 Terminal time as parameter

Consider now the fixed-time optimal control problem (4.1), considering the terminal time T as a parameter. The problem can be rewritten with a modified differential equation:

$$\dot{\mathbf{x}}(t) = m(x(t), u(t), t)\pi(T - t) \qquad (0 \le t \le T');\ \mathbf{x}(0) = \mathbf{x}_0$$

where T' is some fixed time, greater than the variable time T, and $\pi(s) = 1$ for $s \ge 0$, $\pi(s) = 0$ for $s < 0$. Then the Hamiltonian involves T in a term $\boldsymbol{\lambda}(t)\pi(T - t)$. Let $I' = [0, T']$. Therefore the derivative of the optimal objective function with respect to T equals

$$\int_{I'} \boldsymbol{\lambda}(t)\delta'(T - t)\, \mathrm{d}t = \dot{\boldsymbol{\lambda}}(T)$$

4.8 EXERCISES

4.1 If the Pontryagin theory is applied to the linear program

$$\text{Minimize}_{x,u}(c^{\mathrm{T}}x + d^{\mathrm{T}}u) \quad \text{subject to} \quad Dx = Ax + Bu + b,\ x \ge 0,\ |u| \le 1$$

in which D, A, B are matrices, and $|u| \le 1$ applies to each component of u, show that there results an 'adjoint equation'

$$-(D^{\mathrm{T}}\lambda)^{\mathrm{T}} = c^{\mathrm{T}} - \lambda^{\mathrm{T}}A - v^{\mathrm{T}}, \quad v \ge 0,\ v^{\mathrm{T}}x = 0$$

and an associated minimization problem:

$$\text{Minimize}_u\, (d - B^{\mathrm{T}}\lambda)^{\mathrm{T}}u \quad \text{subject to} \quad |u| \leq 1$$

It follows that u_j has the sign of $(d - B^{\mathrm{T}}\lambda)_j$. How does this formulation relate to the dual of the given linear program?

4.2 Discuss the optimal control problem (*regulator with hard constraints*):

$$\text{Minimize } F(x, u) = (1/2)\int_0^T x(t)^{\mathrm{T}}Qx(t)\,\mathrm{d}t$$
$$\text{subject to} \quad x(0) = x_0, \dot{x}(t) = Ax(t) + Bu(t), Lu(t) \leq c$$

Here L, A, B are constant matrices and c is a constant vector. (Only a partial answer is obtainable without numbers and computation. A useful qualitative conclusion follows from the associated problem. Quadratic–linear problems, such as this one, often give useful approximations to engineering problems. Sometimes then a term $u(t)^{\mathrm{T}}Ru(t)$ is added to the integrand.)

4.3 Discuss the optimal control problem

$$\text{Maximize } (1/2)\int_0^1 x(t)^2\,\mathrm{d}t$$
$$\text{subject to} \quad x(0) = 0, \dot{x}(t) = u(t), |u(t)| \leq 1 \quad (0 \leq t \leq 1)$$

Include a statement of what function spaces are appropriate for this problem. The Pontryagin theory leads to an analytic solution for this problem.

4.4 Discuss the inventory model in section 1.2 from the Pontryagin viewpoint. Here the inventory vector forms the *state*, the amounts m_t manufactured from the *control*, and the demands d_t are assumed given. (This model is far too simple for the real world! Note, in particular, that there would be some random variation in the demands d_t.)

4.5 For the problem:

$$\text{Maximize } J(u) = \int_0^1 x(t)^2\,\mathrm{d}t$$
$$\text{subject to} \quad \dot{x}(t) = u(t), x(0) = 0, |u(t)| \leq 1, x(t) \leq 3/4$$

calculate the gradient $J'(u)$ when $u(t) = 0$ $(0 < t < 1/8)$, 1 $(1/8 < t < 3/4)$, 0 $(3/4 < t < 1)$.

4.6 In Figure 4.2, sufficient conditions for a minimum of the control problem are shown by arrows requiring convex hypotheses. Show that *convex* can be weakened here to a suitable *invex* hypothesis (section 3.6).

(*Hint*: Consider the control problem as

$$\min F(\mathbf{x}, \mathbf{u}) \quad \text{subject to} \quad D\mathbf{x} - M(\mathbf{x}, \mathbf{u}) = 0, \mathbf{u} \in \Gamma$$

with variable $\mathbf{z} = (\mathbf{x}, \mathbf{u})$ replacing variable $\mathbf{x}$ in section 3.6. The constraint $\mathbf{u} \in \Gamma$ can be treated as in the second exercise in section 3.6.2. Note that, for an equality constraint, say $h(\mathbf{z}) = 0$, the cone is $\{0\}$, so invex requires

$$h(\mathbf{z}) - h(\mathbf{a}) = h'(\mathbf{a})\boldsymbol{\kappa}(\mathbf{z} - \mathbf{a}, \mathbf{a})$$

for appropriate $\boldsymbol{\kappa}$. Unlike convex, h is then not always linear.)

4.9 REFERENCES

Craven, B. D. (1978) *Mathematical Programming and Control Theory*, Chapman & Hall, London.

Craven, B. D. (1989) Boundary conditions in optimal control, *J. Austral. Math. Soc., Ser. B*, **30** 343–9.

Craven, B. D. (1994) Convergence of discrete approximations for constrained minimization, *J. Austral. Math. Soc., Ser. B*, **35** 1–12.

Luenberger, D. G. (1969) *Optimization by Vector Space Methods*, Wiley, New York.

Teo, K. L., Jepps, G., Moore, E. J. and Hayes, S. (1987) A computational method for free time optimal control problems, with application to maximizing the range of an aircraft-like projectile, *J. Austral. Math. Soc., Ser. B*, **28** 343–9.

5

Worked examples of control problems

5.1 OUTLINE OF PROCEDURE

Consider an optimal control problem, either fixed-time or time-optimal, described by differentiable functions $f(.,.,.)$ and $m(.,.,.)$, and with state function $x(.)$ and control function $u(.)$. The theoretical approach is based on the Hamiltonian function, whose integral in turn represents a Lagrangian, omitting certain terms. Among ordinary differential equations, only a minority possess analytic solutions in closed terms. This limitation holds, even more emphatically, for optimal control problems, which involve two ordinary differential equations, as well as other things relating to optimality.

This chapter presents a number of examples of optimal control problems which permit analytic solutions. Some of the concepts that arise in these examples, notably *switching locus, singular arc, steady state solution, present value Hamiltonian*, may still be significant, even when (as typically) the numbers must be found using a computer. It must be made clear what spaces of functions are being considered for state and control.

Important steps in the analytic approach are as follows:

(i) Formulate the Hamiltonian for the problem. (For some problems, where the time interval is $[0, \infty)$, a present value Hamiltonian is appropriate – see section 5.6 for illustration.)
(ii) From the Hamiltonian, construct the adjoint differential equation, with its boundary condition (typically at the further endpoint of the time range), and the associated problem.
(iii) Consider solutions of the adjoint differential equation, backwards in time from the endpoint boundary condition.
(iv) The associated problem gives information about the value of the control function and of possible switching times.

(v) Possible singular arcs must be considered (see section 5.7 for an example of the procedure).

The conditions provided by the Pontryagin theory are necessary for a constrained minimum, but not generally sufficient. In fact, a *Pontryagin point* (where Pontryagin's necessary conditions hold) corresponds closely to a *Karush–Kuhn–Tucker point* of a constrained optimization problem. For some classes of control problems (section 5.10), an argument using line integrals can show that a minimum has, in fact, been obtained. For some other control problems, the given functions describing the problem may have convex (or invex) properties, which imply a minimum point.

5.2 OSCILLATOR

The movement of a particle along a line is described by the differential equation $\ddot{y}(t) + y(t) = u(t)$, given initial values for $y(0)$ and $\dot{y}(0)$ and a control function $u(t)$. It is required to find a control function $u(.)$, satisfying $(\forall t)\ |u(t)| \leq 1$, which will minimize the time T at which $y(0) = 0$ and $\dot{y}(0) = 0$. This is a *time-optimal* problem; T here is a variable.

Define $x(t) = (x_1(t), x_2(t))$ by $x_1(t) = y(t)$ and $x_2(t) = \dot{y}(t)$. The differential equation then becomes $\dot{x}(t) = Ax(t) + bu(t)$, where

$$A = \begin{bmatrix} 0 & 1 \\ -1 & 0 \end{bmatrix} \quad \text{and} \quad b = \begin{bmatrix} 0 \\ 1 \end{bmatrix}$$

(This is the standard process of reducing an ordinary differential equation of order greater than one to a first-order ordinary differential equation in a vector variable.) The variables $x_1(t)$ and $x_2(t)$, representing position and velocity of the particle, are the coordinates of the *phase plane* for this motion. So it is required to start at a given point in the phase plane, and travel to the origin in minimum time. The time to reach the origin is

$$F(x, u) = \int_0^T 1\, dt$$

so the Hamiltonian is

$$h(.) = 1 + \lambda(t)[Ax(t) + bu(t)] \qquad (0 \leq t < T)$$

The costate $\lambda(t) = [\lambda_1(t), \lambda_2(t)]$ has two components.

From the theory of section 4.6, Pontryagin's theory applies to such time-optimal problems. (Indeed, Pontryagin proved this kind first, by a different road.) Here $\lambda(t)$ satisfies the adjoint differential equation:

$$-[\dot{\lambda}_1(t), \dot{\lambda}_2(t)] = [\lambda_1(t), \lambda_2(t)]\, A$$

and $\lambda(t)bu(t) = \lambda_2(t)u(t)$ is minimized with respect to $u(t) \in [-1, 1]$ when $u(t) = \bar{u}(t)$, the optimal control. It happens that the boundary conditions for the adjoint differential equation are not needed to optimize this problem. From

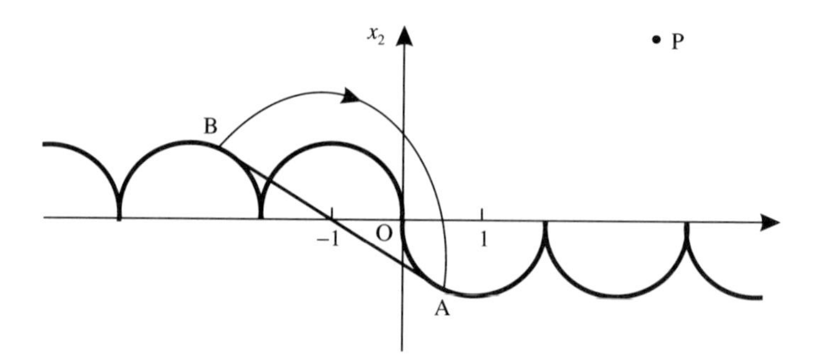

Figure 5.1 The switching locus.

the adjoint equation, $-\dot{\lambda}_1(t) = \lambda_2(t)$ and $-\dot{\lambda}_2(t) = -\lambda_1(t)$; differentiating one and substituting in the other gives $\ddot{\lambda}_2(t) + \lambda_2(t) = 0$. So $\lambda_2(t)$ (also similarly $\lambda_1(t)$) is a sinusoid, with period 2π; and the optimum control function is $\bar{u}(t) = -\operatorname{sgn} \lambda_2(t)$ (where sgn means algebraic sign). So $\bar{u}(t)$ changes sign at times $t_1 + n\pi$ ($n = 0, 1, 2, \ldots$) for some time t_1.

When $u(t) \equiv 1$, integrating $\ddot{y}(t) + y(t) = 1$ shows that the point $(y(t), \dot{y}(t))$ in the phase plane lies on a circle $\dot{y}^2 + y^2 = 2y + c_1$, where c_1 is constant. When $u(t) \equiv -1$, the point in the phase plane lies on a circle, given by

$$\dot{y}^2 + y^2 = -2y + c_2$$

where c_2 is constant. So the paths in the phase plane are made up from two families of concentric circles, with centres at $(1, 0)$ and $(-1, 0)$. Exactly one circle from each family passes through the origin; so the optimum path must reach the origin by one of these. Suppose it is the circle $\dot{y}^2 + (y-1)^2 = 1$; then the optimum path can be traced backwards from its termination (backwards from O to A in the diagram). The path must have switched onto this circle from one of the other family (in Figure 5.1, backwards by the semicircle from A to B); and so on. The points where the control $u(t)$ switches between -1 and $+1$ lie on a curve called the *switching locus* (see Figure 5.1 for its construction from semicircles). If the initial point P in the phase plane lies above [below] the switching locus, the optimum control starts with $u(t) \equiv -1$ [$u(t) \equiv +1$] until the switching locus is reached. Thereafter, the path proceeds by semicircles, as illustrated above.

5.3 REMARK ON OPEN-LOOP AND CLOSED-LOOP CONTROL

(See also section 2.9).

With this problem – and other optimal control problems where a switching locus exists in a phase plane – the control switches when the current position $(y(t), \dot{y}(t))$ reaches the switching locus. Thus the control switches (or not) at time t, depending on $(y(t), \dot{y}(t))$, but not on the previous part of the path. In general, the Pontryagin theory gives an optimum *open-loop control*; that is, the optimum $\bar{u}(.)$ is completely specified in advance. This may be quite unstable in practice, since small perturbations of the system may make $\bar{u}(.)$ quite unsuitable. However, when the switching depends on the current position only, this leads to a *closed-loop control*, in which $\bar{u}(t)$ is not completely specified in advance, but switching occurs when the point $(y(t), \dot{y}(t))$, including any perturbations, reaches a switching locus. This is a much more stable situation.

Note, however, that time lags in the system often cause instability. A differential equation of higher order than 2 may be needed to describe them (section 2.9). Alternatively, if the system is linear, then stability against oscillations relates to the position of eigenvalues of the linear system. (See any book on linear systems theory. Some relevant material is in Rosenbrock (1970) and Barnett and Storey (1970)).

5.4 SINGULAR ARCS

These examples (also those in sections 4.4.1 and 4.4.2) show that bang-bang control, where the control $\bar{u}(t)$ jumps between two (or a larger finite number) of extreme values, is likely to happen when the equations are linear in the control $u(t)$. However, as mentioned in section 4.4.1, the coefficient of $u(t)$ in the associated problem may happen to be identically zero for some interval of time t, say $\gamma \leq t \leq \delta$. The optimum path for such an interval (γ, δ) is called a *singular arc*; on such an arc, the associated problem gives no information, but $\dot{\lambda}(t) \equiv 0$ often does.

The *turnpike problem* illustrates this. A traveller wishes to get from A to B by car, as quickly as possible. Part of the journey is by a main road (turnpike), where there is a high speed limit. But the journey involves local roads, with a much lower speed limit, from A to the turnpike (at a point P, say), and from the turnpike (at a point Q, say) to B, at the other end of the journey. If this is formulated as an optimal control problem, then typically AP and QB are optimized by bang-bang, whereas PQ (on the turnpike) is a singular arc.

5.5 THE 'DODGEM CAR' PROBLEM

This problem is described in Fleming and Rishel (1973).

A vehicle moves in a plane, at unit speed. At time t, its position is given by cartesian coordinates $(x(t), y(t))$ and also $\theta(t)$, the direction in which it is steering. The equations of motion are

$$\dot{x}(t) = \cos\theta(t), \quad \dot{y}(t) = \sin\theta(t), \quad \dot{\theta}(t) = u(t)$$

where $u(t)$, the control function, is restricted by $|u(t)| \leq 1$. It is required to get from $(x(0), y(0), \theta(0)) = (4, 0, \pi/2)$ to $(x, y) = (0, 0)$ in minimum time. (Thus, the problem describes a 'dodgem car' in an amusement park for children. But mathematically, it is interesting.)

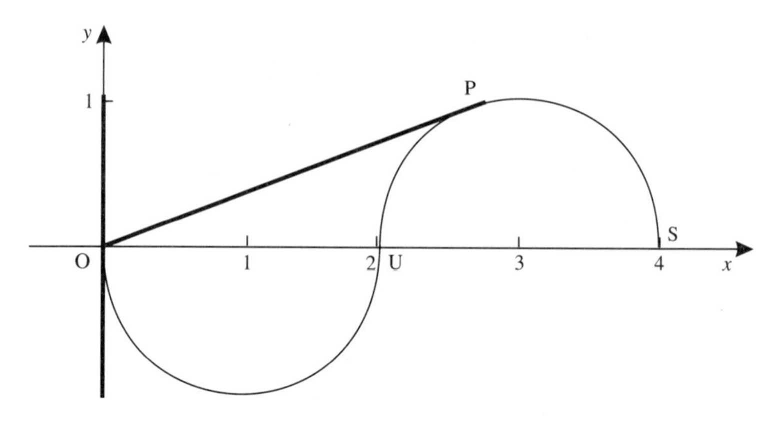

Figure 5.2 Example of a singular arc.

The bang-bang paths in (x, y) space are arcs of circles, as shown in Figure 5.2 (first the semicircle SPU, then the semicircle from U to O); but this is not the shortest-time path from S to O. The optimal path follows the first semicircle from S to P (with $u(t) \equiv 1$), and then a straight line singular arc from P to O. Along this singular arc, $\dot{\theta}(t) \equiv 0$ and $u(t) \equiv 0$ (not +1 or −1 as for bang-bang).

The *state* is $x(t) := [x(t), y(t), \theta(t)]$. An optimum solution has a switching time at $t = t_1$ and a singular arc for $t \in (t_1, t^*)$, and takes the form:

$$(0 < t < t_1) \quad u(t) \equiv 1,\ x(t) = 3 + \cos t,\ y(t) = \sin t,\ \theta(t) = (\pi/2) + t$$

$$\lambda(t) \equiv [\lambda_1(t), \lambda_2(t), \lambda_3(t)] = [c \cos \beta, c \sin \beta, -c + c \sin (t - \beta)]$$

$$(t_1 < t < t^*) \quad u(t) \equiv 0,\ x(t) = (t^* - t) \cos \beta,\ y(t) = (t - t^*) \sin \beta,\ \theta(t) = (\pi/2) + t_1$$

$$\lambda(t) \equiv [\lambda_1(t), \lambda_2(t), \lambda_3(t)] = [c \cos \beta, c \sin \beta, 0]$$

Here c is a positive constant, $\beta = t_1 - \pi/2$, $t_1 \approx 1.9806$ and $t^* \approx 4.7386$. The switching point P (Figure 5.2) is calculable since OP is tangent to the semicircle SPU. The Pontryagin theory requires $\lambda_3(t) < 0$ when $u(t) = 1$, which holds since $-1 + \sin (t - (t_1 - \pi/2)) < 0$ when $0 < t < t_1$; and $\lambda_3(t) = 0$ when $t_1 \leq t < t^*$. Since this problem is symmetric in sgn $y(t)$, there is another optimum solution with the opposite sign of $y(t)$. Any positive constant c will do here.

Consider now the boundary condition formula (4.15), taking

$$q(x(t)) := [x(t)^2 + y(t)^2]^{1/2}$$

For $t^* - t$ small and positive, $x(t) \approx -(t^* - t) \cos \beta$ and $y(t) \approx -(t^* - t) \sin \beta$, where $\theta(t) = \pi + \beta$. (These approximations represent an informal use of l'Hôpital's rules for differentiating a quotient.) Hence

$$q_x^* \approx [(t^* - t) \cos \beta, (t^* - t) \sin \beta, 0] / [(t^* - t)(\cos^2 \beta + \sin^2 \beta)]$$
$$= \{\cos \beta, \sin \beta, 0\}]$$

Then $q_x^* m^* \approx [\cos\beta, \sin\beta, 0]^T [-\cos\beta, -\sin\beta, 0] = -1$. From (4.15), since $R(x(t^*)) \equiv 0$ and $f(.,.,.) \equiv 1$, the terminal condition for the adjoint equation is

$$\lambda^* = 0 - (1+0)[\cos\beta, \sin\beta, 0]/(-1) = [\cos\beta, \sin\beta, 0]$$
$$= [-\cos\theta(t^*), -\sin\theta(t^*), 0]$$

But if instead $q(\mathbf{x}(t)) = (1/2)[x(t)^2 + y(t)^2]$ then, since $x(t^*) = 0$ and $y(t^*) = 0$, the limiting case $q_x^* = 0$, $\lambda^* = 0$ arises. Although this choice of $q(x(t))$ specifies the same terminal point to the path as does the previous $q(x(t))$, they do not appear to be computationally equivalent.

Although these terminal conditions apply to the optimal solution, they may not be appropriate for approximations to it obtained as iterations of some computational algorithm (section 6.4).

5.6 VIDALE–WOLFE MODEL

In the Vidale–Wolfe advertising model (section 2.2), the functional

$$J(u) = \int_0^T e^{-\delta t}[px(t) - u(t)]\,dt$$

is maximized, subject to the constraints

$$x(0) = x_0, \quad \dot{x}(t) = \rho u(t)[1 - x(t)] - \beta x(t) \ (0 \le t \le T), \quad 0 \le u(t) \le c \ (0 \le t \le T)$$

The Hamiltonian, for the equivalent minimizing problem, is

$$-e^{-\delta t}[px(t) - u(t)] + \lambda(t)[\rho u(t)(1 - x(t)) - \beta x(t)]$$

where $\lambda(.)$ is the costate.

It is convenient to work instead with the present value costate $\Lambda(.)$, given by $\lambda(t) = \Lambda(t)\,e^{-\delta t}$. From this,

$$\dot{\lambda}(t) = [\dot{\Lambda}(t) - \delta\Lambda(t)]e^{-\delta t}$$

Substituting into the Hamiltonian gives the present value Hamiltonian:

$$e^{-\delta t}[-px(t) + u(t) + \Lambda(t)[\rho u(t)(1 - x(t)) - \beta x(t)]]$$

in which the term $e^{-\delta t}$ that depends explicitly on time t appears only as a positive multiplier for the whole expression.

The adjoint differential equation is

$$-\dot{\lambda}(t) = -e^{-\delta t}p - \lambda(t)[\rho u(t) + \beta]$$

by differentiating the Hamiltonian with respect to $x(t)$. Substituting for $\lambda(t)$ in terms of $\Lambda(t)$ then gives

$$-\dot{\Lambda}(t) + \delta\Lambda(t) = -p - \Lambda(t)[\rho u(t) + \beta]$$

after cancelling the positive factor $e^{-\delta t}$. Thus

$$\dot{\Lambda}(t) - (\delta + \beta + \rho u(t))\Lambda(t) = p$$

Note that the boundary condition $\lambda(T) = 0$ does not give any boundary condition for $\Lambda(T)$.

Can this control problem have a steady state (equivalently, a singular arc)? If so, then for some interval of t there hold $\dot{\Lambda}(t) \equiv 0$ and $\dot{x}(t) \equiv 0$. Denote by x^*, Λ^*, u^* (constant) steady state values for $x(.)$, $\Lambda(.)$, $u(.)$. Since the associated problem minimizes a linear function of $u(t)$, namely

$$[1 + \rho\Lambda(t)\,(1 - x(t))]u(t)$$

the coefficient of $u(t)$ must vanish along a singular arc. Hence x^*, Λ^*, u^* must satisfy

$$\rho u^*(1 - x^*) - (\beta x^* = 0, -(\Delta + \beta + \rho u^*) = p, 1 + \rho\Lambda^*(1 - x^*) = 0$$

Eliminating Λ^* and u^*, by $\Lambda^* = -1/[\rho(1 - x^*)]$ and $u^* = \beta x^*/[\rho(1 - x^*)]$, gives

$$(\delta + \beta + \beta x^*/(1 - x^*)) = -p/\Lambda^* = p\rho(1 - x^*)$$

Setting $\sigma := 1 - x^*$, and simplifying, gives the quadratic equation in σ:

$$-p\rho\sigma^2 + \delta\sigma + \beta = 0 \tag{5.1}$$

The positive root of this quadratic (a negative σ would make no sense in the problem) gives

$$\sigma = [\delta - (\delta^2 + 4\beta p\rho)^{1/2}]/[2p\rho] = 2\kappa - (\kappa^2 + \theta)^{1/2}$$

where $\kappa := \delta/(2p\rho)$ and $\theta := \beta/(p\rho)$ are parameters. (Observe that the expression has been simplified, so to involve as few parameters as possible.)

If $\sigma < 0$, let $\sigma^* := 0$; if $\sigma > 1$, let $\sigma^* := 1$; otherwise let $\sigma^* := \sigma$. Then the Pontryagin necessary conditions are satisfied. So there is a singular arc, on which $x^* = 1 - \sigma^*$, and u^* and Λ^* are calculable as above from x^*.

However, it is not obvious whether this Pontryagin point (equivalent to a Karush–Kuhn–Tucker point) is, in fact, a minimum, of the control problem. Some information on this can be obtained by a technique using Green's theorem (Sethi, 1974, 1978; Sethi and Staats, 1978). This technique is detailed in section 5.10. For the present problem, observe that $J(u)$ can be expressed as a line integral along some plane curve C defined by $x(.)$, by

$$J(u) = \int_0^T e^{-\delta t}[px(t) - (\dot{x}(t) + \beta x(t))/(\rho(1 - x(t)))]\,dt$$

$$= \int_0^T e^{-\delta t}\{[px(t) - \beta x(t)/(\rho(1 - x(t)))]\}\,dt - 1/[\rho(1 - x(t))]\,dx$$

$$\equiv \int_C \{P(x, t)\,dx - Q(x, t)\}\,dt$$

say, after substituting for $u(t)$ in terms of $\dot{x}(t)$.

Consider now a closed plane curve Γ in the (t, x) plane, consisting of the points $(t, x(t))$ for some interval of t. By Green's theorem for line integrals in the plane, the line integral around Γ equals a double integral over the region E enclosed by Γ. Thus

$$\int_{\Gamma} \{P(x, t)\, \mathrm{d}x + Q(x, t)\}\, \mathrm{d}t = \iint_{E} \{\partial P/\partial t - \partial Q/\partial x\}\, \mathrm{d}x\, \mathrm{d}t$$

This integrand simplifies to

$$\mathrm{e}^{-\delta t}[\delta\zeta(t) - p\rho + \beta\zeta(t)^2] \tag{5.2}$$

in which $\zeta(t) := 1/(1 - x(t))$. Apart from the (positive) exponential term, the integrand corresponds closely to the quadratic expression (5.1) obtained above from the Pontryagin theory.

Consider now $J(u) - J(u^*)$, where u^* is the possible optimum obtained above from the Pontryagin theory. The Green's theorem construction expresses this difference as the difference of two double integrals, over different regions, E_1 and E_2 say, bounded by closed Γ_1 and Γ_2.. (See Figures 5.4 and 5.5 for the detailed construction.) By construction of σ^*, and hence u^*, it follows that E_1 lies on one side of the curve defined by σ^* and E_2 lies on the other side, with opposite signs for the integrand in E_1 and in E_2. This results in $\iint_{E_1} > 0$ and $\iint_{E_2} <$ 0, which proves that $J(.)$ is minimized at u^*. This holds under some restriction on the problem; $u^* < c$ is required.

5.7 INVESTMENT MODEL

In the investment model of section 2.4, there are two state variables $P(t)$ and $E(t)$ and two control variables $u_r(t)$ and $u_s(t)$. The Hamiltonian, $h(.)$ say, as stated in section 2.4, can now be simplified by defining *present value costate variables* $\Lambda_1(t) := \lambda_1(t)\,\mathrm{e}^{\rho t}$ and $\Lambda_2(t) := \lambda_2(t)\,\mathrm{e}^{\rho t}$; then $h(.)\mathrm{e}^{\rho t}$ takes the form

$$A + Bu_r + Cu_s$$

where

$$B = rE + c\Lambda_1 rE - \Lambda_2 rE, \quad C = -\Lambda_2 rE(1 - \tau E/P)$$

and A does not depend on the controls. Here the dependence on t is suppressed to simplify the notation. The adjoint differential equations are

$$-\dot{\Lambda}_1 = -(1 + c)\rho\Lambda_1 + \Lambda_2 rE\tau u_s/P^2$$

$$-\dot{\Lambda}_2 = (-\rho + ru_r + ru_s(1 - 2\tau\varepsilon/P))\,\Lambda_2 - r(1 - u_r) + cr(1 - u_r)\,\Lambda_1$$

The associated problem requires the minimization of $Bu_r + Cu_s$ over the triangular region shown in Figure 5.3. Depending on the values of B and C, which depend on the states, the minimum of the associated problem can be at a vertex of the triangle or the whole of an edge can be minimal. The criteria for minimality are shown in the figure.

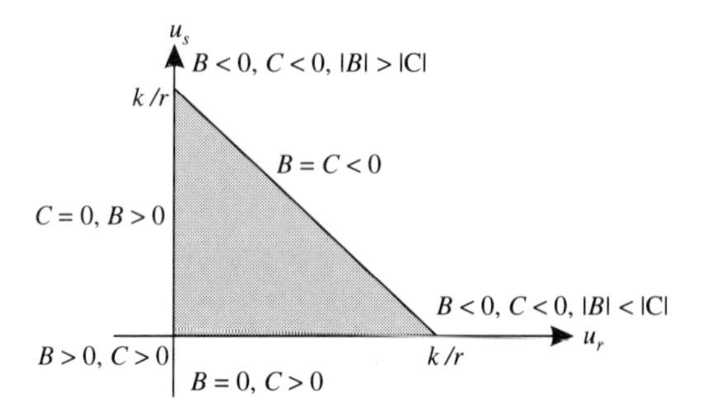

Figure 5.3 Minimization of $Bu_r + Cu_s$.

At time T, $\Lambda_1(T) = 1$ (in consequence of the $\delta(t - T)$ term in the Hamiltonian), $P(T) \equiv s_P > 0$, $\Lambda_2(T) = 0$, $E(T) \equiv s_E > 0$. Consider a solution, backwards in time, thus starting at $t = T$. Initially, $B > 0$ and $C = 0$, with $u_r = 0$. This is a singular condition; can it be continued for some interval $(T - \varepsilon, T)$ of t? This would require $\Lambda_2[1 - \tau E/P) \equiv 0$ on this interval of time. If $\Lambda_2 \equiv 0$ then $0 \equiv \dot{\Lambda}_2 = cE\Lambda_1 - r$ since $u_r \equiv 0$, so $c\Lambda_1 = 1$. But then $0 \equiv -\dot{\Lambda}_1$ implies $\Lambda_1 \equiv 0$, a contradiction. So consider the remaining case, when $1 - \tau\varepsilon/P \equiv 0$ on $(T - \varepsilon, T)$. Differentiation gives $\tau\dot{E} = \dot{P}$; substitution in the differential equations for $\dot{E}$ and $\dot{P}$ shows that $crE \equiv \rho P$. This, with $E/P \equiv 1/\tau$, shows that $r/\rho = \tau$. So the singular condition can only be maintained under the restrictive condition $r/\rho = \tau$.

Consider the case when $r/\rho > \tau$ and $P(T)/E(T) < \tau$. Then $C > 0$ and $B > 0$ on $(T - \varepsilon, T)$ for some $\varepsilon > 0$. So consider a vertex of the triangle, with $u_r \equiv 0$, $u_s \equiv 0$ (thus, not a singular arc). The state equations now give $\dot{P} = crE - c\rho P$, $\dot{E} = 0$. Integrating from t to T leads to

$$P(t)/E(t) = \{P(T)/E(T)\} - (r/\rho)] \, e^{c\rho(T-t)} + (r/\rho)$$

Solving the adjoint differential equations gives

$$\Lambda_1(t) = e^{\rho(1+c)(t-T)} \quad \text{and} \quad \Lambda_2(t) = (r/\rho)[1 - e^{-\rho(1+c)(T-t)}]$$

Calculation from these shows that $C > 0$ continues so long as $B > 0$ and $(P/E) < \tau$. This happens when $T - t = [\rho(c + 1)]^{-1} \ln[(\rho c + r)/(r - \rho)]$. At this time, $B = 0$ and $C > 0$, and it is again necessary to check whether this singular state can be continued.

By continuing in this manner, regarding s_P and s_E as parameters, and considering at each possible singular point whether a singular state can be continued, and if not, considering both possible vertices of the triangle as possible solutions, it is possible (at some length) to construct the entire system of switching curves for this problem. This will not be detailed here, since the intention is only to present the method of solution; refer to Davis and Elzinga (1971) for the details.

5.8 FISH MODEL

Consider the fish model of section 2.7. In dimensionless form, this model (with a quadratic profit function) is as follows:

$$\text{Minimize } J(u) = \int_0^T e^{-\delta t}[p_1 u(t) - p_2(u(t))^2 - m(x(t))u(t)]\,dt$$

subject to the differential equation (with initial condition)

$$x(0) = x_0, \quad \dot{x}(t) = x(t)[1 - x(t)] - u(t) \quad (0 \leq t \geq T)$$

Here the symbols t, T,$x(t)$, $q(t)$, $u(t)$ now refer to the scaled variables τ etc. constructed in section 2.7. The state function $x(t)$ now describes the fish population at a given time as a fraction of the environmental capacity, and the control function $u(t) = q(t)x(t)$, where $q(t)$ measures the effort put into catching fish, and $m(x(t)) = c/x(t)$, where c is constant. Denote $F(x(t)) = x(t)[1 - x(t)]$.

For this model, the adjoint differential equation is

$$-\dot{\lambda}(t) = e^{-\delta t} m'(x(t))u(t) + \lambda(t)F'(x(t))$$

where $\lambda(t)$ is the costate function. In terms of the present value costate function $\Lambda(t) = \lambda(t)e^{\delta t}$, for which $\dot{\Lambda}(t) = (\dot{\lambda}(t) + \delta\lambda(t))e^{\delta t}$, the adjoint differential equation becomes

$$-\dot{\Lambda}(t) = m'(x(t))\,u(t) + \Lambda(t)[F'(x(t)) - \delta]$$

Because of the $e^{-\delta t}$ term, no steady state is possible for $\lambda(t)$; but a steady state may be reached for $\Lambda(t)$. Assume, for the present, that the price coefficient $p_2 = 0$. Consider then a possible steady state, in which $\Lambda(t) = \Lambda^\#$, $x(t) = x^\#$ and $u(t) = u^\#$ have constant values, Then $\dot{\Lambda}(t) \equiv 0$ and $\dot{x}(t) \equiv 0$, so that the differential equations give

$$m'(x^\#)u^\# + \Lambda^\#[F'(x^\#) - \delta] = 0$$

$$F(x^\#) - u^\# = 0$$

In the associated problem, when $p_2 = 0$, the function being minimized is linear in $u(t)$. Hence a steady state requires that the coefficient of $u(t)$ shall vanish over the time interval to which the steady state applies (thus generating a singular arc). This gives

$$-p_1 + m(x^\#) - \Lambda^\# = 0$$

These three equations then solve to give:

$$\begin{aligned} F'(x^\#) - \delta &= -m'(x^\#)\,u^\#/\Lambda^\# \\ &= m'(x^\#)u^\#/[p_1 - m(x^\#)] \\ &= m'(x^\#)F(x^\#)/[p_1 - m(x^\#)] \end{aligned}$$

This provides an equation that determines $x^\#$ implicitly, given the functions $F(.)$ and $m(.)$.

While this solution is not restricted to the case considered where $F(x) = x(1-x)$ and $m(x) = c/x$, an explicit solution can be obtained in that case. From

$$(1-2x^{\#}-\delta) = (p_1 - c/x^{\#})/[-cx^{\#-2}x^{\#}(1-x^{\#})]$$

there results the quadratic in $x^{\#}$:

$$(1-2x^{\#}-\delta)(\sigma x^{\#}-1) = -x^{\#}(1-x^{\#})$$

where $\sigma = p_1/c$. The positive root of this quadratic is

$$x^{\#} = \tfrac{1}{4}[1-\delta+\sigma+((1-\delta+\sigma)^2+8\sigma\delta)^{1/2}]$$

In this case (and some others), the solution for $x^{\#}$ is unique.

With some other functions $F(.)$ there may be more than one solution for $x^{\#}$. It can then happen that one such solution is *stable* (the state returns towards $x^{\#}$ after a small deviation from it), and another solution may be *unstable* (if $x(t)$ moves a small distance away from this $x^{\#}$, then $x(t)$ will continue to move further away). With the present quadratic $F(.)$, the solution is stable.

There is an initial interval of time during which the steady state is approached, starting with the given initial condition for $x(0)$. During this interval, bang-bang control applies. From the associated problem, assuming that $u(t)$ must be nonnegative, and not exceed some maximum level u_{max},

$$u(t) = \begin{cases} u_{max} & \text{while } x(t) > u^{\#} \\ 0 & \text{while } x(t) < u^{\#} \end{cases}$$

If it happens that $m(0) < p_1 < \infty$, then the unit value of a fish exceeds the cost of catching the last fish. So then unrestricted fishing may be expected to lead to extinction of the fish population; thus $x(t) \to 0$ as $t \to \infty$, instead of to a (nonzero) steady state. This has happened to various biological populations in the past.

This model has made some serious assumptions. The assumed linear price function $p_1u - p_2u^2$ with p_2 taken as 0 has led to initial bang-bang control. But (unrealistically) no cost has been attached to switching the fishing effort on and off. Moreover, if $p_2 > 0$, then the associated problem minimizes

$$-p_1u + p_2u^2 + mu + \Lambda u$$

in which $u \equiv u(t)$, $m \equiv m(x(t))$, $\Lambda = \Lambda(t)$ and $0 \le u \le u_{max}$. Denote by $\tilde{u}$ the value of u obtained by equating the gradient to zero; thus

$$-p_1 + 2p_2u + m + \Lambda = 0$$

Then set $u(t) = \tilde{u}$ if $0 < \tilde{u} < u_{max}$, 0 if $\tilde{u} \le 0$ and u_{max} if $\tilde{u} \ge u_{max}$. With this $u(t)$, dependent on $x(t)$ and $\Lambda(t)$, the pair of coupled differential equations

$$\dot{x}(t) = x(t)[1-x(t)] - u(t); \quad \dot{\Lambda}(t) = m'(x(t))u(t) + \Lambda(t)[F'(x(t)) - \delta]$$

would have to be solved. Then there would be no bang-bang control, and no steady state after a finite time – those are consequences of linearity of the model in $u(t)$. Instead, it will often happen that $x(t) \to x_\infty$ and $\Lambda(t) \to \Lambda_\infty$ as $t \to \infty$, where the asymptotic values x_∞ and Λ_∞ are positive and finite.

In order to study the behaviour in a nonlinear case, consider the simpler (nonlinear) example:

$$\max J(u) = \int_0^T e^{-\delta t} g(u(t))\,dt \quad \text{subject to}$$

$$x(0) = x_0, \quad \dot{x}(t) = F(x(t)) - u(t) \quad (0 \le t \le T), \quad x(T) = x_T,$$

$$u(t) \ge 0 \quad (0 \le t \le T)$$

For this problem, with $g(.)$ nonlinear, the adjoint differential equation is

$$\dot{\lambda}(t) = \lambda(t) F'(x(t))$$

The associated problem minimizes $-e^{-\delta t} g(u(t)) + \lambda(t)u(t)$ with respect to $u(t)$. If, subject to later verification, this minimum happens when $u(t) > 0$, and so is unconstrained, then the gradient $-e^{-\delta t} g'(u(t)) + \lambda(t) = 0$. Hence

$$\lambda(t) = e^{-\delta t} g'(u(t))$$

Differentiation gives

$$\dot{\lambda}(t) = e^{-\delta t}[-\delta g'(u(t)) + g''(u(t))\dot{u}(t)]$$

Comparison with $\dot{\lambda}(t)$ from the adjoint differential equation shows that

$$-\delta g'(u(t)) + g''(u(t))\dot{u}(t) = \lambda(t)F'(x(t)) = e^{-\delta t} g'(u(t))F'(x(t))$$

Hence

$$\dot{u}(t) = [\delta - F'(x(t))]\, g'(x(t))/g''(x(t))$$

must be solved (assuming, for $g(.)$ nonlinear, that $g''(x(t)) \ne 0$), together with the given differential equation

$$\dot{x}(t) = F(x(t)) - u(t)$$

This pair of coupled differential equations may have an equilibrium point (x^*, u^*), at which both $\dot{x}(.)$ and $\dot{u}(.)$ vanish; but it will only be approached asymptotically.

5.9 EPIDEMIC MODELS

Consider the first of Sethi's epidemic models, presented in section 2.8. When $I(t)$ denotes the number of infectives at time t, out of a total population of N people, and $v(t)$ denotes the intensity of medicare at time t, then the model is as follows:

$$\min J(v) = \int_0^T e^{-\delta t}\,[kv(t) + KI(t)]\,dt$$

subject to

$$\dot{I}(t) = bI(t)[N - I(t)] - v(t)I(t), \quad I(0) = I_0, \quad I(T) = I_T$$

$$(\forall t)\ 0 \le I(t) \le N, \quad 0 \le v(t) \le V$$

Here $v(.)$ is the control function and $I(.)$ is the state function. For now, the state constraint $0 \le I(t) \le N$ will be neglected; if it turns out not to be satisfied, then additional terms will have to be added. The resulting Hamiltonian is then

$$e^{-\delta t}\{kv(t) + KI(t)] + \lambda(t)\{bI(t)[N - I(t)] - v(t)I(t)\}$$

Then the associated problem is

$$\min_{v(t)} e^{-\delta t} kv(t) - \lambda(t)I(t)v(t) \quad \text{subject to} \quad 0 \le v(t) \le V$$

for each separate $t \in [0, T]$. In terms of the present value costate function $\Lambda(t) = e^{\delta t}\lambda(t)$, the associated problem simplifies to

$$\min_{v(t)} [k - \Lambda(t)I(t)]\, v(t) \quad \text{subject to} \quad 0 \le v(t) \le V$$

since the factor $e^{-\delta t}$ is always positive.

The adjoint differential equation is

$$-\dot{\lambda}(t) = Ke^{-\delta t} + \lambda(t)\{bN - 2bI(t) - v(t)\}$$

Hence

$$\dot{\lambda}(t) + \{bN - 2bI(t) - v(t)\}\lambda(t) = -Ke^{-\delta t}$$

Since $I(0)$ and $I(T)$ are fixed, there are no boundary conditions on $\lambda(0)$ or $\lambda(T)$. Since $\dot{\Lambda}(t) = e^{\delta t}[\dot{\lambda}(t) + \delta\lambda(t)]$, the adjoint differential equation becomes:

$$\dot{\Lambda}(t) + [-\delta + bN - 2bI(t) - v(t)]\Lambda(t) = -K$$

Note that the time-dependent term $e^{-\delta t}$ has dropped out. One may also write the present value Hamiltonian as

$$e^{-\delta t}\{[kv(t) + KI(t)] + \Lambda(t)[bI(t)[N - I(t)] - v(t)I(t)]\}$$

in which $e^{-\delta t}$ appears only as an external multiplying factor.

From the associated problem, there follows:

$$v(t) = 0 \quad \text{if } \Lambda(t)I(t) < k$$

$$v(t) = V \quad \text{if } \Lambda(t)I(t) > k$$

However, if $I(t)$ hits a boundary, whether 0 or N, then $v(t)$ must be modified to prevent $I(t)$ crossing the boundary.

Consider then the given differential equation for $I(t)$, when $v(t) \equiv 0$ on some interval of time t, and when $v(t) \equiv V$ on some interval of time t. When $v(t) \equiv 0$, $I(t) = N(1 + \alpha e^{-Nbt})$, where α is a constant of integration. When $v(t) \equiv V$, then

$$\dot{I}(t) = (bN - V)I(t) - b(I(t))^2$$

Consider first $b < V/N$; then

$$\dot{I}(t) = -b(gI(t) + (I(t))^2)$$

with $g = (V/b) - N > 0$. This solves to

$$I(t) = g/(\beta e^{gbt} - 1)$$

where β is a constant of integration. If, instead, $b > V/N$, then

$$\dot{I}(t) = -bI(t)(-h + I(t))^2$$

with $h = -(V/b) + N > 0$. This solves to

$$I(t) = h/(1 + \gamma e^{-hbt})$$

where γ is a constant of integration.

Observe that, when $v(t) \equiv V$ and $b < V/N$, $\dot{I}(t) < 0$. When $v(t) \equiv V$ and $b > V/N$, then $\dot{I}(t) > 0$ when $I(t) < b$. So in this case, $I(t)$ cannot be controlled to a level less than h, in the long run. (The condition $b > V/N$ means that the rate of growth of the epidemic is too fast for the health services to keep up with it.)

What might an optimum look like? The above theory leads to bang-bang control, except for a possible *singular interval*, (T^*, T) say, in which $v(t)$ is adjusted (within $(0, V)$) to fulfil the terminal constraint $I(T) = I_T$. (There is some resemblance to the turnpike model.) If, initially $I(t)$ is 'small', there would be an initial interval on which $v(t) \equiv 0$ and $I(t)$ is increasing. There would then be a switch to an interval with $v(t) \equiv V$ and (provided that $b < V/N$)) $I(t)$ decreasing. This is consistent with the associated problem, provided that $\Lambda(t) > 0$. The solutions for $I(t)$ involve exponentials, which suggest that there may be only one switching time. (If the differential equation for $I(t)$ had, instead, involved oscillatory solutions, then several switching times might have been expected, and the optimum might then, perhaps, be calculated by assuming either $\Lambda(0)$ or T^* as a parameter, to be adjusted later to fit the terminal condition.)

Consider a possible solution with just one switching time $T^\#$. Assume, subject to verification that, when $t < T^\#$, $v(t) = 0$ if $I(0) < I^\#$ or $v(t) = V$ if $I(0) > I^\#$. Let $I^\# = I(T^\#)$. When $t > T^\#$, consider a possible singular interval solution, for which $I(t) \equiv I^\#$, $\Lambda(t) \equiv \Lambda^\#$, where $I^\#\Lambda^\# = k$ and $v(t) \equiv v^\#$. This solution would represent a steady state, during which the time derivatives $\dot{I}(t) \equiv 0$ and $\dot{\Lambda}(t) \equiv 0$.

Since these derivatives are zero,

$$(-r + bN - bI^\# - v^\#)\Lambda^\# = -K$$

and

$$0 = bI^\#[N - I^\#] - v^\# I^\#$$

Hence

$$-r + bN - 2bI^\# - v^\# = -K/\Lambda^\# = -KI^\#/k$$

and

$$0 = bN - v^\# - bI^\#$$

The first of these two equations solves to

$$I^{\#} = rk/(K - bk)$$

and then the second gives

$$v^{\#} = b(N - I^{\#})$$

(In these steady state solutions, if $I^{\#} > N$ then $I^{\#}$ must be replaced by N and $v^{\#}$ by 0 to fulfil the constraints.)

This steady state solution needs adjustment over a terminal interval, say (T^*, T), in order to fulfil the terminal constraint $I(T) = I_T$. This is done by setting $v(t) = 0$ (if $I^{\#} < I_T$) or V (if $I^{\#} > I_T$), and then taking the shortest time interval $T - T^*$ that will bring $I(t)$ from $I^{\#}$ to I_T; this is calculable by solving the differential equation for $I(t)$ over this interval.

So far, this calculation has used only *necessary conditions* for an optimum. So it is not yet clear whether the optimum obtained is in fact a minimum. By following a different approach, due to Sethi (1978), it may be shown that the solution obtained is optimal if $b \geq v^{\#}$. It is remarked that if $b > k/K$, then $I^{\#} < 0$, so that $v(t) \equiv 0$ ('no medicare') is optimal.

5.10 SUFFICIENT CONDITIONS FOR A MINIMUM

Under convex (or invex) hypotheses, a Pontryagin point, namely a solution satisfying the necessary conditions of Pontryagin's theorem, is in fact a minimum point. This follows because a Pontryagin point is a Karush–Kuhn–Tucker point of the control problem when expressed as a mathematical programming problem. In addition, in some circumstances when $x(t)$ and $u(t)$ are real-valued (thus, not vector-valued), a Pontryagin point can be proved to be, in fact, a minimum, by a technique based on Green's theorem for line integrals in the plane.

Consider the following optimal control problem:

$$\max_u \int_0^T [f(x(t), t) + k(x(t), t)u(t)]\, dt \quad \text{subject to}$$

$$x(0) = x_0, \quad x(T) = x_T, \quad \dot{x}(t) = m(x(t), t) + q(x(t), t)u(t) \quad (0 \leq t \leq T)$$

$$u_{\min} \leq u(t) \leq u_{\max} \quad (0 \leq t \leq T)$$

Note that the functions defining this problem are linear in the control $u(t)$. Assume that $q(x(t), t)$ is not zero. Then $u(t)$ can be expressed in terms of $\dot{x}(t)$; thus $u(t) = [\dot{x}(t) - m(x(t), t)]/q(x(t), t)$. Substituting this expression for $\dot{x}(t)$ into the objective function leads to an optimization problem (in fact a calculus of variations problem) of the form:

$$\max_x J(x) := \int_0^T [g(x(t), t) + h(x(t), t)\dot{x}(t)]\, dt \tag{5.3}$$

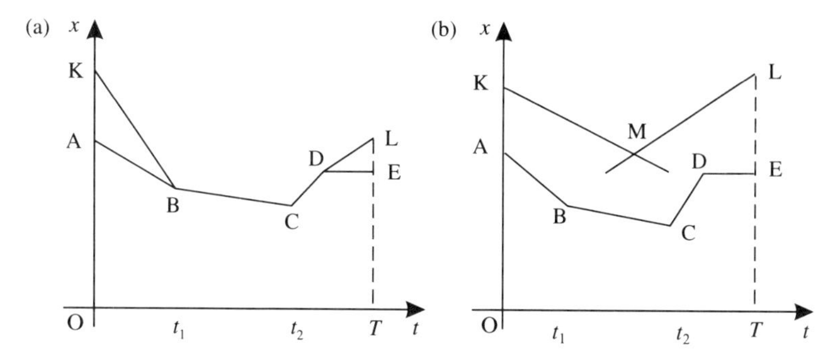

Figure 5.4 Pontryagin solution.

subject to bounds

$$u_l(t) \leq \dot{x}(t) \leq u_u(t) \quad (0 \leq t \leq T)$$

The objective function can be expressed as a line integral in the (x, t)-plane by

$$J(x) = \int_{\Gamma} [g(x(t), t)\,\mathrm{d}t + h(x(t), t)\,\mathrm{d}x]$$

integrated along an appropriate path Γ.

Since the problem involves $u(t)$ linearly, the Pontryagin necessary conditions lead to a singular solution, say during a time interval $[t_1, t_2]$, with a bang-bang solution on $[0, t_1]$ and again on $[t_2, T]$, to fulfil the initial and terminal conditions. Figure 5.4(a) illustrates this; the solution is shown as KBCDL, with BC as the singular arc. If the singular arc were extended to times before t_1 or after t_2 the path ABCDE would be obtained, but the parts AB and DE are not optimal. It can happen that the singular arc is never reached; this is illustrated in Figure 5.4(b), where the Pontryagin solution consists of just the two bang-bang parts KM and ML.

Note that, if the integrand of (5.3) is written as $F(x(t), \dot{x}(t), t)$, then

$$J'(u)z = \int_0^T (F_x z + F_{\dot{x}}\dot{z})\,\mathrm{dt} = \int_0^T (F_x - (\mathrm{d}/\mathrm{d}t)\,F_{\dot{x}})z\,\mathrm{d}t$$

by integrating by parts and using the boundary conditions on $x(.)$, which imply that $z(0) = 0 = z(T)$. Hence, if the constraints on $\dot{x}(t)$ are not active, then

$$0 = (F_x - (\mathrm{d}/\mathrm{d}t)F_{\dot{x}}) = [g_x + h_x\dot{x}] - (\mathrm{d}/\mathrm{d}t)h = g_x - h_t$$

so that $g_x = h_t$ describes the singular arc.

Figure 5.5 shows an optimal path (say for $u = \bar{u}$) in the (x, t)-phase plane, represented by KBCDL, and some non-optimal path KCQL (say for a control $u \neq \bar{u}$). (The arcs are drawn straight for convenience; they may in fact be curved.) In terms of line integrals of $[g(x(t), t)\,\mathrm{d}t + h(x(t), t)\,\mathrm{d}x]$,

$$J(u) - J(\bar{u}) = \int_{\mathrm{KBCDL}} - \int_{\mathrm{KCQL}} = \int_{\mathrm{KBCK}} - \int_{\mathrm{CDLQC}}$$

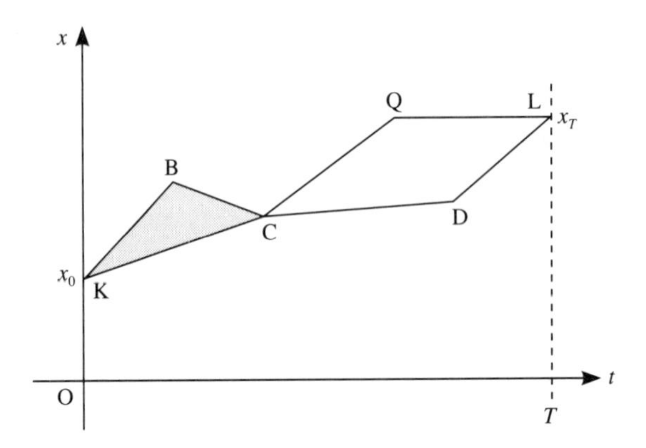

Figure 5.5 Sufficient conditions for optimality.

the difference of two line integrals around closed regions. By Green's theorem,

$$\int_{\text{KBCK}} [g(x(t), t)\, \mathrm{d}t + h(x(t), t)\, \mathrm{d}x] = \iint [g_x(x(t), t)\, h_t(x(t), t)]\, \mathrm{d}x\, \mathrm{d}t$$

Now the singular arc BCD is described by $g_x = h_t$. If it happens that $g_x \le h_t$ whenever $x(t) \le \bar{x}(t)$, then the integrand is nonnegative for the (shaded) area KBC. If also $g_x \ge h_t$ whenever $x(t) \ge \bar{x}(t)$, then the integrand is nonpositive for the area CDLQ. It follows then that $J(u) \ge J(\bar{u})$; thus, $\bar{u}$ is a minimum point for this control problem.

5.11 EXERCISES

5.1 Discuss the control problem:

$$\text{Minimize} \int_I [u(t)x(t) + \tfrac{1}{2}x(t)^2 - \tfrac{1}{2}u(t)]\, \mathrm{d}t$$

subject to $x(0) = \frac{1}{2}$, $(\forall t \in I = [0, 1])\ \dot{x}(t) = u(t)$, $0 \le u(t) \le 2$. What function spaces are chosen for state $x(.)$ and control $u(.)$?

5.2 Consider a *resource depletion* model, where $x(t)$ denotes the amount of resource remaining at time t and $u(t)$ is the rate of extraction of the resource, so that $\dot{x}(t) = -u(t)$ for $t \ge 0$, and $p(u(t))$ denotes the sale price for the resource at time t, where $p(.)$ is an increasing concave function, representing a *law of diminishing returns*. An appropriate profit function to be maximized could be

$$\int_0^T e^{-\delta t}[p(u(t)) - cu(t)]\, \mathrm{d}t$$

where c is the unit cost of extracting the resource and T is the (variable) time during which extraction can take place (so $x(t) = 0$ when $x > T$). Apply the Pontryagin theory to discuss what an optimal policy might be.

5.3 Show in detail how the Pontryagin theory can be applied to the dodgem car problem (section 5.5); in particular, show how the associated problem is used.

5.4 What happens to the oscillator problem if there is a damping term, thus if $\ddot{y}(t) + 2\beta\dot{y}(t) + y(t) = u(t)$, with $\beta > 0$?

5.5 A biological resource, of amount $x(t)$ at time t, is exploited at a rate $u(t)$. It is required to maximize a profit function

$$\int_0^T e^{-\delta t}\,[1 - e^{-u(t)}]\,dt$$

subject to the constraints

$$\dot{x}(t) = \alpha x(t) - u(t), \quad x(0) = a, \quad x(T) = 0, \quad u(t) \geq 0 \quad (0 \leq t \leq T)$$

Assuming that $0 < \alpha < \beta$ and $0 < a < T$, show that the optimal $u(t)$ takes the form $u(t) = [k - (\beta - \alpha)t]_+$ for some constant k.

5.12 REFERENCES

Barnett, S. and Storey, C. (1970) *Matrix Methods in Stability Theory*, Nelson, London.

Clark, C. (1976) *Mathematical Bioeconomics: the Optimal Management of Renewable Resources*, Wiley, New York.

Davis, B. E. and Elzinga, D. J. (1971) The solution of an optimal control problem in financial modeling, *Operations Research*, **19** 1419–73.

Fleming, W. H. and Rishel, R. W. (1975) *Deterministic and Stochastic Optimal Control*, Springer, Berlin.

Rosenbrock, H. H. (1970) *State-space and Multivariable Theory*, Nelson, London.

Sethi, S. P. (1974) Quantitative guidelines for communicable disease control programs: a complete synthesis, *Biometrics*, **30** 681–91.

Sethi, S. P. (1978) Optimal quarantine programmes for controlling an epidemic spread, *J. Operational Research Society*, **29** 265–8.

Sethi, S. P. and Staats, P. W. (1978) Optimal control of some simple deterministic epidemic models, *J. Operational Research Society*, **29** 129–36.

6

Algorithms for control problems

6.1 INTRODUCTION

If an optimization problem is linear, then an optimum can be computed by linear programming methods. Variants of Dantzig's *simplex method* allow optimization of linear problems with large numbers of variables and constraints. However, many optimization problems – including most optimal control problems – involve nonlinear functions. This brings up additional questions, which are discussed in the following sections. They include:

- The distinction between local and global optima.
- Convergence and convergence rate (since an optimum is only reached in a finite number of steps for linear programs and quadratic programs with linear constraints).
- Discretization – the approximation of a control problem in continuous time by a related problem with only finitely many variables and constraints, so that it may be computed.
- Using the special structure of an optimal control problem.
- Sensitivity analysis.

The descriptions of algorithms in this chapter are directed towards an understanding in principle of how the algorithms work, what limitations they have, and what results may be expected from them. For every detailed descriptions, as may be required to write a reliable computer program, reference may by made in particular to the books by Fletcher (1969, 1987) and by Gill, Murray and Wright (1981). See also Bertsekas (1982) for augmented Lagrangian algorithms and Schittkowski (1985) for sequential quadratic algorithms. The computation for a control problem in continuous time requires in addition some discretization of the time-scale (sections 6.4.4 and 6.4.5), leading to a mathematical program in finite dimensions, for which methods are available.

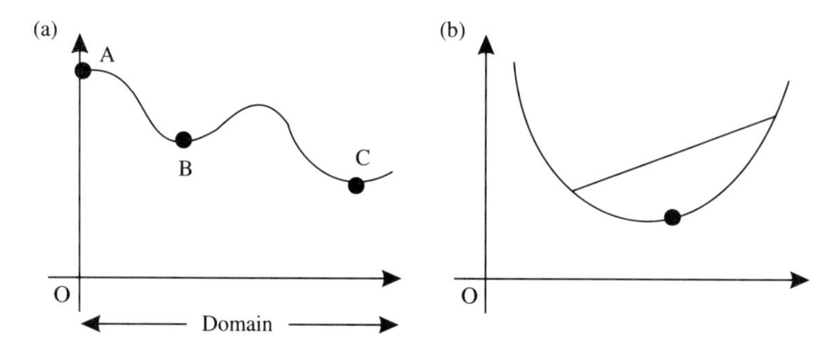

Figure 6.1 (a) Nonconvex and (b) convex problems.

The sensitivity of an optimum to a small change of a parameter is as relevant to a control problem as it is to a problem in finite dimensions; the required details are given in section 6.5. For examples see section 6.6.

Figure 6.1 illustrates the fact that a nonconvex problem (Figure 6.1(a)) can have a local minimum B which is not a global minimum. The global minimum (over the indicated domain) is C. Note that a maximum or minimum can occur on the boundary of the domain (e.g. A); of course, this is usual in linear programming, but can happen also in nonlinear problems. For a convex (e.g. Figure 6.1(b)) or invex problem, a local minimum is always a global minimum (section 3.5.5).

Suppose that the function $f: \mathbf{R}^n \to \mathbf{R}$ has an unconstrained local minimum at the point $\mathbf{x} = \bar{\mathbf{x}}$. Note that the local minimum need not be unique. So, in order to compute it, it is usually necessary to start the computation at an initial point $\mathbf{x}_0$ which is not too far away from $\bar{\mathbf{x}}$. It may, in fact, be necessary to try several different initial points, since there is no way to tell, in advance, how near to $\bar{\mathbf{x}}$ is 'near enough'. Each algorithm considered constructs a sequence of vectors $\mathbf{x}_0, \mathbf{x}_1, \mathbf{x}_2, \ldots, \mathbf{x}_n, \ldots$, called *iterates*. (If the components of a vector $\mathbf{x}$ are required, they will be denoted by $(x_1, x_2, \ldots, x_n)$.)

The simplex method for a linear program reaches an optimum in a finite, though unpredictable, number of iterations. This is not generally so for a nonlinear problem. A sequence of iterates $\mathbf{x}_1, \mathbf{x}_2, \ldots, \mathbf{x}_k, \ldots$ is generated. This sequence of vectors converges (hopefully) to some local optimum $\bar{\mathbf{x}}$. If an algorithm is be computationally useful, the rate of convergence to $\bar{\mathbf{x}}$ must be reasonably fast. Of the many possibilities for rate of convergence, three often arise in practice, described below.

6.1.1 Definition – convergence rates

Linear convergence: $(\forall k)\ \|\mathbf{x}_{k+1} - \bar{\mathbf{x}}\| \le \rho \|\mathbf{x}_k - \bar{\mathbf{x}}\|$, where ρ is a constant, with $0 < \rho < 1$.

Superlinear convergence: $(\forall k)\ \|\mathbf{x}_{k+1} - \bar{\mathbf{x}}\| \le \rho_k \|\mathbf{x}_k - \bar{\mathbf{x}}\|$, where the sequence $\rho_1, \rho_2, \ldots, \rho_k, \ldots, \to 0$.

Second-order convergence: $(\forall k)\ \|\mathbf{x}_{k+1}-\bar{\mathbf{x}}\| \le \rho\|\mathbf{x}_k - {}^{-}\|^2$, where ρ is constant.

Most optimization algorithms of practical use show at least linear convergence. Some show superlinear convergence, which is clearly better. Newton's method gives second-order convergence, at the considerable cost of computing a Hessian matrix of second derivatives at each iteration.

6.2 ALGORITHMS FOR UNCONSTRAINED MINIMIZATION

Although it is usually required to minimize subject to constraints, most algorithms for doing so rely on algorithms for minimization without constraints. The latter are accordingly considered first. Newton's method provides a great deal of motivation for other methods.

6.2.1 Newton's method

Let $f: \mathbf{R}^n \to \mathbf{R}$ be twice continuously differentiable; let $\mathbf{g}(\mathbf{x}) := f'(\mathbf{x})^{\mathrm{T}}$. Suppose that $\bar{\mathbf{x}}$ is a local minimum point of f; then $\bar{\mathbf{x}}$ is a zero of $\mathbf{g}$. Now

$$\mathbf{g}(\mathbf{x}+\mathbf{h}) \approx \mathbf{g}(\mathbf{x}) + \mathbf{g}'(\mathbf{x})\mathbf{h} = 0 \quad \text{when } \mathbf{h} = -\mathbf{g}'(\mathbf{x})^{-1}\mathbf{g}(\mathbf{x})$$

Newton's method, applied to $\mathbf{g}(\mathbf{x})$, constructs a sequence of iterates (given $\mathbf{x}_0$) by:

$$\mathbf{x}_{k+1} := \mathbf{x}_k - \mathbf{g}'(\mathbf{x}_k)^{-1}\mathbf{g}(\mathbf{x}_k) \quad (k = 0, 1, 2, \ldots) \tag{6.1}$$

To analyse the convergence, assume now that f is three times continuously differentiable. Denote $A_k := \mathbf{g}'(\mathbf{x}_k)^{-1}$. Then

$$\mathbf{g}_{k+1} = \mathbf{g}(\mathbf{x}_k - A_k^{-1}\mathbf{g}_k) = A_k(-A_k^{-1}\mathbf{g}_k) + o(\|\mathbf{g}_k\|^2)$$

which implies second-order convergence.

An estimate of the constant in the convergence rate formula can be obtained as follows. From Taylor's theorem, if the components of $\mathbf{g}$ are (for now) written as g_i, then (since $g_i = 0$ at the minimum)

$$g_i(\bar{\mathbf{x}}+\mathbf{z}) = g_i'(\bar{\mathbf{x}})\mathbf{z} + \mathbf{z}^{\mathrm{T}}g_i''(\bar{\mathbf{x}})\mathbf{z} + \cdots \quad (i = 1, 2, \ldots, n)$$

where the Hessian $g_i''(\bar{\mathbf{x}})$ is the matrix of second derivatives of the component $g_i(.)$. This will be summarized by the notation

$$\mathbf{g}(\bar{\mathbf{x}}+\mathbf{z}) = \mathbf{g}'(\bar{\mathbf{x}})\mathbf{z} + \mathbf{z}^{\mathrm{T}}\mathbf{g}_{.}''(\bar{\mathbf{x}})\mathbf{z} + \cdots$$

where $\mathbf{g}''(\bar{\mathbf{x}})$ stands for the set of Hessian matrices $g_i''(\bar{\mathbf{x}})$ $(i = 1, 2, \ldots, n)$. Set $\mathbf{z}_k := \mathbf{x}_k - \bar{\mathbf{x}}$, $\bar{\mathbf{g}}' := \mathbf{g}'(\bar{\mathbf{x}})$, $\bar{\mathbf{g}}_{.}'' := \mathbf{g}_{.}''(\bar{\mathbf{x}})$. Then, substituting into (6.1),

$$\begin{aligned}\mathbf{z}_{k+1} &\approx \mathbf{z}_k - [\bar{\mathbf{g}}' + \mathbf{z}_k^{\mathrm{T}}\bar{\mathbf{g}}_{.}'']^{-1}[\bar{\mathbf{g}}'\mathbf{z}_k + \tfrac{1}{2}\mathbf{z}_k^{\mathrm{T}}\bar{\mathbf{g}}_{.}''\mathbf{z}_k] \\ &\approx \tfrac{1}{2}\mathbf{z}_k^{\mathrm{T}}M_{.}\mathbf{z}_k\end{aligned}$$

retaining terms up to quadratic in $\mathbf{z}_k$, where $M_. := \mathbf{g}'^{-1}\bar{\mathbf{g}}''_.$; thus, $M_.$ stands for a family of Hessian matrices, with component i of the form $\Sigma Q_{il}(\mathbf{z}_k^{\mathrm{T}} R_l \mathbf{z}_k)$, where $Q := \mathbf{g}'^{-1}$ and $R_l = \bar{\mathbf{g}}_l''$.

Consequently,

$$\|\mathbf{z}_{k+1}\| \leq \kappa \|\mathbf{z}_k\|^2$$

where the constant κ depends on first and second derivatives of the gradient, thus on second and third derivatives of the function f. This proves the second-order convergence.

It is noted that the algorithm requires only first and second derivatives to compute it, whereas third derivatives are required in order to obtain some estimate of the rate of convergence. The requirement for third derivatives can be weakened a little, to require that the second derivative $\mathbf{g}(.)$ satisfies a Lipschitz condition. It is commonly the case that if an algorithm uses derivatives up to order r, then estimates of convergence rate require derivatives of order $r + 1$, or at any rate Lipschitz conditions on the derivatives of order r.

6.2.2 Descent methods

Suppose that the function $f: \mathbf{R}^n \to \mathbf{R}$ has an unconstrained local minimum at the point $\mathbf{x} = \bar{\mathbf{x}}$. Assume that f is differentiable. Let $\mathbf{g}_k := f'(\mathbf{x}_k)^{\mathrm{T}}$. (Thus, $\mathbf{g}_k$ is a column vector.) If $\mathbf{x}$ moves away from $\mathbf{x}_k$ in the direction $\mathbf{t}$, then the rate of increase of $f(.)$ at $\mathbf{x}_k$ in direction $\mathbf{t}$ is

$$(\partial/\partial\alpha) f(\mathbf{x}_k + \alpha t)|_{\alpha=0} = f'(\mathbf{x}_k)\mathbf{t} = \mathbf{g}_k^{\mathrm{T}}\mathbf{t}$$

Hence the direction of *steepest descent* is $\mathbf{t} = -\mathbf{g}_k$. The *steepest descent algorithm*, starting at $\mathbf{x}_0$, constructs the iterates by

$$\mathbf{x}_{k+1} := \mathbf{x}_k - \alpha_k \mathbf{g}_k \quad (k = 0, 1, 2, \ldots)$$

where $\alpha_k :=$ argmin $f(\mathbf{x}_k - \alpha\mathbf{g}_k)$. Here, *argmin* means the value of α that minimizes the function. A *linesearch* subroutine (described in section 6.2.9) is required to find (exactly or approximately) α_k, such that $f(.)$ decreases along the halfline $\mathbf{x} = \mathbf{x}_k - \alpha\mathbf{g}_k$ until α_k, and thereafter increases.

Unfortunately, the convergence of steepest descent is slow. It behaves well when descending the steep sides of the 'valley' formed by the graph of f, where f is well approximated by a tangent plane (or hyperplane). But in the 'bottom' of the valley, near the minimum, steepest descent often zigzags about the minimum, instead of moving directly to it. In this region, a quadratic approximation to f would be more appropriate instead of the linear approximation which steepest descent uses.

In order to evaluate and compare minimization algorithms, the quadratic function

$$f(\mathbf{x}) := \tfrac{1}{2}\mathbf{x}^{\mathrm{T}} A\mathbf{x} + \mathbf{b}^{\mathrm{T}}\mathbf{x} \quad (\mathbf{x} \in \mathbf{R}^n)$$

is used as a test function; A is a positive definite symmetric matrix and $\mathbf{b}$ is a constant vector. (Recall that *positive definite* means that

$$(\forall \mathbf{v} \neq \mathbf{0}) \quad \mathbf{v}^{\mathrm{T}} A \mathbf{v} > 0$$

An equivalent property is that all eigenvalues of A are strictly positive.)

An algorithm that performs well for quadratic functions is also suitable for non-quadratic functions f, provided that f satisfies the *quadratic approximation property*:

$$(\forall \mathbf{x} \in E)\, (\forall \mathbf{v} \neq \mathbf{0}) \quad 0 < m_1 \mathbf{v}^{\mathrm{T}} \mathbf{v} \leq \mathbf{v}^{\mathrm{T}} f''(\mathbf{x}) \mathbf{v} \leq m_2 \mathbf{v}^{\mathrm{T}} \mathbf{v}$$

where E is some open region containing the minimum point $\bar{\mathbf{x}}$ (so that the iterations are supposed to happen in E), and m_1 and m_2 are finite constants, with $0 < m_1 < m_2$. Equivalently, when $\mathbf{x} \in E$, all the eigenvalues of $f''(\mathbf{x})$, the *Hessian* matrix of f, are assumed to lie in the interval $[m_1, m_2]$.

Two general approaches are commonly used to obtain much faster convergence that steepest descent can give. A *quasi-Newton* algorithm avoids the computation of second derivatives (in $\mathbf{g}_k'$) by iterating

$$\mathbf{x}_{k+1} := \mathbf{x}_k - \alpha_k H_k \mathbf{g}_\kappa \quad (k = 0, 1, 2, \ldots)$$

in which the H_k are suitable matrices, with $H_0 = I$ (the identity matrix), so that the algorithm begins as steepest descent. Then some rule is used for constructing H_k that makes $\{H_k\} \rightarrow f''(\bar{\mathbf{x}})^{-1}$ (the inverse Hessian), as $k \rightarrow \infty$, so that the algorithm eventually resembles Newton's method. The several such methods in use differ in how they generate the H_k.

An alternative approach uses

$$\mathbf{x}_{k+1} := \mathbf{x}_k + \alpha_k \mathbf{t}_{k+1} \quad (k = 0, 1, 2, \ldots)$$

with search direction $\mathbf{t}_k$ constructed as some compromise between $-\mathbf{g}_{k+1}$ and the previous search direction $\mathbf{t}_k$; thus

$$\mathbf{t}_{k+1} := -\mathbf{g}_{k+1} + \beta_{k+1} \mathbf{t}_k$$

for some coefficient β_k to be chosen. The start of this process (for $k = 0$) is illustrated in Figure 6.2.

In either approach, a *linesearch* is required to find the step length α_k.

6.2.3 Theorem – convergence rate inequality

When the quasi-Newton algorithm

$$\mathbf{x}_{k+1} := \mathbf{x}_k - \alpha_k H_k \mathbf{g}_k \quad (k = 0, 1, 2, \ldots);\ \mathbf{x}_0 \text{ given}$$

is applied to the quadratic objective function

$$f(\mathbf{x}) := \tfrac{1}{2} \mathbf{x}^{\mathrm{T}} A \mathbf{x} + \mathbf{b}^{\mathrm{T}} \mathbf{x}$$

with A and each H_k positive definite, and the linesearch is exact,thus

$$\alpha_k := \operatorname{argmin}_{\alpha > 0} f(\mathbf{x}_k - \alpha H_k \mathbf{g}_k), \quad \text{where } \mathbf{g}_k := f'(\mathbf{x}_k)^{\mathrm{T}} = A\mathbf{x}_k + \mathbf{b}$$

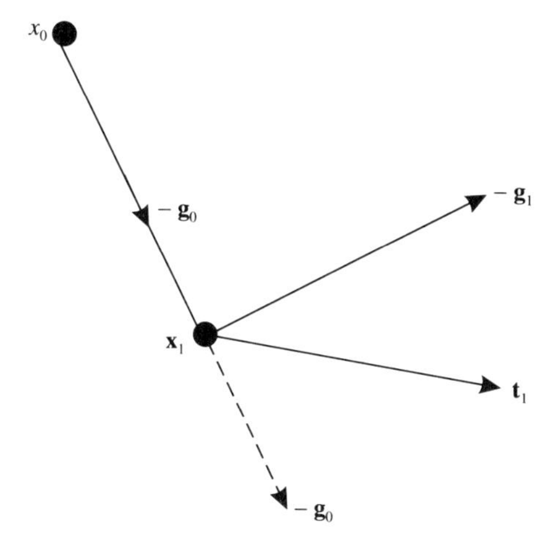

Figure 6.2 The start of a quasi-Newton algorithm.

then the *convergence rate inequality* holds:

$$\|\mathbf{x}_{k+1} - \bar{\mathbf{x}}\| / \|\mathbf{x}_k - \bar{\mathbf{x}}\| \leq \sigma(1 - \rho_k)/(1 + \rho_k) \tag{6.2}$$

where $\bar{\mathbf{x}} = -A^{-1}\mathbf{b}$ is the minimum point, σ is constant and ρ_k is the ratio of least to greatest eigenvalue for the matrix $H_k A$. Thus the convergence rate is *linear*. If also the matrices H_k are so constructed that $\{H_k\} \to A^{-1}$ as $k \to \infty$, then the convergence is *superlinear*.

Proof

Since $f(\mathbf{x}_k - \alpha H_k \mathbf{g}_k)$ is quadratic in α, its minimum point with respect to α satisfies a linear equation, given by differentiating with respect to α. Its solution gives the step length α_k, given below in (6.3).

Define the *modified distance* for vector $\mathbf{x}$ to $\bar{\mathbf{x}}$ as

$$d_A(\mathbf{x}, \bar{\mathbf{x}}) := [(\mathbf{x} - \bar{\mathbf{x}})^{\mathrm{T}} A (\mathbf{x} - \bar{\mathbf{x}})]^{1/2}$$

Then, on substituting

$$\mathbf{x}_{k+1} - \bar{\mathbf{x}} = \mathbf{x} - \bar{\mathbf{x}} - \alpha_k H_k \mathbf{g}_k$$

and

$$\alpha_k = (\mathbf{g}_k^{\mathrm{T}} H_k \mathbf{g}_k)/(\mathbf{g}_k^{\mathrm{T}} H_k^{\mathrm{T}} A H_k \mathbf{g}_k) \tag{6.3}$$

$$d_A(\mathbf{x}_{k+1}, \bar{\mathbf{x}})/d_A(\mathbf{x}_k, \bar{\mathbf{x}}) = 1 - (\mathbf{g}_k^{\mathrm{T}} H_k \mathbf{g}_k)^2/[(\mathbf{g}_k^{\mathrm{T}} H_k^{\mathrm{T}} A H_k \mathbf{g}_k)\,(\mathbf{g}_k^{\mathrm{T}} H_k^{-1} \mathbf{g}_k)]$$

$$= 1 - (\mathbf{w}_k^{\mathrm{T}} \mathbf{w}_k)/[(\mathbf{w}_k^{\mathrm{T}} B_k \mathbf{w}_k)(\mathbf{w}_k^{\mathrm{T}} B_k^{-1} \mathbf{w}_k) \tag{6.4}$$

after substituting $H_k = S_k^{\mathrm{T}} S_k$, $\mathbf{w}_k = S_k \mathbf{g}_k$ and $B_k = S_k^{\mathrm{T}} A S_k$.

Note here that, by a theorem in matrix algebra, a real symmetric positive definite matrix M can be diagonalized by an orthogonal matrix L; thus,

$$ML = LD$$

where D is a diagonal matrix of the eigenvalues $\mu_1, \mu_2, \ldots, \mu_n$ say, of M, and L consists of the (column) eigenvectors $\xi_1, \xi_2, \ldots, \xi_n$ of M, which can be arranged so that $\xi_i^T \xi_j = 0$ if $i \neq j$ and $= 1$ if $i = j$; then $L^{-1} = L^T$, which defines L as an orthogonal matrix. Then $M = LDL^T$. Since M is positive definite, all the μ_i are > 0. Define $D^{1/2}$ as the diagonal matrix whose diagonal elements are the positive square roots $\mu_1^{1/2}, \mu_2^{1/2}, \ldots, \mu_n^{1/2}$ and let $S := LD^{1/2}L^T$. Then

$$S^T S = M = SS^T$$

This construction is applied to $M = H_k$, yielding $S = S_k$, as required in (6.4) above.

Moreover, $M = LDL^T$ expands to give $M = \sum \mu_i(\xi_i \xi_i^T)$. Now, substituting B_k for M (and λ_i for μ_i) shows that $\mathbf{w}^T B_k \mathbf{w} = \sum \lambda_i v_i^2$ where $v_i := \mathbf{w}^T \xi_i$. Since B_k^{-1} has eigenvalues λ_i^{-1} and the same eigenvectors as B, $\mathbf{w}^T B_k^{-1} \mathbf{w} = \sum \lambda_i^{-1} v_i^2$. It also follows that

$$(\forall \mathbf{w}) \quad (\min \mu_i)\mathbf{w}^T\mathbf{w} \leq \mathbf{w}^T M \mathbf{w} \leq (\max \mu_i)\mathbf{w}^T\mathbf{w}$$

Hence, if $M = A$, and γ and γ' are the smallest and largest eigenvalues of A,

$$(\forall \mathbf{x} \neq \bar{\mathbf{x}}) \quad 0 \leq \gamma \leq d_A(\mathbf{x}, \bar{\mathbf{x}})/\|\mathbf{x} - \bar{\mathbf{x}}\| \leq \gamma' < \infty$$

Hence, setting $t_i := v_i^2/(\sum_s v_s^2)$, so that $\sum_i t_i = 1$,

$$\begin{aligned} d_A(\mathbf{x}_{k+1}, \bar{\mathbf{x}})/d_A(\mathbf{x}_k, \bar{\mathbf{x}}) &= 1 - \left(\sum_i v_i^2\right)^2 \Big/ \left[\left(\sum_i \lambda_i v_i^2\right)\left(\sum_j \lambda_j^{-1} v_j^2\right)\right] \\ &= 1 - \left[\left(\sum_i \lambda_i t_i\right)\left(\sum_j \lambda_j^{-1} t_j\right)\right]^{-1} \\ &\leq 1 - 4\lambda_1\lambda_n/(\lambda_1 + \lambda_n)^2 \\ &= [(\lambda_1 - \lambda_n)/(\lambda_1 + \lambda_n)]^2 \end{aligned}$$

subject to proof of *Kolmogorov's inequality*, which asserts that

$$\tfrac{1}{4}(\lambda_1 + \lambda_n)^2/(\lambda_1\lambda_n) = \max_{t \in \mathbf{R}^n}\left(\sum_i \lambda_i t_i\right)\left(\sum_j \lambda_j^{-1} t_j\right) \quad \text{subject to} \quad (\forall i) t_i \geq 0, \ \sum t_i = 1$$

Denote $\kappa := \tfrac{1}{2}[(\lambda_1/\lambda_n) + (\lambda_n/\lambda_1)]$. If $t_2 = \ldots = t_{n-1} = 0$, the function to be maximized is $t_1^2 + t_n^2 + 2\kappa t_1 t_n$. For this problem the Karush–Kuhn–Tucker conditions require that, for some Lagrange multiplier β,

$$t_1 + \kappa t_n = \beta, \quad \kappa t_1 + t_n = \beta, \quad t_1 + t_n = 1$$

These solve to $t_1 = t_n = \tfrac{1}{2}$, objective $= \tfrac{1}{2}(1 + \kappa) = \tfrac{1}{4}(\lambda_1 + \lambda_n)^2/(\lambda_1\lambda_n)$. Now, if additional constraints $t_2 = \ldots = t_{n-1} = 0$ are not imposed, the Karush–Kuhn–Tucker conditions require that

$$(\forall j) \quad \lambda_j b + \lambda_j^{-1} a \geq \beta$$

where $a := \Sigma\lambda_j t_j$ and $b := \Sigma\lambda_j^{-1} t_j$, and complementary slackness requires equality whenever $t_j > 0$. But equality can only hold for two different λ_j; so the present case reduces to the previous special case.

Therefore

$$\|\mathbf{x}_{k+1} - \bar{\mathbf{x}}\|/\|\mathbf{x}_k - \bar{\mathbf{x}}\| \leq (\gamma'/\gamma) d_A(\mathbf{x}_{k+1}, \bar{\mathbf{x}})/d_A(\mathbf{x}_k, \bar{\mathbf{x}})$$
$$\leq (\gamma'/\gamma)(1 - \rho_k)/(1 + \rho_k)$$

where $\rho_k := \lambda_n/\lambda_1$. Since

$$B\xi = \lambda\xi \Leftrightarrow H_k A(S_k\xi) = S_k S_k^{\mathrm{T}} A(S_k\xi) = \lambda(S_k\xi)$$

the eigenvalues of B are also those of $H_k A$. This proves (6.2), with $\sigma = \gamma'/\gamma$.

If $\{H_k\} \to A^{-1}$ as $k \to \infty$, then $\{H_k A\} \to I$, the unit matrix, all of whose eigenvalues are 1. For this $H_k A$, $\{\rho_k\} \to 1$, thus the right-hand side of (6.2) tends to 0, proving superlinear convergence. □

6.2.4 Conjugate gradient algorithms

Consider the test function

$$f(\mathbf{x}) = \tfrac{1}{2}\mathbf{x}^{\mathrm{T}} A\mathbf{x} + \mathbf{b}^{\mathrm{T}}\mathbf{x} \quad (\mathbf{x} \in \mathbf{R}^n)$$

A set $\{\mathbf{t}_1, \mathbf{t}_2, \ldots\}$ of search directions is called *conjugate* with respect to A if $\mathbf{t}_i^{\mathrm{T}} A\mathbf{t}_j = 0$ for all i and j with $i \neq j$. (This is a property that eigenvectors have, though it does not apply only to eigenvectors.)

If the symmetric matrix A is positive definite, and the vectors $\mathbf{e}_1, \mathbf{e}_2, \ldots, \mathbf{e}_n$ are linearly independent, then the construction $\mathbf{t}_1 := \mathbf{e}_1$, then

$$\mathbf{t}_i := \mathbf{e}_i + c_{i1}\mathbf{t}_1 + c_{i2}\mathbf{t}_2 + \ldots + c_{i,i-1}\mathbf{t}_{i-1} \tag{6.5}$$

with the coefficients c_{ij} chosen to make $\mathbf{t}_i^{\mathrm{T}} A\mathbf{t}_j = 0$ whenever $j = 1, 2, \ldots, i - 1$, will make the set of vectors $\mathbf{t}_i$ conjugate. Now $\mathbf{t}_1$ alone is trivially conjugate. If $\mathbf{t}_1, \mathbf{t}_2, \ldots, \mathbf{t}_{i-1}$ are assumed to be conjugate, then

$$0 = \mathbf{t}_i^{\mathrm{T}} A\mathbf{e}_j = \mathbf{e}_i^{\mathrm{T}} A\mathbf{t}_j + c_{ij}(\mathbf{t}_j^{\mathrm{T}} A\mathbf{t}_j) \quad (j = 1, 2, \ldots, i - 1)$$

since the other terms vanish. With this choice of c_{ij}, $\mathbf{t}_1, \ldots, \mathbf{t}_i$ are conjugate. So, by induction, a set of n conjugate vectors $\mathbf{t}_1, \ldots, \mathbf{t}_n$ is obtained. Although one would never compute conjugate directions in this way, the construction throws light on a class of algorithms called *conjugate gradient* methods.

For the quadratic test function, start at the point $\mathbf{x}_0$, and construct iterates $\mathbf{x}_2, \mathbf{x}_3, \ldots$ as follows, denoting the (transposed) gradient at $\mathbf{x}_i$ by $\mathbf{g}_i = A\mathbf{x}_i + \mathbf{b}$. Note that the initial search direction $\mathbf{t}_1$ must satisfy $\mathbf{t}_1^{\mathrm{T}}\mathbf{g}_0 < 0$, in order that $\mathbf{t}_1$ is a decreasing direction for f. Now construct successively:

$$\mathbf{e}_i := -\mathbf{g}_{i-1}$$

$$\mathbf{t}_i \text{ from (6.5) with } c_{ij} := \mathbf{g}_{i-1}^{\mathrm{T}}(\mathbf{g}_j - \mathbf{g}_{j-1})/(\mathbf{t}_j^{\mathrm{T}}(\mathbf{g}_j - \mathbf{g}_{j-1}))$$

$$\mathbf{x}_i := \mathbf{x}_{i-1} + \alpha_i\mathbf{t}_i$$

$$\text{where } \alpha_i := \operatorname{argmin} f(\mathbf{x}_{i-1} + \alpha\mathbf{t}_i)$$

Since after j exact linesearches along directions $\mathbf{t}_1, \ldots, \mathbf{t}_i$ the function f is minimized in a hyperplane spanned by these directions, $\mathbf{g}_j$ must be orthogonal to this hyperplane. Since the search directions are linear combinations of $\mathbf{g}_1, \ldots, \mathbf{g}_{j-1}$, it follows that $\mathbf{g}_i^{\mathrm{T}}\mathbf{g}_j = 0$ for $0 < i < j$. Hence the $\mathbf{e}_i$ are linearly independent, as (6.5) has assumed; and $c_{ij} = 0$ for $j = 2, 3, \ldots, i-2$. Hence

$$\mathbf{t}_i = -\mathbf{g}_{i-1} + \mathbf{c}_{i1}\mathbf{t}_i + \mathbf{c}_{i,i-1}\mathbf{t}_{i-1} \quad (\text{where } c_{10} := 0)$$

In the usual conjugate gradient algorithms, the initial search direction $\mathbf{t}_1 := -\mathbf{g}_0$; then $c_{i1} = 0$, and

$$\mathbf{t}_i = -\mathbf{g}_{i-1} + \beta_i'\mathbf{t}_{i-1}$$

where

$$\beta_i' := \mathbf{g}_{i-1}^{\mathrm{T}}(\mathbf{g}_{i-1} - \mathbf{g}_{i-2})/[\mathbf{t}_{i-1}^{\mathrm{T}}(\mathbf{g}_{i-1} - \mathbf{g}_{i-2})]$$

After the algorithm is applied for n iterations to a quadratic function (of $\mathbf{x} \in \mathbf{R}^n$), the resulting gradient is orthogonal to n previous gradients, which span $\mathbf{R}^n$. Thus a zero gradient has been reached, and hence a minimum of the convex objective function (convex since A is positive definite).

For a quadratic objective function, there are several other equivalent formulae for the coefficient of $\mathbf{t}_{i-1}$. These include:

$$\beta_i := \mathbf{g}_{i-1}^{\mathrm{T}}\mathbf{g}_{i-1}/(\mathbf{g}_{i-2}^{\mathrm{T}}\mathbf{g}_{i-2}) \qquad \text{(Fletcher–Reeves)}$$

$$\beta_i'' := \mathbf{g}_{i-1}^{\mathrm{T}}(\mathbf{g}_{i-1} - \mathbf{g}_{i-2})/(\mathbf{g}_{i-2}^{\mathrm{T}}\mathbf{g}_{i-2}) \qquad \text{(Polak–Ribière)}$$

However, these different formulae are not equivalent when the algorithm is applied to a function f which is not convex, and the computational behaviour may be different. The Polak–Ribière formula has often been recommended.

When such a conjugate gradient algorithm is applied to a nonconvex function it will not reach a minimum in n iterations. Since the above theory does not hold for more than n iterations, it is necessary then to restart the process. Thus, for a nonquadratic function, a *reset* after each n iterations consists of a steepest descent step from the value of $\mathbf{x}$ that has been reached.

6.2.5 Implementation of Fletcher–Reeves

A computer program does not store the iterations $\mathbf{x}_0, \mathbf{x}_1, \mathbf{x}_2, \ldots$ as separate vectors; instead it overwrites $\mathbf{x}_{i-1}$ with $\mathbf{x}_i$. A program requires an *initialization* part, where initial values are set for various variables, followed by an *iteration* part, which is then repeated a number of times, until the optimum point is approached, within some small preset *tolerance*.

The iteration must contain an *exit test*, so that the sequence of iterations can come to an end. The iteration will call subroutines to generate function values and gradient values and to perform a linesearch (section 6.2.9).

The following example outlines how this may be done for the Fletcher–Reeves algorithm. Three n-dimensional vectors $\mathbf{x}$, $\mathbf{g}$ and $\mathbf{t}$ are required, and scalar variables r, β, c, tol and α^*. It is more efficient *not* to require a special version of *iteration* for its first occurrence. Usually this can be arranged; it is done here by suitable use of **t** and c.

Initialization Enter *tol* (= tolerance) and $\mathbf{x}_0$ (initial point)
Let $\mathbf{x} := \mathbf{x}_0$, $\mathbf{t} = \mathbf{0}$, $c := 1$
Iteration Let $\mathbf{g} := f'(\mathbf{x})$ (calling *gradient* subroutine)
Let $r := \mathbf{g}^T\mathbf{g}$
If $r < tol$ then EXIT
Let $\beta := r/c$; $\mathbf{t} := \beta\mathbf{t} - \mathbf{g}$; $c := r$
Let $\alpha^* := \text{argmin}\, f(\mathbf{x} + \alpha\mathbf{t})$ (calling *linesearch* subroutine, which in turn calls *function* subroutine)
Let $\mathbf{x} := \mathbf{x} + \alpha^*\mathbf{t}$
GOTO *Iteration*

This outline is to show the flow of logic, not the details relating to a particular computing language. Often, the roles of

If $r < tol$ then EXIT

and

GOTO *Iteration*

are replaced by a WHILE statement (or something similar), as in the following alternative version:

Initialization Enter *tol* (= tolerance) and $\mathbf{x}_0$ (initial point)
Let $\mathbf{x} := \mathbf{x}_0$, $\mathbf{t} = \mathbf{0}$, $c := 1$
Let $\mathbf{g} := f'(\mathbf{x})$; let $r := \mathbf{g}^T\mathbf{g}$ (calling *gradient* subroutine)
WHILE $r \geq tol$ DO the *Iteration* subroutine.
Iteration Let $\beta := r/c$; $\mathbf{t} := \beta\mathbf{t} - \mathbf{g}$; $c := r$
Let $\alpha^* := \text{argmin}\, f(\mathbf{x} + \alpha\mathbf{t})$ (calling *linesearch* subroutine, which in turn calls *function* subroutine)
Let $\mathbf{x} := \mathbf{x} + \alpha^* t$; let $\mathbf{g} := f'(\mathbf{x})$; let $r := \mathbf{g}^T\mathbf{g}$

6.2.6 Quasi-Newton algorithms

Consider the quasi-Newton algorithm described by $\mathbf{x}_0$ given and

$$\mathbf{x}_{k+1} := \mathbf{x}_k + \alpha_k \mathbf{t}_k; \quad \mathbf{t}_k := -H_k \mathbf{g}_k \quad (k = 0, 1, 2, \ldots)$$

where $\mathbf{g}_k := f'(\mathbf{x}_k)^T$ and H_{k+1} is a specified function of H_k, $\mathbf{p}_k := \mathbf{x}_{k+1} - \mathbf{x}_k$ and $\mathbf{q}_k := \mathbf{g}_{k+1} - \mathbf{g}_k$. The step length α_k may be exact, for constructing the theory:

$$\alpha_k := \text{argmin}_{\alpha>0}\, f(\mathbf{x}_k + \alpha_k \mathbf{t}_k)$$

or some approximation to this, as nearly always in practical computing.

If $f(\mathbf{x}) := \frac{1}{2}\mathbf{x}^T A\mathbf{x} + b^T\mathbf{x}$, then

$$\mathbf{q}_k = (A\mathbf{x}_{k+1} + \mathbf{b}) - (A\mathbf{x}_k + \mathbf{b}) = A\mathbf{p}_k$$

thus $\mathbf{p}_k = A^{-1}\mathbf{q}_k$. For other objective functions than quadratic, assumed twice differentiable,

$$\mathbf{q}_k = A\mathbf{p}_k + o(\|\mathbf{p}_k\|^2)$$

For a quasi-Newton algorithm, it is desired that H_k should approximate A^{-1} when k is large, where now A denotes the Hessian $f''(\mathbf{x}_k)$. It is usual to impose the *quasi-Newton condition*

$$\mathbf{p}_k = \mathbf{H}_{k+1}\mathbf{q}_k$$

modelled on $\mathbf{p}_k = A^{-1}\mathbf{q}_k$, in constructing a quasi-Newton algorithm.

Consider a possible update formula:

$$H_{k+1} := H_k + c_1'\mathbf{p}_k\mathbf{p}_k^{\mathrm{T}} + c_2'(H_k\mathbf{q}_k)(H_k\mathbf{q}_k)^{\mathrm{T}} + c_3'(H_k\mathbf{q}_k\mathbf{p}_k^{\mathrm{T}} + \mathbf{p}_k\mathbf{q}_k^{\mathrm{T}}H_k)$$

in which c_1', c_2' and c_3' are constants to be chosen. Note that $\mathbf{p}_k\mathbf{p}_k^{\mathrm{T}}$ is an $n \times n$ matrix of rank 1, which must not be confused with the 1×1 matrix $\mathbf{p}_k^{\mathrm{T}}\mathbf{p}_k$ (which may be identified with a scalar). The term that c_3' multiplies is a symmetric matrix. Thus H_k is symmetric, given that H_k is symmetric. Desirably the coefficients should be dimensionless, and thus should not change when there is a scale change, $\mathbf{x} \mapsto \sigma\mathbf{x}$ (for some positive constant σ). This is achieved by setting

$$c_1' = c_1/(\mathbf{p}_k^{\mathrm{T}}\mathbf{q}_k), \quad c_2' = c_2/(\mathbf{q}_k^{\mathrm{T}}H\mathbf{q}_k), \quad c_3' = c_3/(\mathbf{p}_k^{\mathrm{T}}\mathbf{q}_k)$$

The update formula can then be written in simpler notation:

$$H := H + c_1\mathbf{p}\mathbf{p}^{\mathrm{T}}/(\mathbf{p}^{\mathrm{T}}\mathbf{q}) + c_2(H\mathbf{q})(H\mathbf{q})^{\mathrm{T}}/(\mathbf{q}^{\mathrm{T}}H\mathbf{q}) + c_3(H\mathbf{q}\mathbf{p}^{\mathrm{T}} + \mathbf{p}\mathbf{q}^{\mathrm{T}}H)/(\mathbf{p}^{\mathrm{T}}\mathbf{q})$$

where the terms on the right relate to iteration k, and H on the left is H_{k+1}.

If the quasi-Newton condition is applied and terms in $\mathbf{p}$ and $H\mathbf{q}$ are equated, then

$$c_1\mathbf{p}^{\mathrm{T}}\mathbf{q} + c_3\mathbf{q}^{\mathrm{T}}H\mathbf{q} = 1 \quad \text{and} \quad -1 = c_3\mathbf{p}^{\mathrm{T}}\mathbf{q} + c_2\mathbf{q}^{\mathrm{T}}H\mathbf{q}$$

The update formula becomes, after rearrangement of terms,

$$H := H + \mathbf{p}\mathbf{p}^{\mathrm{T}}/(\mathbf{p}^{\mathrm{T}}\mathbf{q}) - H\mathbf{q}\mathbf{q}^{\mathrm{T}}H/(\mathbf{q}^{\mathrm{T}}H\mathbf{q}) + \theta\mathbf{z}\mathbf{z}^{\mathrm{T}} \tag{6.6}$$

where $\mathbf{z} := (\mathbf{q}^{\mathrm{T}}H\mathbf{q})^{1/2}[\mathbf{p}/(\mathbf{p}^{\mathrm{T}}\mathbf{q}) - H\mathbf{q}/(\mathbf{q}^{\mathrm{T}}H\mathbf{q})]$ and $\theta := -c_3\mathbf{p}^{\mathrm{T}}\mathbf{q}$. If $\theta = 0$, the *Davidson–Fletcher–Powell* (DFP) update formula is obtained:

$$H := H + \mathbf{p}\mathbf{p}^{\mathrm{T}}/(\mathbf{p}^{\mathrm{T}}\mathbf{q}) - H\mathbf{q}\mathbf{q}^{\mathrm{T}}H/(\mathbf{q}^{\mathrm{T}}H\mathbf{q}) \tag{6.7}$$

If $\theta = 1$, the *Broyden–Fletcher–Goldfarb–Shanno* (BFGS) update formula is obtained; the terms rearrange to

$$H := H + [1 + (\mathbf{q}^{\mathrm{T}}H\mathbf{q})/(\mathbf{p}^{\mathrm{T}}\mathbf{q})]\mathbf{p}\mathbf{p}^{\mathrm{T}}/\mathbf{p}^{\mathrm{T}}\mathbf{q} - (H\mathbf{q}\mathbf{p}^{\mathrm{T}} + \mathbf{p}\mathbf{q}^{\mathrm{T}}H)/(\mathbf{p}^{\mathrm{T}}\mathbf{q}) \tag{6.8}$$

Various properties are common to all updates (6.6), with $0 \le \theta \le 1$.

In order that the objective function shall decrease in iteration k, it is required that $\mathbf{g}_k^{\mathrm{T}}\mathbf{t}_k < 0$. An exact linesearch requires that $f(\mathbf{x}_k + \alpha\mathbf{t}_k)$ is minimized when

$\alpha = \alpha_k$, hence that

$$\mathbf{g}_{k+1}^{\mathrm{T}}\mathbf{t}_k = f'(\mathbf{x}_k + \alpha\mathbf{t}_k)\mathbf{t}_k = 0$$

Combining these shows that

$$\mathbf{p}_k^{\mathrm{T}}\mathbf{q}_k = \alpha_k\mathbf{t}_k^{\mathrm{T}}(\mathbf{g}_{k+1} - \mathbf{g}_k) = -\alpha_k\mathbf{t}_k^{\mathrm{T}}\mathbf{g}_k > 0$$

unless $\alpha_k = 0$, at which the iterations terminate.

Assume then that $\mathbf{p}_k^{\mathrm{T}}\mathbf{q}_k > 0$, and that $H \equiv H_k$ is symmetric and positive definite; thus $(\forall \mathbf{z} \neq 0)\ \mathbf{z}^{\mathrm{T}}H\mathbf{z} > 0$. Factorizing $H = S^{\mathrm{T}}S$ (as in the proof of the convergence rate inequality), and setting $\mathbf{a} = L\mathbf{z}$ and $\mathbf{b} = L\mathbf{q}$,

$$\mathbf{z}^{\mathrm{T}}[H - H\mathbf{q}\mathbf{q}^{\mathrm{T}}H/(\mathbf{q}^{\mathrm{T}}H\mathbf{q})]\mathbf{z} = \mathbf{a}^{\mathrm{T}}\mathbf{a} - (\mathbf{a}^{\mathrm{T}}\mathbf{b})^2/(\mathbf{b}^{\mathrm{T}}\mathbf{b}) \geq 0 \tag{6.9}$$

by the Cauchy–Schwarz inequality. Since $\mathbf{p}^{\mathrm{T}}\mathbf{q} > 0$,

$$\mathbf{z}^{\mathrm{T}}[\mathbf{p}\mathbf{p}^{\mathrm{T}}/(\mathbf{p}^{\mathrm{T}}\mathbf{q})]\mathbf{z} = (\mathbf{z}^{\mathrm{T}}\mathbf{p})^2/(\mathbf{p}^{\mathrm{T}}\mathbf{q}) \geq 0 \tag{6.10}$$

Since $\mathbf{a} \neq \mathbf{0}$ (from $\mathbf{z} \neq \mathbf{0}$), equality holds in (6.9) only if $\mathbf{a}$ is parallel to $\mathbf{b}$ (thus if $\mathbf{z}$ is parallel to $\mathbf{q}$) and then strict inequality holds in (6.10). Hence, for the DFP update, h_k is positive definite. The same conclusion holds for any update in the *Broyden family* (6.6), because an additional nonnegative term $\theta(\mathbf{z}^{\mathrm{T}}\mathbf{v})^2$ is added to $\mathbf{z}^{\mathrm{T}}H_{k+1}\mathbf{z}$.

It follows then, from the discussion following the convergence rate inequality, that the convergence for a quadratic objective function is superlinear, provided that it is shown that $\{H_k\} \to f''(\bar{x})^{-1}$, the inverse Hessian at the minimum point $\bar{x}$.

Assume now that $H_0 := I$ (or κI for some positive constant κ). Consider a quadratic objective function, with Hessian matrix A. For any update of the Broyden family (6.9), it can be shown by induction that the successive search directions $\mathbf{t}_i$ are conjugate directions with respect to A, and that

$$(\forall j \leq i) \quad H_{i+1}\mathbf{q}_j = \mathbf{p}_j$$

(The quasi-Newton condition, which was assumed in constructing the update formula, is the case $j = i$.) From conjugacy, it follows that the successive directions $\mathbf{t}_i$ are linearly independent, up to the stage (say at iteration m) when the step length is zero, and so the process terminates. From the linear independence, $m \leq n$ (the dimension of the space). If $m = n$, then

$$(\forall j \leq n) \quad \mathbf{p}_j = H_{n+1}\mathbf{q}_j = H_{n+1}A\mathbf{p}_j$$

for a quadratic objective. The $\mathbf{p}_j$ are linearly independent since the $\mathbf{t}_j$ are, and the step lengths are positive. Hence $H_{n+1}A = I$, so $H_{n+1} = A$, the inverse Hessian. So $\{H_k\} \to f''(\bar{x})^{-1}$.

To prove the statements

$$(\forall j \leq i) \quad H_{i+1}\mathbf{q}_j = \mathbf{p}_j \tag{6.11}$$

$$(\forall j < i) \quad \mathbf{t}_i^{\mathrm{T}}A\mathbf{t}_j = 0 \tag{6.12}$$

note that they hold for $i = 1$. If they hold for some $i < m$, then

$$(\forall j \leq i) \quad \alpha_j \mathbf{t}_{i+1}^{\mathrm{T}} A \mathbf{t}_j = (-\mathbf{g}_{i+1}^{\mathrm{T}} H_{i+1}) \mathbf{q}_j = -\mathbf{g}_{i+1}^{\mathrm{T}} \mathbf{p}_j \quad \text{by (6.11)}$$
$$= 0$$

for an exact linesearch. So (6.12) holds for $i + 1$ replacing i. Substituting in the update formula (6.6), $H_{i+2}\mathbf{q}_{i+1} = \mathbf{p}_{i+1}$. For $j \leq i$, $H_{i+2}\mathbf{q}_j = H_{i+1}\mathbf{q}_j$ plus terms which vanish, from (6.12) with $i + 1$ and (6.11). So (6.11) holds, by induction, provided that each step length α_j is > 0, thus until the process terminates. (It may, of course, terminate with $m < n$).

This theory applies for exact linesearches. In computing practice, DFP works well with fairly exact linesearch, but not so well with very approximate linesearches. However, BFGS continues to work well in practice with approximate linesearches.

6.2.7 Partitioned matrix inverse

The following construction for the inverse of a matrix that is partitioned into suitable submatrices is required for the theory of some of the algorithms, and is also of computational use.

Let A be a square matrix, partitioned into four submatrices, with A_{11} and A_{22} square matrices. Then

$$\begin{bmatrix} A_{11} & A_{12} \\ A_{21} & A_{22} \end{bmatrix} \begin{bmatrix} I & P \\ 0 & Q \end{bmatrix} = \begin{bmatrix} A_{11} & 0 \\ A_{21} & I \end{bmatrix} \Leftrightarrow A \begin{bmatrix} P \\ Q \end{bmatrix} = \begin{bmatrix} 0 \\ I \end{bmatrix} \tag{6.13}$$

Then

$$A^{-1} = \begin{bmatrix} I & P \\ 0 & Q \end{bmatrix} \begin{bmatrix} A_{11} & 0 \\ A_{21} & I \end{bmatrix}^{-1} = \begin{bmatrix} I & P \\ 0 & Q \end{bmatrix} \begin{bmatrix} A_{11}^{-1} & 0 \\ -A_{21}A_{11}^{-1} & I \end{bmatrix}$$

Multiplying out (6.13) gives

$$A_{11}P + A_{12}Q = 0, \quad \text{so } P = -A_{11}^{-1}A_{12}Q$$

and

$$A_{21}P + A_{22}Q = I, \quad \text{so } (-A_{21}A_{11}^{-1}A_{12} + A_{22})Q = I$$

Hence

$$Q = (A_{22} - A_{21}A_{11}^{-1}A_{12})^{-1}$$

and

$$A^{-1} = \begin{bmatrix} A_{11}^{-1} + A_{11}^{-1}A_{12}QA_{21}A_{11}^{-1} & -A_{11}^{-1}A_{12}Q \\ -QA_{21}A_{11}^{-1} & Q \end{bmatrix}$$

This construction has assumed that A and A_{11} are invertible matrices, or equivalently that A_{11} and $A_{22} - A_{21}A_{11}^{-1}A_{12}$ are invertible matrices.

6.2.8 Modification of an invertible matrix by matrix of lower rank

If one or more columns of an invertible matrix are replaced by other columns, then the inverse of the modified matrix can be calculated without a complete reinversion. The case where just one column is altered is used in the update step of the simplex method for linear programming.

Let A be an invertible matrix, partitioned as in the following formula. Consider the replacement of columns(s)

$$\begin{bmatrix} A_{12} \\ A_{22} \end{bmatrix}$$

by column(s)

$$\begin{bmatrix} U \\ V \end{bmatrix}$$

Then

$$\begin{bmatrix} A_{11} & A_{12} \\ A_{21} & A_{22} \end{bmatrix}\begin{bmatrix} I & P \\ 0 & Q \end{bmatrix} = \begin{bmatrix} A_{11} & U \\ A_{21} & V \end{bmatrix} \Leftrightarrow A\begin{bmatrix} P \\ Q \end{bmatrix} = \begin{bmatrix} U \\ V \end{bmatrix}$$

Consequently

$$\begin{bmatrix} A_{11} & U \\ A_{12} & V \end{bmatrix}^{-1} = \begin{bmatrix} I & -PQ^{-1} \\ 0 & Q^{-1} \end{bmatrix} A^{-1}$$

A different formula applies when a matrix of lower rank is added to A, not necessarily by replacing one or more columns of A. Consider the modification of an invertible $n \times n$ matrix A by adding to it RSU^{T}, where R is $n \times m$ with rank m, S is $m \times m$ invertible, and U is $n \times m$ with rank m, with $m < n$. The *Sherman–Morrison formula* states that

$$(A + RSU^{\mathrm{T}})^{-1} = A^{-1} - A^{-1}RW^{-1}U^{\mathrm{T}}A^{-1}$$

where

$$W = S^{-1} + U^{\mathrm{T}}A^{-1}R$$

This formula may be proved by changes of coordinates, which reduce R and U^{T} to the respective forms $\{I \quad 0]^{\mathrm{T}}$ and $\{I \quad 0]$.

6.2.9 Linesearch

If $f: \mathbf{R}^n \to \mathbf{R}$ is the differentiable function to be minimized and an iterate $\mathbf{x}_k$ has been reached, then a *linesearch* is required in some direction $\mathbf{t}_k$, to minimize $\varphi(\alpha) := f(\mathbf{x}_k + \alpha\mathbf{t}_k)$ with respect to $\alpha \geq 0$, either exactly (for theory), or to a sufficient approximation (in computing practice). There are numerous methods for linesearch. Note first that $\mathbf{t}_k$ must be a descent direction along which $f(.)$ decreases, hence the directional derivative of $f(.)$ at $\mathbf{x}_k$ in direction $\mathbf{t}_k$ must satisfy $\Delta := f'(\mathbf{x}_k)\mathbf{t}_k < 0$. If $\hat{\alpha}$ is the step length obtained by the linesearch, then the new iterate is $\mathbf{x}_{k+1} := \mathbf{x}_k + \hat{\alpha}\mathbf{t}_k$. For brevity, denote $f_k := f(\mathbf{x}_k)$ and $f_{k+1} := f(\mathbf{x}_{k+1})$. Let $\delta_k := \mathbf{x}_{k+1} - \mathbf{x}_k$. Assume that $f(.)$ reaches a (local) minimum at some point $\bar{\mathbf{x}}$, and denote $\bar{f} := f(\bar{\mathbf{x}})$.

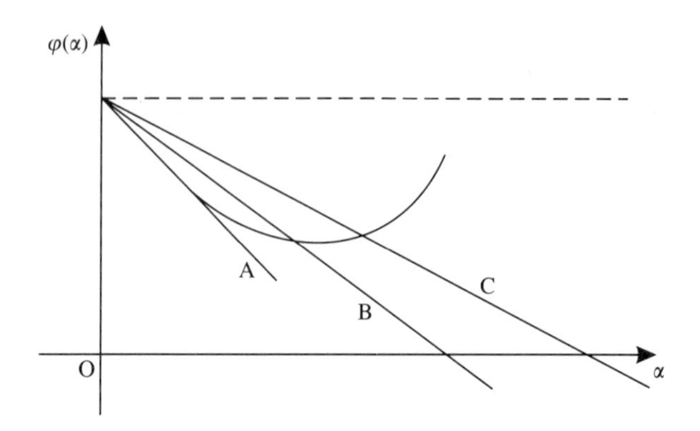

Figure 6.3 Linesearches: the Goldstein–Armarijo criterion.

A linesearch makes $f_{k+1} < f_k$; hence the sequence $f_1, f_2, \ldots, f_k, \ldots$ converges to some limit $\tilde{f}$, assuming that there is some lower bound to the values of $f(.)$. But if the steps $f_k - f_{k+1}$ are too small, it may happen that $\tilde{f} > \bar{f}$, so the minimum has not been reached. Now $\varphi(.)$ has initially negative slope (see line A in Figure 6.3), so $\varphi(.)$ decreases to some minimum, say at $\alpha = \tilde{\alpha}$, and then increases again, as α increases. A linesearch finds some approximation $\hat{\alpha}$ to $\alpha^{\#}$.

The Goldstein–Armarijo criterion requires that

$$\rho(-\Delta) \leq f_k - f_{k+1} \leq (1-\rho)(-\Delta) \tag{6.14}$$

for some constant ρ in $0 < \rho < \frac{1}{2}$. This requirement puts the step length between two bounds, illustrated by the lines B and C in Figure 6.3; the exact minimum is not always included! If an approximate linesearch algorithm satisfies these conditions, and if also the angle between $\mathbf{t}_k$ and $-\ f'(\mathbf{x}_k)^{\mathrm{T}}$ is bounded away from $\frac{1}{2}\pi$, then (Fletcher, 1980) either there is *finite termination* ($f'(\mathbf{x}_k) = \mathbf{0}$ for some k), or $f'(\mathbf{x}_k) \to 0$ as $k \to \infty$, or $f_k \downarrow -\infty$. The second case is the interesting one; various linesearch algorithms build in a test that (6.14) is satisfied.

A basic concept is *bracketing*. A *bracket* is an interval of α that contains the minimum point $\alpha^{\#}$. The approach is to choose some initial bracket, generally $[0, \alpha^*]$ where α^* is chosen large enough (often $\alpha^* = 1$) to ensure (almost certainly) a bracket, and then to look for ways to reduce the length of the bracket.

If $\varphi(.)$ is assumed to have just one minimum, then a bracket can be found by a systematic search (though this is not often done in practice because it may involve too many function evaluations). Choosing an initial step β_0, $\varphi(0)$ and $\varphi(\beta_0)$ are computed. If $\varphi(\beta_0) < \varphi(\alpha_0)$ then $\varphi(\beta_0 + 2\beta_0)$ is computed. If $\varphi(3\beta_0) < \varphi(\beta_0)$ then $\varphi(3\beta_0 + 4\beta_0)$ is computed. Observe that the steps are doubling: β_0, $2\beta_0$, $4\beta_0$, $8\beta_0$, ... The process stops when $\varphi(.)$ increases again – in the diagram $\varphi(15\beta_0) > \varphi(7\beta_0)$. Then $\varphi(.)$ is evaluated after a reverse step of half the current

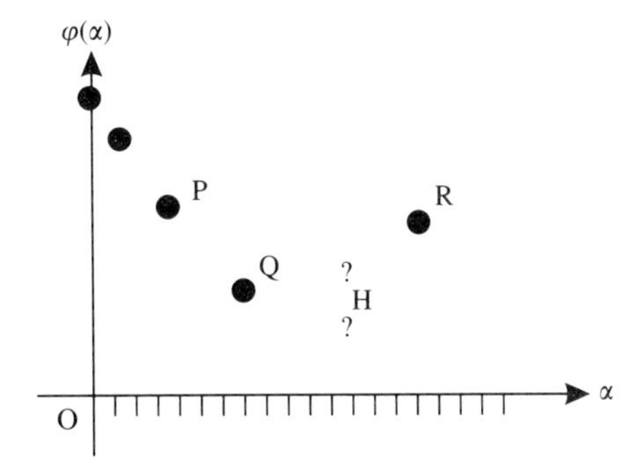

Figure 6.4 Finding a bracket.

increment. In Figure 6.4, P, Q, R and H label ordinates, H being that after the reverse step. Call the corresponding α values α_P etc.

Two cases now arise. Since there is only one minimum for $\varphi(.)$, there follow:

If H < Q then the minimum of $\varphi(.)$ lies in $[\alpha_Q, \alpha_R]$

If H > Q then the minimum of $\varphi(.)$ lies in $[\alpha_P, \alpha_H]$

In either case, a *bracket* has been found, containing the minimum point. Note that the computation has available only the points indicated on the graph, *not* the whole course of the graph of $\varphi(.)$.

If $[\alpha_L, \alpha_U]$ is a bracket, with midpoint $\alpha = \frac{1}{2}(\alpha_L + \alpha_U)$ and half-length $\beta = \frac{1}{2}(\alpha_U - \alpha_L)$, then a quadratic $q(.)$ could be fitted to the three points

$$(\alpha_L, \varphi(\alpha_L)), (\alpha, \varphi(\alpha)), (\alpha_U, \varphi(\alpha_U))$$

The minimum point of $q(.)$ is then calculated to be at $\alpha^{\#}$, where

$$\alpha^{\#} - \alpha = (\tfrac{1}{2}\beta)\,[\varphi(\alpha_L) - \varphi(\alpha_U)]/[\varphi(\alpha_U) - 2\varphi(\alpha) + \varphi(\alpha_L)]$$

Then $\alpha^{\#}$ provides an approximation to the exact minimum point $\hat{\alpha}$ of $\varphi(.)$.

Consider first an 'exact' linesearch (thus, exact up to the tolerance within which the computing is done) – although such methods are not usually used in practical computing. Assume here that the function $\varphi(.)$ has just one minimum.

An approximate linesearch due to Powell (1978) is based on the idea of successively fitting a quadratic approximation $q(.)$ to $\varphi(.)$, for which $q(0) = \varphi(0)$, $q'(0) = \varphi'(0) \equiv \Delta < 0$ and $q(\alpha) = \varphi(\alpha)$ for some chosen $\alpha > 0$. The method proceeds as follows:

Initialization	$\alpha := 1$; compute $\varphi(0)$; $b := 0.1$
Iteration	Compute $\varphi(\alpha)$. If $\varphi(\alpha) \leq \varphi(0) + b\alpha\Delta$ then EXIT linesearch Otherwise compute the minimum point of the fitted $q(.)$ as $s := -\frac{1}{2}\Delta\alpha^2/[\varphi(\alpha) - \varphi(0) - \Delta\alpha]$ Let $\alpha := \text{MAX}\{s, 0.1\alpha\}$ GOTO *Iteration*

This is a very approximate linesearch, taking very little computing time, so that it can be done many times. The EXIT test demands some minimal reduction in $\varphi(.)$, based on the initial negative slope Δ. The method does not quite work as it stands, because cumulative errors in step lengths can eventually produce a search direction in which $\varphi(.)$ does not sufficiently decrease, and then this linesearch may not terminate. One could count the number N of iterations within the linesearch, and then if $N > 2$ (say), terminate the linesearch, go back to the previous $\mathbf{x}_K$, and do a steepest descent step from there (which is certain to reduce $f(.)$.)

There are many variations of these ideas. One common approach approximates $\varphi(.)$ by a cubic instead of a quadratic.

Exercise

For $\varphi(\alpha) := -\alpha + \alpha^2 + 0.1\alpha^3$, compute an approximate minimum by the Powell method. (The exact minimum is $\hat{\alpha} = 0.467\,251\,3\,\ldots$). For comparison, find a bracket and an estimate $\alpha^{\#}$ by the search method described, with initial step $\beta_0 = 0.1$. Then start the search method again with new initial point $\alpha^{\#}$ and new initial step $= \frac{1}{2}\beta_0$. It is then not obvious whether α should increase or decrease, and both directions may have to be tried. Also, if no decrease of $\varphi(.)$ is found, the step must be decreased, say $\beta := \sqrt{10}\beta$. If $\hat{\alpha}$ is to be found within a stated tolerance τ say, then the computation must reduce the half-bracket β to less than τ.

6.3 CONSTRAINED MINIMIZATION

Consider now a minimization problem subject to both inequality and equality constraints:

$$\min_{\mathbf{x}} f(\mathbf{x}) \quad \text{subject to} \quad \mathbf{g}(\mathbf{x}) \leq \mathbf{0},\ \mathbf{h}(\mathbf{x}) = \mathbf{0} \tag{6.15}$$

As before, $\mathbf{x} \in \mathbf{R}^n$. Assume that a local minimum is reached at a point $\bar{\mathbf{x}}$. Some algorithms are outlined as follows.

6.3.1 Penalty methods

One classical approach replaces (6.15) by an unconstrained problem which approximates to (6.15). If the objective function $f(.)$ is considered as a cost to be minimized, then additional *penalty costs* are added to $f(.)$ when $\mathbf{x}$ does not satisfy the constraints. Define $t_+ := t$ if $t \geq 0$, 0 if $t < 0$, and, for a vector $\mathbf{v} = (v_1, v_2, \ldots, v_n)$, define $[\mathbf{v}]_+ := (v_{1+}, v_{2+}, \ldots, v_{n+})$. Consider then the unconstrained problem:

$$\min_{\mathbf{x}} f(\mathbf{x}) + \mu\|[\mathbf{g}(\mathbf{x})]_+\|^2 + \mu\|\mathbf{h}(\mathbf{x})\|^2 \tag{6.16}$$

The terms in $\mathbf{g}$ and $\mathbf{h}$ constitute the *penalty function*; note that they are zero when $\mathbf{x}$ satisfies all the constraints. The positive parameter μ could, more generally, be replaced by different positive parameters for each component of $[\mathbf{g}(\mathbf{x})]_+$ and $\mathbf{h}(\mathbf{x})$.

This penalty approach has been given different names by various authors, including *Sequential Unconstrained Minimization Technique* (SUMT) (Fiacco and McCormick, 1968). The approach is directly meaningful if the constraints describe physical limitations on some process, with a monetary penalty if they are violated. Such constraints may be called *soft constraints*. However, if the constraints are so-called *hard constraints* that must be satisfied exactly, then the penalty method gives only an approximate solution.

Under some fairly unrestrictive assumptions, the problem (6.16) reaches an unconstrained minimum at a point $\hat{\mathbf{x}}(\mu)$, where $\hat{\mathbf{x}}(\mu) \to \bar{\mathbf{x}}$ as $\mu \to \infty$. This means that a fairly large value of μ may be needed in order to obtain an accurate solution. But that may be computationally difficult, because if μ is large, the Hessian of the unconstrained objective function may be *ill-conditioned*, meaning that the ratio of largest to smallest eigenvalue is large. That results in slow convergence – see the convergence rate inequality (section 6.2.3).

If (6.15) has only inequality constraints, then the region of $\mathbf{x}$ satisfying the constraints may be pictured as a region with interior points, with a *boundary fence* described by $\mathbf{g}(\mathbf{x}) = \mathbf{0}$ (in fact, a boundary surface). The minimum point of the penalty problem lies typically a little outside the boundary fence. However, if the fence is moved in by a suitable small shift, then the unconstrained minimum can be made to lie exactly on the original boundary. The required shift turns out to depend on the Lagrange multipliers for the given problem (6.15). This idea forms the basis for the *augmented Lagrangian* algorithm, described in section 6.3.2.

6.3.2 Augmented Lagrangian algorithm

Consider first a problem with equality constraints:

$$\min_{\mathbf{x}} f(\mathbf{x}) \quad \text{subject to} \quad \mathbf{h}(\mathbf{x}) = \mathbf{0} \tag{6.17}$$

in which f and $\mathbf{h}$ are twice continuously differentiable, $\mathbf{x} \in \mathbf{R}^n$, $\mathbf{h}(\mathbf{x}) \in \mathbf{R}^m$, $m < n$, the minimum is reached at $\mathbf{x} = \bar{\mathbf{x}}$, and the $m \times n$ matrix $\mathbf{h}'(\bar{\mathbf{x}})$ has full rank (thus has rank m, since $m < n$). Then the constraint $\mathbf{h}(\mathbf{x}) = \mathbf{0}$ is locally solvable (sections 3.5.2 and 3.5.7) at $\bar{\mathbf{x}}$, and Karush–Kuhn–Tucker necessary conditions hold, with optimal Lagrange multiplier $\bar{\boldsymbol{\lambda}}$ say; to each direction $\mathbf{d} \in S := \mathbf{h}'(\bar{\mathbf{x}})^{-1}(\mathbf{0})$ there corresponds a local solution $\mathbf{x} = \bar{\mathbf{x}} + \alpha\mathbf{d} + \mathbf{o}(\alpha)$ $(\alpha \downarrow 0)$ to $\mathbf{h}(\bar{\mathbf{x}}) = \mathbf{0}$. Define now the Lagrangian as $L(\mathbf{x}, \boldsymbol{\lambda}) := f(\mathbf{x}) - \lambda\mathbf{h}(\mathbf{x})$, noting that for an equality constraint the components of $\boldsymbol{\lambda}$ can have either sign. Denote by M the Hessian matrix of $L(., \bar{\boldsymbol{\lambda}})$ at $(\bar{\mathbf{x}}, \bar{\boldsymbol{\lambda}})$, namely the matrix of second partial derivatives with respect to $\mathbf{x}$.

Then, for feasible $\mathbf{x}$, Taylor's theorem gives

$$f(\mathbf{x}) = L(\mathbf{x}, \bar{\boldsymbol{\lambda}}) = L(\bar{\mathbf{x}}, \bar{\boldsymbol{\lambda}}) + L_{\mathbf{x}}(\bar{\mathbf{x}}, \bar{\boldsymbol{\lambda}})\,(\alpha\mathbf{d} + \mathbf{o}(\alpha)) + \tfrac{1}{2}\alpha^2\mathbf{d}^{\mathrm{T}}M\mathbf{d} + o(\alpha^2)$$

where $L_{\mathbf{x}}$ denotes partial derivative with respect to $\mathbf{x}$. Now $L_{\mathbf{x}}(\bar{\mathbf{x}}, \bar{\boldsymbol{\lambda}}) = \mathbf{0}$ by Karush–Kuhn–Tucker conditions. Assume that M is positive definite on S, meaning that $\mathbf{d}^{\mathrm{T}}M\mathbf{d} > 0$ whenever $\mathbf{0} \neq \mathbf{d} \mapsto \in S$. Then

$$f(\mathbf{x}) = f(\bar{\mathbf{x}} + \alpha\mathbf{d} + o(\alpha)) > L(\bar{\mathbf{x}}, \bar{\boldsymbol{\lambda}}) = f(\bar{\mathbf{x}})$$

whenever α is positive and sufficiently small. Thus the problem (6.17) has an *isolated local minimum* at $\bar{\mathbf{x}}$.

Consider now a related unconstrained minimization problem:

$$\min_{\mathbf{x}} F(\mathbf{x}, \boldsymbol{\lambda}) \tag{6.18}$$

where

$$F(\mathbf{x}, \boldsymbol{\lambda}) := f(\bar{\mathbf{x}}) + \tfrac{1}{2}\mu\|\mathbf{h}(\mathbf{x})\|^2 - \boldsymbol{\lambda}\mathbf{h}(\mathbf{x})$$
$$= f(\bar{\mathbf{x}}) + \tfrac{1}{2}\mu\|\mathbf{h}(\mathbf{x}) - \mu^{-1}\boldsymbol{\lambda}\|^2 - \tfrac{1}{2}\mu^{-1}\|\boldsymbol{\lambda}\|^2$$

with μ a fixed positive parameter. Note that omitting the term $-\tfrac{1}{2}\mu^{-1}\|\boldsymbol{\lambda}\|^2$ does not affect minimization with respect to $\mathbf{x}$, and that the *augmented Lagrangian* $F(\mathbf{x}, \boldsymbol{\lambda})$ equals the ordinary Lagrangian $L(\mathbf{x}, \boldsymbol{\lambda})$ plus a *penalty term* $\tfrac{1}{2}\mu\|\mathbf{h}(\mathbf{x})\|^2$ for the constraint $\mathbf{h}(\mathbf{x}) = \mathbf{0}$. The term $-\mu^{-1}\boldsymbol{\lambda}$ represents the shift of boundary, mentioned in section 6.3.1.

Differentiating with respect to $\mathbf{x}$,

$$F_{\mathbf{x}}(\mathbf{x}, \boldsymbol{\lambda}) = f_{\mathbf{x}}(\mathbf{x}) + \mu\mathbf{h}(\mathbf{x})^{\mathrm{T}}\mathbf{h}'(\mathbf{x}) - \boldsymbol{\lambda}\mathbf{h}'(\mathbf{x})$$

Then, from the Karush–Kuhn–Tucker conditions,

$$F_{\mathbf{x}}(\bar{\mathbf{x}}, \bar{\boldsymbol{\lambda}}) = \mathbf{0} + \mu\mathbf{h}(\bar{\mathbf{x}})^{\mathrm{T}}\mathbf{h}'(\bar{\mathbf{x}}) = \mathbf{0}$$

Also, using $\mathbf{h}(\bar{\mathbf{x}}) = \mathbf{0}$,

$$\mathbf{F}_{xx}(\bar{\mathbf{x}}, \bar{\boldsymbol{\lambda}}\mathbf{h}) = (f - \bar{\boldsymbol{\lambda}})''(\bar{\mathbf{x}}) + \mu\mathbf{h}'(\bar{\mathbf{x}})^{\mathrm{T}}\mathbf{h}'(\bar{\mathbf{x}}) = M + \mu\mathbf{h}'(\bar{\mathbf{x}})^{\mathrm{T}}\mathbf{h}'(\bar{\mathbf{x}})$$

Since $\mathbf{h}'(\bar{\mathbf{x}})$ has been assumed to have full rank, elementary row and column operations (section 1.5.1) change this matrix into the *partitioned matrix* form $[I, 0]$, where I denotes an $m \times m$ unit matrix, and 0 denote an $m \times (n-m)$ matrix of zeros. The change amounts to choosing suitable new coordinates in both the domain and range spaces. Assume that the other matrices in the calculation are partitioned similarly. Then

$$\mathbf{F}_{xx}(\bar{\mathbf{x}}, \bar{\boldsymbol{\lambda}}) = \begin{bmatrix} H_{11} & H_{12} \\ H_{21} & H_{22} \end{bmatrix} = \begin{bmatrix} M_{11} & M_{12} \\ M_{21} & M_{22} \end{bmatrix} + \mu \begin{bmatrix} I \\ 0 \end{bmatrix} [I \ \ 0] = \begin{bmatrix} M_{11} + \mu I & M_{12} \\ M_{21} & M_{22} \end{bmatrix} \equiv P$$

in which M_{22} is positive definite, since M is assumed positive definite on S. After some calculation, it is found that, for any vector $\mathbf{v} \neq \mathbf{0}$,

$$\mathbf{v}^{\mathrm{T}}\mathbf{F}_{xx}(\bar{\mathbf{x}}, \bar{\boldsymbol{\lambda}})\mathbf{v} > 0$$

thus $\mathbf{F}_{xx}(\bar{\mathbf{x}}, \bar{\boldsymbol{\lambda}})$ is positive definite, provided that μ is sufficiently large, namely just large enough for the off-diagonal submatrices to be dominated by the positive definite diagonal submatrices M_{22} and $M_{11} + \mu I$. Thus $F(., \bar{\boldsymbol{\lambda}})$ is convex, for $\mathbf{x}$ near $\bar{\mathbf{x}}$. This, with $F_{\mathbf{x}}(\bar{\mathbf{x}}, \bar{\boldsymbol{\lambda}}) = \mathbf{0}$, shows that $F(., \bar{\boldsymbol{\lambda}})$ is minimized at $\bar{\mathbf{x}}$.

Moreover, by continuity, $F_{\mathbf{xx}}(\bar{\mathbf{x}}, \bar{\boldsymbol{\lambda}})$ remains positive definite for $\mathbf{x}$ near $\bar{\mathbf{x}}$ and $\boldsymbol{\lambda}$ near $\bar{\boldsymbol{\lambda}}$. Then the usual *implicit function theorem* shows that the equation $F_{\mathbf{x}}(.,\bar{\boldsymbol{\lambda}}) = \mathbf{0}$ has a unique solution $\mathbf{x} = \hat{\mathbf{x}}(\boldsymbol{\lambda})$, for all $\boldsymbol{\lambda}$ near $\bar{\boldsymbol{\lambda}}$. Denote $q(\boldsymbol{\lambda}) := F(\hat{\mathbf{x}}(\boldsymbol{\lambda}), \boldsymbol{\lambda})$. Then the gradient $q'(\bar{\boldsymbol{\lambda}}) = \mathbf{0}$. (For a convex problem, the stationary point at $\bar{\boldsymbol{\lambda}}$ of $q(.)$ is actually a maximum, since $L(\mathbf{x}, \boldsymbol{\lambda})$ has a saddle point at $(\bar{\mathbf{x}}, \bar{\boldsymbol{\lambda}})$; see e.g. Craven (1978) for the saddle point theorem.)

Calculations with the chain rule show that

$$F_{\mathbf{x}}(\hat{\mathbf{x}}(\boldsymbol{\lambda}), \boldsymbol{\lambda}) \equiv \mathbf{0} \Rightarrow F_{\mathbf{xx}}\hat{\mathbf{x}}' + F_{\mathbf{x}\lambda} = \mathbf{0} \Rightarrow \hat{\mathbf{x}}' = F_{\mathbf{xx}}^{-1}\mathbf{h}_{\mathbf{x}}^{\mathrm{T}} \tag{6.19}$$

in which arguments $\mathbf{x}(\boldsymbol{\lambda})$ and $\boldsymbol{\lambda}$ have been suppressed, for brevity, and $\hat{\mathbf{x}}'$ denotes the gradient of $\hat{\mathbf{x}}(\boldsymbol{\lambda})$. Similarly,

$$q(\boldsymbol{\lambda}) := F(\hat{\mathbf{x}}(\boldsymbol{\lambda}), \boldsymbol{\lambda}) \Rightarrow q_{\lambda} = F_{\mathbf{x}}\hat{\mathbf{x}}' + F_{\lambda} = \mathbf{0} - \mathbf{h}^{\mathrm{T}} \quad (\text{since } F_{\mathbf{x}} \equiv \mathbf{0})$$

Hence, since $F_{\mathbf{x}} \equiv \mathbf{0}$, and substituting for $\hat{\mathbf{x}}'$ from (6.19),

$$q_{\lambda\lambda} = F_{\lambda\mathbf{x}}\hat{\mathbf{x}}' + F_{\mathbf{x}\lambda} = -\mathbf{h}_{\mathbf{x}}\hat{\mathbf{x}}' + \mathbf{0} = -\mathbf{h}_{\mathbf{x}}F_{\mathbf{xx}}^{-1}\mathbf{h}_{\mathbf{x}}^{\mathrm{T}}$$

In terms of the new coordinates for the matrix P above, $F''(\bar{\mathbf{x}}, \bar{\boldsymbol{\lambda}})$, partition

$$P^{-1} = \begin{bmatrix} D_{11} & D_{12} \\ D_{21} & D_{22} \end{bmatrix}$$

Then

$$P^{-1} = [I \quad 0]\begin{bmatrix} D_{11} & D_{12} \\ D_{21} & D_{22} \end{bmatrix}\begin{bmatrix} I \\ 0 \end{bmatrix} = D_{11}$$

By the *partitioned matrix inverse* formula (section 6.2.7),

$$D_{11} = [(M_{11} + \mu I) - M_{12}M_{22}^{-1}M_{21}]^{-1}$$

which has the form $(\mu I + C)^{-1}$, for a constant matrix C. Hence

$$-q_{\lambda\lambda}(\bar{\boldsymbol{\lambda}}) = \mu^{-1}(I + \mu^{-1}C)^{-1} = \mu^{-1}(I - \mu^{-1}C + O(\mu^{-2}))$$

expanding in geometric series. So, if μ is not too small,

$$-q_{\lambda\lambda}(\bar{\boldsymbol{\lambda}}) \approx \mu^{-1}I$$

Now Newton's method to solve for the zero $\bar{\boldsymbol{\lambda}}$ of $q_{\mathbf{x}}(.)$ would iterate

$$\boldsymbol{\lambda} \mapsto \boldsymbol{\lambda} - \mathbf{q}_{\lambda\lambda}(\boldsymbol{\lambda})^{-1}q_{\mathbf{x}}(\boldsymbol{\lambda})$$

Consider instead a quasi-Newton algorithm, approximating $q_{\lambda\lambda}(\boldsymbol{\lambda})$. This gives the iteration

$$\boldsymbol{\lambda} \mapsto \boldsymbol{\lambda} - \mu\mathbf{h}(\hat{\mathbf{x}}(\boldsymbol{\lambda}))^{\mathrm{T}}$$

This construction gives the *Powell–Hestenes* algorithm, in which each iteration consists of the following two steps:

(i) Given $\boldsymbol{\lambda}$, find the unconstrained minimum of $F(., \boldsymbol{\lambda})$, say at $\tilde{\mathbf{x}}$.
(ii) Update $\boldsymbol{\lambda}$ by

$$\boldsymbol{\lambda} \mapsto \boldsymbol{\lambda} - \mu\mathbf{h}(\tilde{\mathbf{x}})^{\mathrm{T}}$$

The whole calculation must start with some assumed initial $\boldsymbol{\lambda}$ (often with $\boldsymbol{\lambda} = \mathbf{0}$). The parameter μ must be 'large enough' to satisfy a positive definite requirement, and to make an $O(\mu^{-2})$ term small enough. However μ need *not* tend to infinity; so the problem of 'ill-conditioned Hessian' in the penalty function method is avoided here. What has been proved is *local convergence*; that is, convergence of $(\mathbf{x}, \boldsymbol{\lambda})$ to $(\bar{\mathbf{x}}, \bar{\boldsymbol{\lambda}})$, provided that the initial point is not too far away from $(\bar{\mathbf{x}}, \bar{\boldsymbol{\lambda}})$, and assuming that μ is over some threshold, depending on the functions in the problem. Convergence is not guaranteed if M is not positive definite on the null space S of $\mathbf{h}'(\bar{\mathbf{x}})^{-1}$. A simple example is $f(x) = -x^2$ (a concave function of $x \in \mathbf{R}$), with constraints $a \leq x \leq b$; for some a and b, convergence only occurs with a special value of μ.

Now consider a problem with inequality constraints:

$$\min_{\mathbf{x}} f(\mathbf{x}) \quad \text{subject to} \quad \mathbf{g}(\mathbf{x}) \geq \mathbf{0} \tag{6.20}$$

The inequality constraint can be replaced by an equality, $\mathbf{g}(\mathbf{x}) - \mathbf{s} = \mathbf{0}$, where $\mathbf{s} \geq \mathbf{0}$ is a vector of slack variables. The above approach for an equality constraint is modified as follows. The term

$$\|\mathbf{g}(\mathbf{x}) - \mathbf{s} - \mu^{-1}\boldsymbol{\lambda}\|^2 = \sum_i [g_i(\mathbf{x}) - s_i - \mu^{-2}\lambda_i]^2$$

If $g_i(\mathbf{x}) > 0$ then $g_i(\mathbf{x}) - s_i = 0$ and $s_i > 0$, so $g_i(\mathbf{x}) - s_i - \mu^{-1}\lambda_i = \mu^{-1}\lambda_i$.
If $g_i(\mathbf{x}) \leq 0$ then $s_i = 0$ and $|g_i(\mathbf{x}) - s_i - \mu^{-1}\lambda_i| = [g_i(\mathbf{x}) - s_i]_-$, where (for real t)

$$[t]_+ := t \text{ if } t \geq 0,\ 0 \text{ if } t < 0; \quad [t]_- := 0 \text{ if } t \geq 0,\ -t \text{ if } t < 0$$

For a vector $\mathbf{v}$, denote by $[\mathbf{v}]_-$ the vector whose components are $[v_i]_-$, and denote by $[\mathbf{v}]_+$ the vector whose components are $[v_i]_+$.

Disregarding terms not involving $\mathbf{x}$, this calculation results in the two steps of each iteration of the *Powell–Hestenes–Rockafellar* (PHR) algorithm:

(i) Given $\boldsymbol{\lambda}$, find the unconstrained minimum $f(\mathbf{x}) + \frac{1}{2}\mu\|[\mathbf{g}(\mathbf{x}) - \mu^{-1}\boldsymbol{\lambda}]_-\|^2$, say at $\tilde{\mathbf{x}}$.
(ii) Update $\boldsymbol{\lambda}$ by

$$\boldsymbol{\lambda} \mapsto [\boldsymbol{\lambda} - \mu\mathbf{g}(\tilde{\mathbf{x}})^{\mathrm{T}}]_+$$

Note that the $\boldsymbol{\lambda}$ update is modified here to fulfil the requirement (from the Karush–Kuhn–Tucker conditions) that $\boldsymbol{\lambda} \geq \mathbf{0}$ for the inequality constraint $\mathbf{g}(\mathbf{x}) \geq \mathbf{0}$. The convergence behaviour is similar to the case of equality constraints. Note that

$$[g_i(\mathbf{x}) - \mu^{-1}\lambda_i]_-^2 = -\lambda_i b_i + \tfrac{1}{2}\mu b_i^2 \quad \text{if } b_i \leq \lambda_i/\mu, \quad \text{or } -\tfrac{1}{2}\lambda_i^2/\mu \quad \text{if } b_i > -\lambda_i/\mu$$

where the b_i are the components of $\mathbf{g}(\mathbf{x})$, and λ_i are the components of $\boldsymbol{\lambda}$.

Clearly this approach can be applied to a problem with both equality and inequality constraints by combining appropriate terms; the details are left to the reader.

The PHR algorithm has a good reliable reputation, provided that its assumptions are satisfied. For a discussion of convergence rate, refer to Fletcher (1987) or Craven (1991). However, many recent computer codes for constrained optimization use instead the *sequential quadratic* algorithm, which builds in the augmented Lagrangian approach.

6.3.3 Quadratic programming

A quadratic program, one with quadratic objective and linear constraints, is often required to be solved, either because it describes, exactly or approximately, some physical problem – for example, many control problems in engineering can be put in this form, or because the quadratic program is needed as a local approximation, in the course of computing an optimum for a more general nonlinear program.

Consider first a quadratic program with equality constraints:

$$\min_{\mathbf{v}} \tfrac{1}{2}\mathbf{v}^{\mathrm{T}}C\mathbf{v} + \mathbf{r}^{\mathrm{T}}\mathbf{v} \quad \text{subject to} \quad M\mathbf{v} = \mathbf{k} \tag{6.21}$$

with $\mathbf{x} \in \mathbf{R}^n$ and $\mathbf{k} \in \mathbf{R}^m$, k where $m < n$. Assume that (6.21) reaches a minimum when $\mathbf{v} = \bar{\mathbf{v}}$. The Karush–Kuhn–Tucker necessary conditions for this minimum require that

$$C\bar{\mathbf{v}} + \mathbf{r} + M^{\mathrm{T}}\bar{\boldsymbol{\lambda}} = \mathbf{0}, \quad M\bar{\mathbf{v}} = \mathbf{k}$$

in terms of a Lagrange multiplier $\bar{\boldsymbol{\lambda}}$. This linear system may be written in partitioned matrix form as

$$\begin{bmatrix} C & M^{\mathrm{T}} \\ M & 0 \end{bmatrix}\begin{bmatrix} \bar{\mathbf{v}} \\ \bar{\boldsymbol{\lambda}} \end{bmatrix} = \begin{bmatrix} -\mathbf{r} \\ \mathbf{k} \end{bmatrix}$$

Hence, using the partitioned matrix inverse formula of section 6.2.7,

$$\begin{bmatrix} \bar{\mathbf{v}} \\ \bar{\boldsymbol{\lambda}} \end{bmatrix} = \begin{bmatrix} C^{-1} - U^{\mathrm{T}}KU & U^{\mathrm{T}}K \\ KU^{\mathrm{T}} & -K \end{bmatrix}\begin{bmatrix} -\mathbf{r} \\ \mathbf{k} \end{bmatrix} \tag{6.22}$$

where $K := [MC^{-1}M^{\mathrm{T}}]^{-1}$ and $U := C^{-1}M^{\mathrm{T}}$. This has assumed that C and $MC^{-1}M^{\mathrm{T}}$ are invertible; note that $MC^{-1}M^{\mathrm{T}}$ is invertible if C is invertible and M has full rank (thus rank $= m$). With these assumptions, both the minimum point $\bar{\mathbf{x}}$ and the Lagrange multiplier $\bar{\boldsymbol{\lambda}}$ are uniquely determined by solving the linear system (6.22).

Suppose now that a nonlinear problem is approximated, in some neighbourhood of an iterate $\mathbf{x}_k$, by a quadratic program, with inequality and equality constraints:

$$\min_{\mathbf{x}} \tfrac{1}{2}\mathbf{x}^{\mathrm{T}}C\mathbf{x} + \mathbf{s}^{\mathrm{T}}\mathbf{x} \quad \text{subject to} \quad A\mathbf{x} \le \mathbf{b},\ A'\mathbf{x} = \mathbf{b}' \tag{6.23}$$

Set $\mathbf{v} := \mathbf{x} - \mathbf{x}_k$. Construct a matrix M, consisting of those rows of A corresponding to inequality constraints active at $\mathbf{x}_k$, together with all the rows of A'; let $\mathbf{l}$ consist of those elements of $\mathbf{b}$ corresponding to active constraints, together with all elements of $\mathbf{b}'$. Consider the related quadratic program:

$$\min_{\mathbf{v}} \tfrac{1}{2}\mathbf{v}^{\mathrm{T}}C\mathbf{v} + (C\mathbf{x}_k + \mathbf{s})^{\mathrm{T}}\mathbf{v} \quad \text{subject to} \quad M\mathbf{v} = \mathbf{k} := \mathbf{l} - M\mathbf{x}_k \tag{6.24}$$

Note that (6.24) has a more restricted feasible set than (6.23), since active inequalities have been changed into equations. However, (6.24) has the form of (6.21), so is readily solvable, by (6.22).

The solution, $\mathbf{v}_k$ say, obtained for (6.24) gives a search direction. Thus, the next iterate is $\mathbf{x}_{k+1} = \mathbf{x}_k + \alpha_k \mathbf{v}_k$, where α_k is a step length, which may be obtained from a linesearch. Note that the list of active constraints must be recorded, since it is likely to change from one iteration to the next.

A quadratic program, with linear constraints that may include inequalities, could be solved by iterating the above approach until the optimal list of active constraints is found.

The simplex algorithm for linear programming does not work for a quadratic program for the following reason. When a new basis element is introduced, its value is increased from zero until some old basis element is reduced to zero. But this can fail for quadratic programming, since a quadratic function may reach a minimum before the old basis element reaches zero. Thus, the number of basis elements increases with successive iterations. This approach is followed in Beale's algorithm (see Cameron (1985) for a detailed analysis), in which the number of basis elements is kept as low as possible. However, this approach does not seem to have been used for control problems.

6.3.4 Sequential quadratic algorithm

Consider the problem:

$$\min f(\mathbf{x}) \quad \text{subject to} \quad g_j(\mathbf{x}) = 0\ (j = 1, 2, \ldots, l),\ g_j(\mathbf{x}) \geq 0\ (j = l+1, \ldots, m) \qquad (6.25)$$

For brevity, the constraints may be written in vector form as $\mathbf{g}(\mathbf{x}) \in S$, where $S := \{0\}^l \times \mathbf{R}_+^{m-l}$. For (6.25), define the Lagrangian

$$L(\mathbf{x}, \boldsymbol{\lambda}) := f(\mathbf{x}) + \boldsymbol{\lambda}\mathbf{g}(\mathbf{x})$$

and the augmented Lagrangian

$$L^*(\mathbf{x}, \boldsymbol{\lambda}) := f(\mathbf{x}) + \tfrac{1}{2}\mu \| [\mathbf{g}(\mathbf{x}) - \mu^{-1}\boldsymbol{\lambda}]_* \|^2$$

where the components of $[\mathbf{v}]_*$ are v_j for $j = 1, 2, \ldots, l$ (namely for equality constraints), and $[v_j]_-$ for $j = l + 1, \ldots, m$ (namely for $\geq$ inequality constraints).

In a region close to the iterate $\mathbf{x}_k$ (with Lagrange multiplier $\boldsymbol{\lambda}_k$), the problem (6.25) is locally approximated by a quadratic problem:

$$\min_{\mathbf{d},\delta} f'(\mathbf{x}_k)\mathbf{d} + \tfrac{1}{2}\mathbf{d}^{\mathrm{T}} B_k \mathbf{d} + \tfrac{1}{2}\rho_k \delta^2$$

subject to (6.26)

$$\mathbf{g}(\mathbf{x}_k) + (I - \delta U_k)\mathbf{g}'(\mathbf{x}_k) \in S$$

Here B_k is some positive definite matrix (to be specified), which approximates the Hessian matrix of the augmented Lagrangian at $(\mathbf{x}_k, \boldsymbol{\lambda}_k)$. If the terms in δ were omitted, the linearization of the constraints in (6.26) could result in discarding the feasible points, so the problem must be slightly perturbed. The diagonal matrix U_k has diagonal elements 1, for equality constraints and for (active or nearly active) inequality constraints where $g_j(\mathbf{x}_k) \leq \varepsilon$ or $\lambda_k > 0$. The positive parameters ε, ρ_k and μ (which may depend on k) must be chosen; obviously ε is small positive, allowing for the fact that an inactive constraint may become active at a nearby point.

The quadratic–linear problem (6.26) approximates (6.25) in some region around the iterate $\mathbf{x}_k$. The optimum $\mathbf{d}_k$ of (6.26) selects a search diection; denote by $\mathbf{u}_k$ its corresponding Lagrange multiplier (row vector). A linesearch is then done, to minimize (approximately) the augmented Lagrangian

$$L^*(\mathbf{x}_k + \alpha\mathbf{d}_k, \boldsymbol{\lambda}_k + \alpha(\mathbf{u}_k - \mathbf{v}_k))$$

with respect to $\alpha > 0$, say at $\alpha = \alpha_k$. Then updates are done as follows:

$$\boldsymbol{\lambda}_{k+1} := \boldsymbol{\lambda}_k + \alpha_k(\mathbf{u}_k - \boldsymbol{\lambda}_k)$$

$$\mathbf{x}_{k+1} := \mathbf{x}_k + \alpha_k\mathbf{d}_k$$

B_k is updated by the BFGS formula.

This algorithm has been implemented by Schittkowski (1985); another variant has been implemented by Zhou and Tits (1991). It requires some method to solve the quadratic programming subproblems. The algorithm works with a wide variety of constrained minimization problems, notably including such problems arising from optimal control problems. It has been shown theoretically to converge, fairly generally, to some Karush–Kuhn–Tucker point of the problem – which need not be unique, so need not be the minimum sought. Since each iteration of *sequential quadratic* generates a decrease of the objective over some local region, the assumption in PHR that the Lagrangian is convex (from positive definite Hessian) on a certain subspace is not required here. Consequently, sequential quadratic can converge to an optimum when augmented Lagrangian fails to do so.

6.3.5 Projected gradient

Consider a problem with linear constraints:

$$\text{Minimize } f(\mathbf{x}) \quad \text{subject to} \quad A\mathbf{x} \leq \mathbf{b}$$

This could arise, for example, from discretizing a control problem with a linear differential equation:

$$\dot{x}(t) = A(t)x(t) + B(t)u(t), \quad x(0) = x_0$$

At a given iterate $\mathbf{x}_k$, the active constraints will consist of some (usually not all) of the rows of $A\mathbf{x} \leq \mathbf{b}$; denote them by $\tilde{A}\mathbf{x} \leq \tilde{\mathbf{b}}$.Then $\tilde{A}\mathbf{x}_k = \tilde{\mathbf{b}}$. Now a descent direction $\mathbf{t}_k$ from $\mathbf{x}_k$ is likely to move outside the feasible region. Consider instead a modified direction $\tilde{\mathbf{t}}_k$ satisfying $f'(\mathbf{x}_k)\tilde{\mathbf{t}}_k < 0$ (so that $\tilde{\mathbf{t}}_k$ is a descent direction for f), and also $\tilde{A}\tilde{\mathbf{t}}_k = \mathbf{0}$ (so that $\tilde{\mathbf{t}}_k$ runs along the boundary of the feasible region). Such a direction $\tilde{\mathbf{t}}_k$ can be found by projecting $\mathbf{t}_k$ onto the boundary of the feasible set – hence the name *projected gradient*. Thus $\tilde{\mathbf{t}}_k = P_k \mathbf{t}_k$, where P_k is the projection matrix

$$P_k = I - \tilde{A}^{\mathrm{T}}(\tilde{A}\tilde{A}^{\mathrm{T}})^{-1}\tilde{A}$$

assuming that $\tilde{A}$ has full rank, so that $\tilde{A}\tilde{A}^{\mathrm{T}}$ is invertible.

The direction $\mathbf{t}_k$ could arise from e.g. a Fletcher–Reeves algorithm or a BFGS algorithm. The step length is obtained, as usual, by a linesearch (usually approximate) in the direction $\tilde{\mathbf{t}}_k$.

The list of active constraints generally changes from one iteration to the next. This requires an update of the projection matrix P_k. Since $\tilde{A}$ is being changed by adding to it a matrix of small rank, some modifications of the formulae in section 6.2.8 are possible, so that the inverse matrix within P_k need not be computed from the beginning at each iteration. Alternatively, since $f(\mathbf{x})$ is being minimized near $\mathbf{x}_k$ subject to the equality constraint $\tilde{A}\mathbf{x} = \tilde{\mathbf{b}}$, it is equivalent to eliminating some components of $\mathbf{x}$ in terms of other components, in the manner of section 6.3.3. This leads to a *generalized reduced gradient* algorithm.

6.4 COMPUTATION OF OPTIMAL CONTROL PROBLEMS

Consider first the fixed-time optimal control problem:

$$\text{Minimize } J(u) \equiv F(x, u) = \int_0^T f(x(t), u(t), t)\,\mathrm{d}t + \Phi(x(T))$$

subject to

$$x(0) = x_0,\ \mathrm{d}x(t)/\mathrm{d}t = m(x(t), u(t), t) \quad (0 \le t \le T)$$

$$u(t) \in \Gamma(t) \quad (0 \le t \le T)$$

The Hamiltonian is

$$h(x(t), u(t), t, \lambda(t) := f(x(t), u(t), t) + \lambda(t)m(x(t), u(t), t)$$

In general, $x(t)$ and $u(t)$ are vector-valued, $\lambda(t)$ is a row vector and $m(x(t), u(t), t)$ is a column vector.

While analytic solutions are possible only in limited cases, the Pontryagin theory can sometimes be used for computation (sections 6.4.1 and 6.4.2). The gradient formulae arising from the Pontryagin theory (in section 6.4.4) are needed in most algorithms.

6.4.1 Bang-bang

Consider such a control problem where the control function is scalar-valued and constrained by $(\forall t)\ a \le u(t) \le b$, and where furthermore the optimum control is bang-bang. (This may happen when the control $u(t)$ enters the problem linearly, but other conditions are also required.) If the (possibly empty) set $\mathbf{S}$ of *switching times* $t_1, t_2, \ldots$ was known, then the optimum control would be determined as one of two possible cases (depending on whether $u(0) = a$ or $u(0) = b$). So the optimization problem then reduces to optimization over a finite set of variables, since $J(u)$ reduces to a function of the variables t_i in **S**. (If $u(t)$ is vector-valued, lying in a polyhedron, then $u(t)$ jumps between vertices.)

The assumptions may need to be relaxed to allow for a possible singular arc for the initial and final time intervals.

6.4.2 Iterations using Pontryagin's principle

For the above fixed-time problem, assume an initial control function $u^0(.)$, and construct iterations $u^1(.)$, $u^2(.)$, $u^3(.)$, ... in the following manner.

Given $u^k(.)$, the given differential equation for $x(.)$, with initial condition, determines $x(.) \equiv u^k(.)$. Then the adjoint differential equation, with terminal condition $\lambda(T) = 0$, determines the costate function $\lambda(.) \equiv \lambda^k(.)$. The Hamiltonian for the current iterations is then calculable, since $x^k(t)$ and $u^k(t)$ are now known. Let

$$v(t) := \operatorname{argmin}_{u(t)\in\Gamma(t)} h(x^k(t), u(t), t, \lambda^k(t))$$

In general, $v(t) \neq u^k(t)$, since $u^k(t)$ is not optimal.

An iterative process is then defined by

$$(\forall t) \quad u^{k+1}(t) := \Psi(u^k(t), v(t))$$

where $\Psi(.,.)$ is some suitable function. Let

$$M_k(t) := h(x^k(t), u^k(t), t, \lambda^k(t)) - h(x^k(t), v(t), t, \lambda^k(t))$$

Then $M_k(t) \geq 0$ for all t, and $M_k(t) > 0$ for some t. Let $t^* := \operatorname{argmax} M_k(t)$. Consider in particular

$$\Psi(u^k(t), v(t)) = \begin{cases} v(t) & t \in E \\ u^k(t) & t \notin E \end{cases}$$

where E is some interval centred on t^*. For suitable choice of E, the sequence

$$\{h(x^k(t), u^k(t), t, \lambda^k(t)) : k = 0, 1, 2, \ldots\} \tag{6.27}$$

converges to $h(\bar{x}(t), \bar{u}(t), t, \bar{\lambda}(t))$, where $\bar{x}(t)$, $\bar{u}(t)$ and $\bar{\lambda}(t)$ are the optimal state, control and costate for the control problem; see references in Teo, Goh and Wong (1991). Note that, in the case of several local minima, $u^0(.)$ is assumed to be sufficiently close to $\bar{u}$ that only a region around $\bar{u}(.)$ that contains no other local minimum point need be considered. The convergence depends on $M_k(.)$ decreasing sufficiently rapidly as k increases, so that the sequence (6.27) does not converge to some other limit.

While this algorithm works, it appears from computational experience that a different kind of discretization (section 6.4.4) is computationally more effective.

6.4.3 Standard formulations

The fixed-time optimal control problem will now be generalized (following Goh and Teo (1987) and Jennings *et al.* (1990, 1991)) to:

$$\text{Minimize } J(u(.,p),p) := F(x(.,p),u(.,p),p)$$
$$:= \int_0^T f(x(t,p),u(t,p),t,p)\,\mathrm{d}t + \Phi(x(T,p),p) \quad (6.28)$$

subject, for $0 \leq t \leq T$, to

$$(\forall t) \quad u(t) \in \Gamma(t) \quad (6.29)$$
$$x(0,p) = x_0(p), \quad (\forall t)\ \partial x(t,p)/\partial t = m(x(t,p),u(t,p),t,p) \quad (6.30)$$
$$(\forall t) \quad \theta^i(u(.,p),p) = 0 \quad (i = 1,2,\ldots,r_1) \quad (6.31)$$
$$(\forall t) \quad \theta^i(u(.,p),p) \geq 0 \quad (i = r_1+1, r_1+2,\ldots,r) \quad (6.32)$$

where

$$\theta^i(u(.,p),p) := \Phi^i(x(T,p),p) + \int_0^T f^i(x(t,p),u(t,p),t,p)\,\mathrm{d}t \quad (6.33)$$

Here $\Gamma(t)$ is a specified interval (or rectangle if $u(t) \in \mathbf{R}^2$); p is a parameter (scalar or vector); and there are r constraints which may involve both the state $x(.)$ and the control $u(.)$. The sets $\Gamma(t)$ could be more general, e.g. with curved boundaries.

It suffices to scale to $T = 1$. If the given interval of t is, instead, (α, β), then the scaling $t = \alpha + (\beta - \alpha)t'$, where t' is the new time variable, achieves this. Some control problems over an infinite time domain, $[0, \infty)$, can also be transformed to the $[0, 1]$ time domain. Consider the transformation of t to a new variable $t' := \tanh t$. Then $t' \in [0, 1)$. In this case, some assumptions are involved, namely that the resulting integrands, as functions of t', have finite integrals, and that $x(t)$ and $u(t)$ tend to finite limits as $t \to \infty$.

For a time-optimal control problem, where the optimization is also over a variable T, the following *Goh–Teo transformation* converts the problem into a fixed-time problem. Let $t := \tau T$, where τ is a scaled time variable and T is taken as the parameter p, or as a component of p if there are already other parameters p_0. Let $p := (T, p_0)$ and

$$\tilde{x}(\tau,p) := x(\tau T,p) = x(t,p_0) \quad \text{and} \quad \tilde{u}(t,p) := u(\tau T,p_0) = u(t,p_0)$$

Then the control problem becomes:

$$\text{Minimize } \tilde{J}(\tilde{u}(.,p),p) := F(\tilde{x}(.,p),\tilde{u}(.,p),p)$$
$$:= \int_0^1 f(\tilde{x}(\tau,p),\tilde{u}(\tau,p),\tau,p)T\,\mathrm{d}\tau + \Phi(\tilde{x}(1,p),p) \quad (6.34)$$

subject to:

$$(\forall \tau \in [0,1]) \quad u(\tau) \in \tilde{\Gamma}(\tau) := \Gamma(T\tau)$$
$$(\forall \tau \in [0,1]) \quad \tilde{x}(0,p) = x_0(p_0), \quad (\partial/\partial\tau)\tilde{x}(\tau,p) = Tm(\tilde{x}(\tau,p),u(\tau,p),\tau,p)$$
$$(\forall \tau \in [0,1]) \quad \tilde{\theta}^i(\tilde{u}(.,p),p) = 0 \quad (i = 1,2,\ldots,r_1)$$
$$(\forall \tau \in [0,1]) \quad \tilde{\theta}^i(\tilde{u}(.,p),p) \geq 0 \quad (i = r_1+1, r_1+2,\ldots,r)$$

where

$$\theta^i(\tilde{u}(.,p),p) := \Phi^i(\tilde{x}(1,p),p) + \int_0^1 f^i(\tilde{x}(\tau,p),\tilde{u}(t\tau,p),\tau,p)T\,\mathrm{d}\tau$$

This is of the fixed-time form (6.28) previously given. (It will appear later that τ will be discretized, but T will not, so that accuracy in T is not lost.)

The many particular cases of this (fixed-time) formulation include the following (the type of inequality, $= 0$ or ≥ 0, is shown at the right):

$$(\forall t) \quad u(t) \leq b \Leftrightarrow \Phi^i(.) \equiv 0, f^i(x(t,p),u(t,p),t,p) = [-b + u(t)]_+^2 := 0$$

$$(\forall t \in E \subset [0,1]) \quad \rho(x(t,p)) \leq 0 \Leftrightarrow \Phi^i(.) \equiv 0, f^i(x(t,p),u(t,p),t,p) = \chi_E(t)\,[\rho(x(t),p)]_+^2 := 0$$

where $\chi_E(t) := 1$ for $t \in E$, 0 for $t \notin E$; or instead (section 6.4.6)

$$f^i(x(t,p),u(t,p),t,p) = \chi_E(t)\,[\rho(x(t),p)]_+ := 0 \tag{6.35}$$

$$\rho(x(1,p),p) \leq 0 \Leftrightarrow \Phi^i(x(1,p),p) = -\rho(x(1,p),p);\ f^i \equiv 0; \geq 0$$

For a time-optimal problem with a terminal constraint $q(x(T)) = 0$, there are two approaches. Either, as in section 4.5, the constraint may be implemented as a boundary condition on the adjoint differential equation, or, following the present approach, this constraint may be implemented as

$$\theta^1(\tilde{u}(.,p),p) := q(\tilde{x}(1,p),p) + \int_0^1 0.T\,\mathrm{d}\tau = 0$$

6.4.4 Discretization and gradients

Computation of an optimal control generally requires discretization of the time variable. A crude approach considers the values of $x(t)$ and $u(t)$ at gridpoints, say $t = j/N$, $j = 0, 1, 2, \ldots, N$, and approximates an integral $\int_0^1 f(.)\,\mathrm{d}t$ by $N^{-1}\sum f_j$, where f_j here denotes the value of the integrand $f(.)$ at $t = j/N$. Then optimization proceeds over a finite number of variables.

A more precise approach approximates the control function $u(.)$ by a *step function*. Figure 6.5 shows, as an example, the interval [0, 1] divided into ten equal subintervals, and a step function constant on each subinterval. Thus $u(t)$ is specified by its values, u_j say, on subintervals $[j/N, (j+1)/N]$ $(j = 0, 1, \ldots, N-1)$. Since $x(t)$ is then obtained by integration, $x(t)$ may be specified to similar accuracy by its values at the gridpoints j/N $(j = 0, 1, 2, \ldots, N)$, and similarly for the costate $\lambda(t)$, specified by values λ_j.

From section 4.7.6, the gradient of $j(.,p)$, subject only to the differential equation for $x(t)$, is given by

$$(\forall w(.)) \quad J'(u(.,p))w = \int_0^1 h_u^0(x(t,p),u(t,p),t,\mu(t,p))w(t)\,\mathrm{d}t \tag{6.36}$$

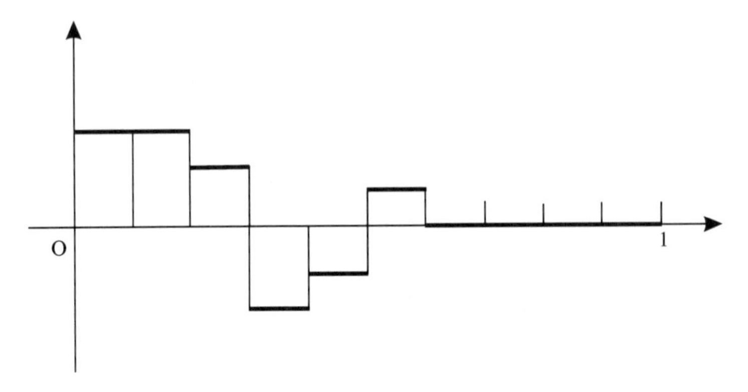

Figure 6.5 Approximating the control function with a step function.

where $h^0(.)$ is the *restricted Hamiltonian*

$$h^0(x(t,p),u(t,p),t,\mu(t,p)) = f(x(t,p),u(t,p),t) + \mu(t,p)m(x(t,p),u(t,p),t) \tag{6.37}$$

and the corresponding costate $\mu(t,p)$ satisfies the corresponding adjoint differential equation:

$$-\dot{\mu}(t,p) = h_x^0(x(t),u(t),t,\mu(t)); \quad \mu(1) = \Phi_x(x(1,p),p) \tag{6.38}$$

The boundary condition on $\mu(.)$ is taken from section 4.5, and $x(.)$ is related to $u(.)$ by the differential equation for $x(t)$ and its initial condition.

Since each $\theta_i(u(t,p),p)$ has the form of another objective function, the gradient of $\theta_i(.,p)$, subject only to the differential equation for $x(.)$, is

$$(\forall w(.)) \quad \theta^{i\prime}(u(.,p),p)w = \int_0^1 h_u^i(x(t),u(t),t,\mu^i(t))w(t)\mathrm{d}t \tag{6.39}$$

where

$$h^i(x(t),u(t),t,\mu^i(t,p)) := \theta^i(t,p) + \mu^i(t)m(x(t,p),u(t,p),t) \tag{6.40}$$

is a new restricted Hamiltonian, and its corresponding costate $\mu^i(.)$ satisfies an adjoint differential equation

$$-\dot{\mu}^i(t,p) = h_x^i(x(t,p),u(t,p),t,\mu^i(t,p)); \mu^i(1) = \Phi_x^i(x(1,p),p) \tag{6.41}$$

If $u(.)$ is a step function with ordinates $u_0, u_1, \ldots$, then (for given p) $J(u(t,p), p)$ becomes a function $\hat{J}(u_0, u_1, \ldots, u_{N-1})$. From section 4.7.7, the gradient of $\hat{J}(.)$, subject only to the given differential equation (6.30) for $x(.)$, is given by

$$\partial \hat{J}/\partial u_j = \int_{I(j)} h_u^0(x(t,p),u(t,p),t,\mu(t,p))\,\mathrm{d}t \tag{6.42}$$

where $I(j)$ denotes the subinterval $(j/N, (j+1)q/N)$. Similarly, $\theta_i(u(t,p),p)$ becomes a function $\hat{\theta}(u_0, u_1, \ldots, u_{N-1})$, and its gradient, subject only to the given differential equation (6.30) for $x(.)$, is given by

$$\partial\hat{\theta}/\partial u_j = \int_{I(j)} h_u^i(x(t,p), u(t,p), t, \mu^i t, p) \tag{6.43}$$

6.4.5 Computing schemes

A local minimum for the optimal control problem (6.28) can therefore be computed approximately – namely, subject to the discretization – by applying some algorithm for constrained minimization to the discretized problem with a finite set of variables, namely $\hat{u}_j$ ($j = 0, 1, \ldots, N$), together with a finite number of parameters p_i. The algorithm could be augmented Lagrangian or sequential quadratic. These or other algorithms require values of the objective and gradient functions as well as their gradients.

Initialization requires the following:

0. Transform the control problem to the standard form, described in section 6.4.3. Choose a number N of subintervals (e.g. $N = 10$ or 20). Choose an initial (step function) control $\hat{u}^0(.)$ (e.g. $\hat{u}^0(.) := 0$). Let $\hat{u}(.) := \hat{u}^0(.)$.

An outline scheme for computing these function and gradient values is as follows.

1. Given $\hat{u}$, compute $\hat{x}$ from the given differential equation (6.30), with initial condition.
2. From $\hat{u}$ and $\hat{x}$, compute $\hat{\mu}$ from the adjoint differential equation (6.38), with boundary condition.
3. Compute the objective value $J(.)$ by integrating (6.28), and the constraint values $\hat{\theta}^i(.)$ by integrating (6.32).
4. For each constraint function $\hat{\theta}^i$, compute the corresponding costate $\hat{\mu}^i$ from the adjoint differential equation (6.41), with boundary condition.
5. Compute the gradient of the objective $\hat{J}(.)$, subject to (6.30), from the restricted Hamiltonian (6.38) and the gradient formula (6.42).
6. Compute the gradient of each constraint function $\hat{\theta}^i$, subject to (6.30), from its Hamiltonian (6.40) and the gradient formula (6.43).

An example of using the gradient formulae is given in section 6.6.1. Note that if the nonsmooth integrand (6.35) in section 6.4.3 is required, then some further approximation, by smoothing, is needed.

6.4.6 Concerning nonsmoothness

The several integrations of differential equations required by this algorithm are usually done by a Runge–Kutta method. A slight modification is required to the usual Runge–Kutta computer subroutines, since they assume a differential equation of the form

$$\dot{\boldsymbol{\varphi}}(t) = \mathbf{g}(\boldsymbol{\varphi}(t), t), \quad \boldsymbol{\varphi}(0) \text{ given}$$

for a vector function $\boldsymbol{\varphi}(.)$, but with a continuous function $\mathbf{g}(\boldsymbol{\varphi}(.).)$ on the right-hand side. But in the present context, since $u(.)$ is a step function, the right-hand side has discontinuities. Consequently, when the Runge–Kutta subroutine calls a function value, it must be a function not only of t, but also of the label j of the current subinterval.

In the integration of (6.30), for example, namely

$$\dot{x}(t, p) = m(x(t, p), u(t, p), t, p)$$

$u(t, p)$ has the constant value $\hat{u}_j$ on the subinterval $I(j)$, but values of $x(., p)$ are known only at the endpoints of $I(j)$. Consequently, some interpolation is required to estimate $x(t, p)$ at those intermediate points t that Runge–Kutta requires.

A further source of nonsmoothness arises when expressions such as

$$\chi_E(t)[\rho(x(t), p)]_+^2 \quad \text{or} \quad \chi_E(t)[\rho(x(t), p)]_+$$

arise as integrands in the formulation of the problem, as in section 6.4.3. It suffices to consider E as consisting of one or more intervals; so $\chi_E(.)$ involves a step function. The function $0 \mapsto [z]_+$ $(z \in \mathbf{R})$ has a jump in gradient, and must be approximated before most minimization algorithms will work. The function $0 \mapsto [z]_+^2$ $(z \in \mathbf{R})$ is twice differentiable, but its second derivative has a jump at the origin, and this may upset a minimization algorithm that assumes a C^2-function.

A nonsmooth function $\sigma : \mathbf{R} \to \mathbf{R}$ may be approximated by a smooth function, $\tilde{\sigma}(.)$, which is uniformly close to $\sigma(.)$. Consider a smooth function

$$\tilde{\sigma}(z) := \int_{\mathbf{R}} \sigma(z - s)\varphi(\varepsilon^{-1}s)\varepsilon^{-1}\, ds$$

where $\varepsilon > 0$ is a small parameter and $\varphi(.)$ is a suitable nonnegative *mollifier function*, satisfying $\int_{\mathbf{R}}\varphi(t)\, dt = 1$ and $\varphi(t) = 0$ when $|t| > 1$. In particular, the *ramp function* $\sigma_R(z) := z$ when $z \geq 0$, 0 when $z < 0$ (thus, $\sigma_R(z) = [z]_+$), can be thus approximated so as to smooth the corner at 0, where $\sigma_R(.)$ is not differentiable. If $\varphi(z) := \frac{1}{2}\chi_{[-1,1]}(z) \equiv 1$ when $0 \leq z \leq 1$, 0 elsewhere, then

$$\tilde{\sigma}_R(z) = \int_{-\infty}^{z} \varphi(\varepsilon^{-1}s)\varepsilon^{-1}\, ds$$

$$= \begin{cases} 0 & z < -\varepsilon \\ \frac{1}{4}((z + \varepsilon)/\varepsilon)^2 & -\varepsilon \leq z \leq \varepsilon \\ z & \varepsilon < z \end{cases}$$

as shown in Figure 6.6. Note that $\sigma(.)$ is differentiable, and $\tilde{\sigma}(z) = \sigma(z)$, $\tilde{\sigma}'(z) = \sigma'(z)$ for $z = \pm\varepsilon$.

If a constraint $h(x) \leq 0$ involving step functions is thus approximated, replacing h by a smoothed $\tilde{h}$, then the constraint may be approximated by $\tilde{h}(x) \leq \varepsilon$ for small positive ε to ensure that a constraint qualification (such as local solvability) holds (Martin and Teo, 1994).

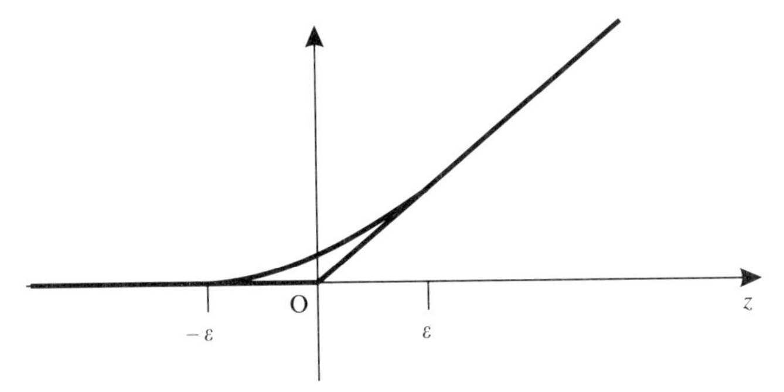

Figure 6.6 Approximating a ramp function.

A similar approximation may be obtained for the function $\sigma_S(z) := [z]_+^2$, namely

$$\tilde{\sigma}_S(z) = \begin{cases} 0 & z < -\varepsilon \\ \frac{1}{6}((z+\varepsilon)/\varepsilon)^3 & -\varepsilon < z < \varepsilon \\ ((z+\varepsilon)/\varepsilon)^2 + ((z+\varepsilon)/\varepsilon) & \varepsilon < z \end{cases}$$

When $\sigma_R(.)$ or $\sigma_S(.)$ arise in representing certain integral constraints, they may appropriately be approximated for computation by the smoother functions $\tilde{\sigma}_R(.)$ or $\tilde{\sigma}_S(.)$, with a smoothing parameter ε which is made to tend to zero as the optimum is approached. The choice $\sigma_R(.)$ is probably better, although less smoothed, because KKT conditions will not generally hold when $\sigma_S(.)$ is involved.

The term $\Phi(x(T, p), p)$ in the objective function (6.28) (section 6.4.3) can be put under the integral sign as a term

$$\Phi(x(t, p), p)\delta(t - T)$$

where $\delta(.)$ is Dirac's delta function. While this formulation is not directly computable, it suggests consideration of a modified control model where the delta function is replaced by a step function with value γ^{-1} on the interval $(T - \gamma, T)$, and zero elsewhere. This replaces the term $\Phi(x(T, p), p)$ by some average involving values of $x(., p)$ on the interval $(T - \gamma, T)$ (which could, for some problems, be a better representation of the real-world problem).

The step function may then require smoothing for computation. Consider a mollifier function $\varphi(z) := 0\ (z < -1),\ 1 + t\ (-1 < z < 0),\ 1 - t\ (0 < z < 1),\ 0\ (z > 1)$ (Figure 6.7). First consider the step function $\sigma(z) := 0\ (z < 0),\ 1\ (z > 0)$. The triangular mollifier is required in order to get sufficient smoothness in the resulting $\tilde{\sigma}(.)$. Since

$$\tilde{\sigma}(z) = \int_{-\infty}^{t} \varphi(s)\, ds$$

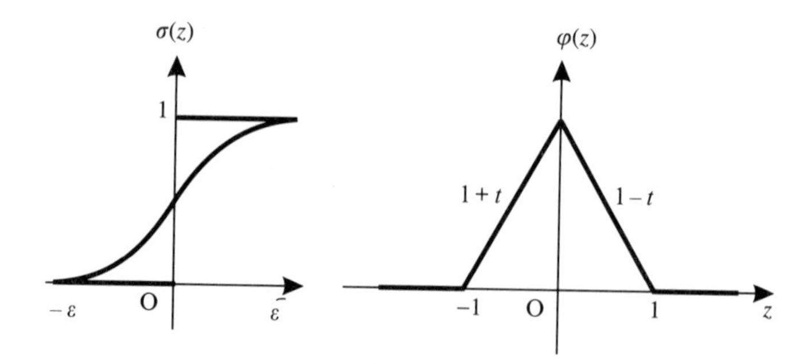

Figure 6.7 Approximating a step function.

there results (in case $\varepsilon = 1$)

$$\tilde{\sigma}(z) = \begin{cases} \tfrac{1}{2} + z + \tfrac{1}{2}z^2 & (-1 \leq z < 0) \\ \tfrac{1}{2} + z - \tfrac{1}{2}z^2 & (0 \leq z \leq 1) \end{cases}$$

and zero elsewhere. Denote this particular $\tilde{\sigma}(.)$ function by $r(.)$. Note that $r(.)$ and $\sigma(.)$ agree, in value and slope, at the joining points -1 and 1, and also that $\int_{-1}^{1}\sigma(z)\,dz = \int_{-1}^{1} r(z)\,dz = 1$.

Now consider a step function $\sigma_B(.)$ with value γ^{-1} on $(T-\gamma, T)$, and zero elsewhere, with a smoothing parameter $\varepsilon < \tfrac{1}{2}\gamma$, a smoothed function could be

$$\tilde{\sigma}_B(z) := r((z - T + \gamma)/\varepsilon) - r((z - T)/\varepsilon)$$

By construction, this function has the same value for its integral over $(T-\gamma-\varepsilon, T+\varepsilon)$ as does the given step function, so that it will give a good representation to the integral when ε is small.

6.5 SENSITIVITY TO PARAMETERS

The approximation of the control $u(.)$ by a step function leads, in many computing instances, to accurate estimates of the optimal control. One may ask why this happens, and under what circumstances?

6.5.1 Sensitivity to the step function approximation

Consider a constrained minimization problem, containing a parameter $\mathbf{q}$. Assume that the problem reaches a *strict* local minimum at a point $\bar{\mathbf{x}}$ when the parameter $\mathbf{q} = \mathbf{0}$, that the objective and constraint functions are uniformly continuous on bounded sets, and that the objective reaches a minimum on each closed bounded set when $\mathbf{q}$ is nonzero and sufficiently small. Then, from sections 4.6.1 and 7.4, the problem reaches a local minimum when $\mathbf{q}$ is nonzero and sufficiently small at a point $\bar{\mathbf{x}}(\mathbf{q})$, where $\bar{\mathbf{x}}(\mathbf{q}) \to \bar{\mathbf{x}}$ as $\mathbf{q} \to \mathbf{0}$. If, in addition, the objective and constraint functions satisfy Lipschitz conditions and two other stability conditions hold – the Robinson stability condition on the constraints and a linear growth condition on the objective – then the function $\bar{\mathbf{x}}(.)$ also satisfies a Lipschitz condition.

The approximation of an optimal control problem by replacing the control function by a step function approximation can be regarded as a Lipschitz perturbation of the problem with Lipschitz constant proportional to $1/N$, where N is the number of equal subintervals for the step function. The assumption that the perturbed objective reaches a minimum on closed bounded sets is fulfilled, since each perturbed (by step function approximation) problem is finite-dimensional. However, the assumption of uniform continuity has to be verified. This is done in section 7.6 by renorming the space of control functions, thus restricting the control function to a suitable subspace in which the optimal control is assumed to lie. This subspace must be chosen so that high-frequency oscillations of control functions become small, at a sufficiently fast rate, as the frequency becomes large. This is a reasonable assumption in practice, since any optimization problem, in order to be computable, entails the assumption that nothing interesting happens below a certain small scale of distance.

6.5.2 Sensitivity to change of parameter

The parametrized fixed-time control problem (6.34) may be expressed as follows, using delta functions to put the additional terms under the integral signs (and considering vector-valued states and controls):

$$\begin{aligned}\text{Minimize } \tilde{J}(\tilde{\mathbf{u}}(.,p),p) &:= F(\tilde{\mathbf{x}}(\tau,p),\tilde{\mathbf{u}}(\tau,p),p)\\ &:= \int_0^1 [f(\tilde{\mathbf{x}}(\tau,\mathbf{p}),\tilde{\mathbf{u}}(\tau,\mathbf{p}),\tau,p)T\\ &\quad + \Phi(\tilde{\mathbf{x}}(t,p),p)\delta(t-1)]\,\mathrm{d}\tau \qquad (6.44)\end{aligned}$$

subject to:

$$\begin{aligned}&(\forall\tau\in[0,1]) \quad \tilde{\mathbf{u}}(\tau)\in\tilde{\Gamma}(\tau)\\ &(\forall\tau\in[0,1]) \quad \tilde{\mathbf{x}}(0,p)=\mathbf{x}_0(\mathbf{p}_0),\ (\partial/\partial\tau)\tilde{\mathbf{x}}(\tau,p)=Tm(\tilde{\mathbf{x}}(\tau,p),\tilde{\mathbf{u}}(\tau,p),\tau,p)\\ &(\forall\tau\in[0,1]) \quad \tilde{\theta}^i(\tilde{\mathbf{u}}(.,p),p)=0 \quad (i=1,2,\ldots,r_1)\\ &(\forall\tau\in[0,1]) \quad \tilde{\theta}^i(\tilde{\mathbf{u}}(.,p),p)\ge 0 \quad (i=r_1+1,r_1+2,\ldots,r)\end{aligned}$$

where

$$\tilde{\theta}_i(\tilde{\mathbf{u}}(.,p),p) := \int_0^1 [f^i(\tilde{\mathbf{x}}(\tau,\mathbf{p}),\tilde{\mathbf{u}}(\mathbf{t}\tau,\mathbf{p}),\tau,p)T + \Phi^i(\tilde{\mathbf{x}}(t,p),p)\delta(t-1)]\,\mathrm{d}\tau$$

Denote by $\tilde{\boldsymbol{\theta}}$ the vector of the $\tilde{\theta}^i$, by $\boldsymbol{\psi}$ the vector of the f^i, and by $\boldsymbol{\Xi}$ the vector of the Φ^i, for $i=1,2,\ldots,r$; denote by S the cone $\{0\}^{r_1}\times\mathbf{R}_+^{(r-r_1)}$. Now consider $\tilde{\boldsymbol{\theta}}$ adjoined to $\tilde{\mathbf{x}}$ as an additional state variable, subject to the constraint $\tilde{\boldsymbol{\theta}}\in S$. In abstract form (as in section 4.3), this problem may be summarized in the form:

$$\text{Minimize}_{\tilde{\mathbf{x}},\tilde{\mathbf{u}},\mathbf{p}} F(\tilde{\mathbf{x}},\tilde{\mathbf{u}},\mathbf{p}) \quad \text{subject to} \quad D\tilde{\mathbf{x}}=M(\tilde{\mathbf{x}},\tilde{\mathbf{u}}),\ \tilde{\mathbf{u}}\in\boldsymbol{\Gamma},\ \boldsymbol{\theta}(\tilde{\mathbf{x}},\tilde{\mathbf{u}},\mathbf{p})\in S$$

This formulation is more general than that of section 4.3, since here $\mathbf{G}$ may depend on $\mathbf{u}$ as well as on $\mathbf{x}$. The integral of the Hamiltonian is then:

$$H(\mathbf{x}, \mathbf{u}, \boldsymbol{\lambda}, \boldsymbol{\omega}, \mathbf{p}) := F + \bar{\boldsymbol{\lambda}} M - \boldsymbol{\omega}\bar{\boldsymbol{\theta}}$$

with Lagrange multipliers $\bar{\boldsymbol{\lambda}}$ and $\bar{\boldsymbol{\omega}}$. Representing these by functions $\boldsymbol{\lambda}(.)$ and $\boldsymbol{\omega}(.)$, with $\boldsymbol{\omega}(t)$ matrix-valued if $\boldsymbol{\psi}(.)$ is vector-valued, the Hamiltonian is

$$\begin{aligned} h(\tilde{\mathbf{x}}(\tau, p), \tilde{\mathbf{u}}(\tau, p), t, \boldsymbol{\lambda}(t), \boldsymbol{\omega}(t)) := & f(\tilde{\mathbf{x}}(\tau, p), \tilde{\mathbf{u}}(\tau, p), \tau, p)T \\ & + \Phi(\tilde{\mathbf{x}}(t, p), p)\delta(t - T) \\ & + \boldsymbol{\lambda}(t)Tm(\tilde{\mathbf{x}}(\tau, p), \tilde{\mathbf{u}}(\tau, p), \tau, p) \\ & - \boldsymbol{\omega}(t)\, [\boldsymbol{\psi}(\tilde{\mathbf{x}}(\tau, p), \tilde{\mathbf{u}}(t\tau, p, \tau, p)T \\ & + \Xi\,(\tilde{\mathbf{x}}(t, p), p)] \end{aligned}$$

The adjoint differential equation is then

$$-\dot{\boldsymbol{\lambda}}(t) = h_{\tilde{\mathbf{x}}}(\tilde{\mathbf{x}}(\tau, p), \tilde{\mathbf{u}}(\tau, p), \tau, p); \quad \boldsymbol{\lambda}(1) = 0; \quad \boldsymbol{\omega}(t) \in S^*$$

(Note that this is different from the adjoint differential equations used in section 6.4.4 to obtain gradients.)

Suppose now that the various functions in the problem also depend on an additional parameter $\mathbf{q}$, with given value $\mathbf{0}$. Unlike $\mathbf{p}$, the parameter $\mathbf{q}$ is not optimized over. From section 4.7.7, the gradient of the optimal objective function with respect to the parameter $\mathbf{q}$ equals

$$H_{\mathbf{q}}(\tilde{\mathbf{x}}, \tilde{\mathbf{u}}, \lambda, \omega, \mathbf{p}, \mathbf{0}) = \int_0^1 h_{\mathbf{q}}(\tilde{\mathbf{x}}(t), \tilde{\mathbf{u}}(t), t, \boldsymbol{\lambda}(t), \boldsymbol{\omega}(t))$$

inserting optimal values for the functions in the integrand.

If the optimum is computed using the augmented Lagrangian algorithm, or using the sequential quadratic algorithm, then discretized versions of the optimal costate functions $\boldsymbol{\lambda}(t)$ and $\boldsymbol{\omega}(t)$ are necessarily computed, as well as the optimal control $\mathbf{u}(t)$ and state $\mathbf{x}(t)$. So estimates of sensitivity to small changes of parameters, analogous to shadow costs, are then calculable.

6.6 EXAMPLES

6.6.1 Gradient calculation for the 'dodgem car' problem

Consider the 'dodgem car' problem of section 5.5. The state functions $x(.)$, $y(.)$, $\theta(.)$ will now be denoted as components $x_2(.)$, $x_3(.)$, $x_1(.)$ of a vector state function $x(.)$. The problem is to minimize the time T to a stated terminal point, subject to various constraints. The derived fixed-time control problem takes the form:

$$\min_{T,\tilde{x}_1(.),\tilde{x}_2(.),\tilde{x}_3(.),\tilde{u}(.)} T \quad \text{subject to} \quad (\forall \tau \in (0,1)) \quad |\tilde{u}(\tau)| \leq 1,$$

$$\dot{\tilde{x}}_1(\tau) = \tilde{u}(\tau), \quad \dot{\tilde{x}}_2(\tau) = \cos \tilde{x}_1(\tau)$$

$$\dot{\tilde{x}}_3(\tau) = \sin \tilde{x}_1(\tau)$$

$$\tilde{x}_1(0) = \tfrac{1}{2}\pi, \quad \tilde{x}_2(0) = 4, \quad \tilde{x}_3(0) = 0$$

$$\theta^1(u) := \tilde{x}_2(1)^2 + \tilde{x}_3(1)^2 = 0$$

An alternative terminal condition would be

$$\theta^1(u) := [\tilde{x}_2(1)^2 + \tilde{x}_3(1)^2]^{1/2} = 0$$

The differential equations required to calculate the gradients of the objective and of θ^1 are:

$$-\dot{\mu}_1 = -\mu_2 T \sin \tilde{x}_1 + \mu_3 T \cos \tilde{x}_1, \quad -\dot{\mu}_2 = 0, -\tilde{\mu}_3 = 0, \quad \mu(1) = (0, 0, 0)$$

$$-\dot{\mu}_1^1 = -\mu_2^1 T \sin \tilde{x}_1 + \mu_3^1 T \cos \tilde{x}_1, \quad -\dot{\mu}_2^1 = 0, -\tilde{\mu}_3^1 = 0$$

$$\mu^1(1) = (0, 2\tilde{x}_2(1), 2\tilde{x}_3(1))$$

(or $\mu^1(1) = [x_2(1)^2 + \tilde{x}_3(1)^2]^{-1/2}(0, \tilde{x}_2(1), \tilde{x}_3(1))$ for the second version of the terminal conditions.)

6.6.2 Computed results for the 'dodgem car' problem

Using the discretization method of section 6.4.4, this control problem has been computed using the MISER3 package (Jennings *et al.*, 1990). Following section 6.6.1, the time to reach the target appears as a parameter T, and the (scaled) time interval [0, 1] is divided into either 10 or 20 subintervals. Figure 6.8(a) plots the trajectory $y(t)$ versus $x(t)$; the theoretical curve is shown, with computed points plotted. Figure 6.8(b) plots the computed optimal control function against actual time. The optimal value of T is 4.73..., in agreement with theory. The theoretical time of switching, from bang-bang to singular arc, is 2.89... . By accident of discretization, the computed switching time agrees closely with this figure; in general, some interpolation between subintervals may be necessary. The theoretical $u(t)$ jumps from 1 to 0; the computed optimal $u(t)$ only approximates the zero portion; but because the state is obtained from $u(t)$ by integration, the computed trajectory is accurate.

6.6.3 A control problem with a time lag

Teo, Wong and Clements (1984) compute the following control problem:

$$\min \tfrac{1}{2}[x_1(2)^2 + x_2(2^2) + \int_0^2 u(t)^2 \, dt$$

subject to $u(t) \geq 0$, $\dot{x}_1(t) = x_2(t)$, $\dot{x}_2(t) = -x_2(t) - x_1(t-1) + u(t)$ $(0 \leq t \leq 2)$, with initial conditions $x_1(t) = 10$, $x_2(t) = 0$ $(-1 \leq t \leq 0)$. (Because of the time lag term, the initial conditions must be given over a time interval.) The computed

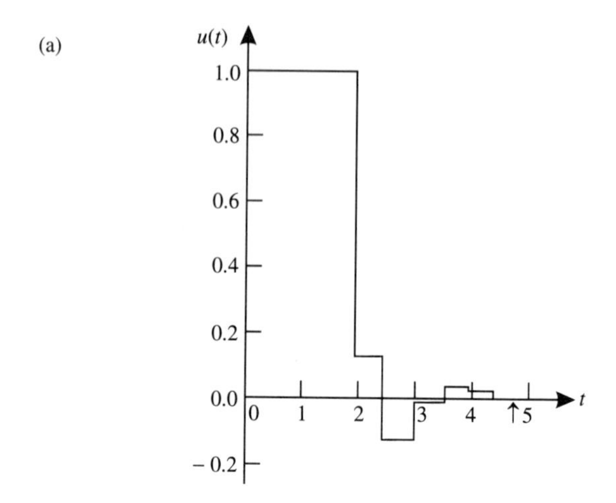

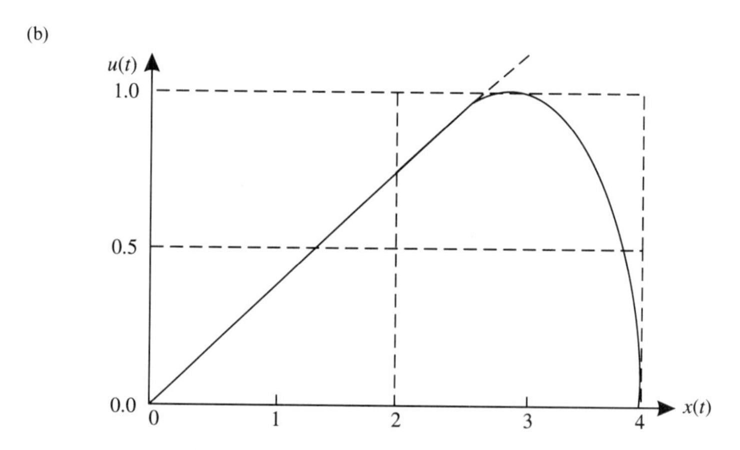

Figure 6.8 The 'dodgem car' problem: (a) computed optimal control; (b) computed optimal path.

optimum control, even for only 10 subintervals, uniformly approximates the exact optimum control – a part is shown in Figure 6.9.

6.6.4 An economic model

Consider an economic model with state functions $x_0(t)$ = production, $x_1(t)$ = level of technology, $x_2(t)$ = capital stock and dynamic equations:

$$x_1(t+1) = \beta_0 x_0(t) - \beta_1 x_1(t), \quad x_2(t+1) = -\beta_2 x_2(t) + [\beta_3 - u(t)]x_0(t),$$
$$x_0(t) = \beta_4 x_1(t) x_2(t)^{\beta_6}$$

where the β_i are given constants, and the control $u(t)$ = consumption/$x_0(t)$. A discounted sum of terms $\beta_7^{-1}[u(t)x_0(t)]^{\beta_7}/[x_0(t)^{\beta_8}]$ is to be maximized. This model

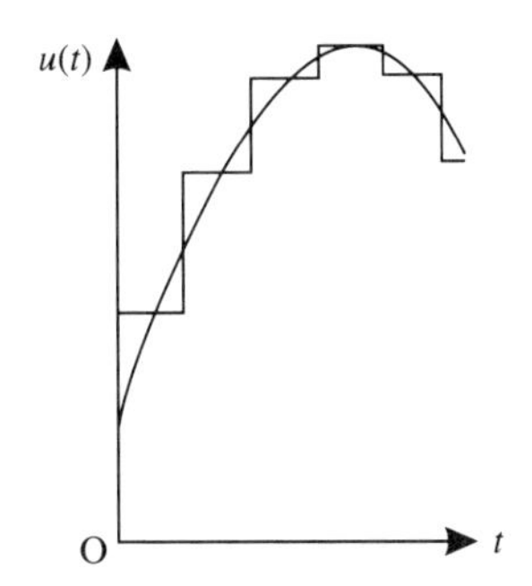

Figure 6.9 Time-lag problem.

(compare section 2.4) is part of Islam's (1994) model of climate and the economy, but here omitting the climate part.

If the difference equations are approximated by differential equations (assuming a small time interval), the following optimal control model is obtained:

$$\max \beta_7^{-1} \int_0^T e^{-\rho t} u(t)^{\beta_7} x_0(t)^{\beta_7 - \beta_8}\, dt \quad \text{subject to} \quad \beta_{10} \le u(t) \le \beta_3,$$

$$\dot{x}_1(t) = \beta_0 x_0(t) - (1 + \beta_1) x_1(t), \quad \dot{x}_2(t) = -(1 + \beta_2) x_2(t) - [\beta_3 - u(t)] x_0(t),$$

$$x_0(t) = \beta_4 x_1(t) x_2(t)^{\beta_6}$$

with initial conditions for $x_1(0)$, $x_2(0)$, and for $x_0(0)$ calculated from $x_1(0)$ and $x_2(0)$. Note that $0 < \beta_{10} < \beta_3 < 1$; $u(t)$ is bounded away from zero (consumption is positive!). Also, if $\beta_7 < 0$ the problem converts to a minimization problem, and the term $u(t)^{\beta_7}$ requires $\beta_{10} > 0$.

The state variable $x_0(t)$ could be eliminated by substituting for $x_0(t)$ into the other equations, yielding an integrand $e^{-\rho t} u(t)^{\beta_7} x_1(t) x_2(t)^{\beta_9}$ and dynamic equations

$$\dot{x}_1(t) = x_1(t)[\beta_0 \beta_4 x_2(t)^{\beta_6} - (1 + \beta_1)]$$

$$\dot{x}_2(t) = -(1 + \beta_2) x_2(t) + \beta_4[\beta_3 - u(t)] x_1(t) x_2(t)^{\beta_6}$$

but computationally it is probably simpler to retain the equality constraint for $x_0(t)$. This constraint has the form $(\forall t)\ h(x_0(t), x_1(t), x_2(t)) = 0$, so it would be approximated by a smooth constraint

$$\int_0^T \varphi \circ h(x_0(t), x_1(t), x_2(t))\, dt \le \varepsilon$$

where φ is a smooth approximation to the absolute-value function $s \mapsto |s|$, as discussed in section 6.4.5.

If the control problem is approximated by linearizing u and x_2 about chosen points $\tilde{u}$ and $\tilde{x}_2$, thus setting $u(t) = \tilde{u}(t) + \eta(t)$ and $x_2(t) = \tilde{x}_2(t) + \xi(t)$ then, neglecting powers of $\eta(t)$ and $\xi(t)$ above the first, an approximate problem is obtained which is linear in the new control function $\eta(t)$. This would give bang-bang control, with a possible singular arc. The nonlinearity of the given problem in $u(t)$ would smooth out the jumps in the bang-bang control.

6.7 EXERCISES

See also the linesearch exercise in section 6.2.9.

6.1 Consider minimization of the function

$$f(x_1, x_2) := (1 - x_1^2)^2 + 4(x_2 - x_1^2)^2$$

and the initial point (0.00, 1.00). Work out (by hand) three iterations of the Fletcher–Reeves algorithm (section 6.2.4). For this exercise, approximate the function near the current iterate by a quadratic; then a linesearch involves minimizing a quadratic in the step length α. (This function is a simplification of the *banana valley function* – which has coefficient 100 in place of 4. This function has its minimum at (1, 1), at the end of a narrow curving valley, which an algorithm must try to follow. The initial steepest descent step from (0.00, 1.00) comes to approximately (0.24, 0.00); this may be assumed for the following hand calculations.)

6.2 Consider the following simple problem, with $\mathbf{x} \in \mathbf{R}^n$, to which an augmented Lagrangian algorithm might be applied:

$$\text{Minimize } \tfrac{1}{2}\mathbf{x}^T\mathbf{x} + \mathbf{b}^T\mathbf{x} \quad \text{subject to} \quad \mathbf{p}^T\mathbf{x} \le \mathbf{q}$$

where $\mathbf{p}^T\mathbf{p} = 1$ and $\mathbf{p}^T\mathbf{b} + q < 0$.

(i) Using the Karush–Kuhn–Tucker theorem, find the optimum point $\bar{\mathbf{x}}$ and the optimum Lagrange multiplier $\bar{\lambda}$ for this problem.

(ii) Given λ, obtain an expression for the vector$\hat{\mathbf{x}}$ that minimizes the augmented Lagrangian

$$\tfrac{1}{2}\mathbf{x}^T\mathbf{x} = \mathbf{b}^T\mathbf{x} + \tfrac{1}{2}\mu\|[\mathbf{p}^T\mathbf{x} - q + \mu^{-1}\lambda]_+\|^2$$

with respect to $\mathbf{x}$. (*Hint*: The inverse of the matrix $I + \varepsilon\mathbf{p}\mathbf{p}^T$ will be required; this is $I + \delta\mathbf{p}\mathbf{p}^T$, where $\varepsilon + \delta + \varepsilon\delta = 0$; multiply the matrices to verify. Note that if the expression $\mathbf{p}^T\hat{\mathbf{x}} - q + \mu^{-1}\lambda$ happens to be positive, then $[.]_+ = [.]$ here, so it suffices to minimize a quadratic form. This supposition must be verified later, using the hypothesis that $\mathbf{p}^T\mathbf{b} + q < 0$.)

(iii) The second part of an augmented Lagrangian iteration updates λ to $\lambda^{\text{new}} := [\lambda + \mu(\mathbf{p}^T\hat{\mathbf{x}} - q)]_+$. Obtain a formula for λ^{new}; then compare $\lambda^{\text{new}} - \bar{\lambda}$ with $\lambda - \bar{\lambda}$. What can be deduced about the convergence rate for this algorithm?

(iv) If λ is set to zero, then these formulae give results on the penalty method (section 6.3.1). For the penalty method, consider$\hat{\mathbf{x}}$ as a function of the penalty parameter μ, and discuss the rate of convergence of $\hat{\mathbf{x}}(\mu)$ to $\bar{\mathbf{x}}$ as $\mu \to \infty$.

6.3 For the control problem

$$\text{Minimize } J(u) := \int_0^1 [-x(t)]\,\mathrm{d}t$$

subject to

$$x(0) = 0 \quad \text{and} \quad (\forall t \in [0, 1]) \quad \dot{x}(t) = u(t), \quad 0 \leq u(t) \leq 1, \quad x(t) \leq \tfrac{3}{4}$$

Starting with an initial control function

$$u(t) = 0\ (0 \leq t \leq \sqrt{8}), \quad 1\ (\tfrac{1}{8} < t \leq \tfrac{3}{4}), \quad 0\ (\tfrac{3}{4} < t \leq 1)$$

examine a steepest descent step. This requires the formula

$$J'(u)z = \int_0^1 (\partial h/\partial u)\, z(t)\,\mathrm{d}t$$

where $h(.) = -x(t) + \lambda(t)u(t)$ is the Hamiltonian, and $\lambda(t)$ must satisfy the adjoint differential equation and boundary conditions.

It is instructive to compare with a problem in, say, three dimensions, where a direction **d** must be chosen to find a suitably negative value of a directional derivative $\mathbf{q}^{\mathrm{T}}\mathbf{d}$. Here a suitable **d** is $-\mathbf{q}$.

This suggests choosing $z(.) = -(\partial h/\partial u)$ for the control problem, and considering $J(u + \alpha z)$ for positive step length α. But $u + \alpha z$ is not feasible, since $u(t) + \alpha z(t) > 1$ for an interval I' of t, violating the control constraint. One could consider a *projected* (negative) *gradient*, which remains in the feasible region. Here, each value of t counts as a dimension, so $u(t) + \alpha z(t)$ gets replaced by 1 for $t \in I'$. What would then be the step length? One could take the largest (positive) α, restricted by (i) $J(.)$ starting to increase again, (ii) $u(.)$ becoming infeasible (as already discussed), or (iii) $x(t)$ becoming infeasible (thus, $> \tfrac{3}{4}$) for some t.

(*Note*: This projected gradient approach is not usually used for control problems. However, the calculation of gradient using the Hamiltonian is needed in all current algorithms.)

6.4 Work out the gradient formulae (section 4.7.6) required to compute (as in section 6.4.5) the 'dodgem car' problem of section 5.5. Hence detail the computing schema in section 6.4.5 for this problem. (If the MISER3 computer package for optimal control is available to you, it would be instructive to run this problem on it.)

6.5 Compare computationally the Fletcher–Reeves and Polak–Ribière updates of section 6.2.4, combined with the Powell approximate linesearch given in section 6.2.9, as applied to the test function

$$f(x) := (1 - x_1)^2 + \sum_{i=1}^{n-1} (x_i^2 - x_{i+1})^2$$

initially for the dimension $n = 5$, but also for other values, and including the starting point $x_{\mathrm{init}} = (2, 2, 2, 2, 2)$. Note that *reset* will be required after each n iterations.

6.8 REFERENCES

See also the extensive bibliographies in Fletcher (1987), Gill, Murray and Wright (1981), Martin and Teo (1994), and Teo, Goh and Wong (1991).

Bertsekas, D. P. (1982) *Constrained Optimization and Lagrange Multiplier Methods*, Academic Press, New York.

Cameron, N. (1985) *Introduction to Linear and Convex Programming*, Australian Mathematical Society Lecture Series 1, Cambridge University Press, Cambridge.

Craven, B. D. (1991) An algorithm for minimax, *ZOR – Methods and Models of Operations Research*, **35** 425–34.

Fiacco, A. V. and McCormick, G. P. (1968) *Nonlinear Programming*, Wiley, New York.

Fletcher, R. (ed.) (1969) *Optimization*, Academic Press, London.

Fletcher, R. (1980) *Practical Methods of Optimization*, Vol. 1 Unconstrained. Optimization Wiley, New York.

Fletcher, R. (1987) *Practical Methods of Optimization*, 2nd edn, Wiley, New York.

Gill, P. E., Murray, W. and Wright, M. H. (1981) *Practical Optimization*, Academic Press, London.

Goh, C. J. and Teo, K. L. (1987) *MISER, An Optimal Control Software*, Dept. of Industrial and Systems Engineering, National University of Singapore.

Islam, S. M. N. (1994) *Australian Dynamic Integrated Model of Climate and the Economy (ADICE)*, Centre for Strategic Economic Studies, Victoria University of Technology, Melbourne, Australia.

Jennings, L. S., Fisher, M. E., Teo, K. L. and Goh, C. J. (1990) *MISER3, Optimal Control Software, Theory and User Manual.*

Jennings, L. S., Fisher, M. E., Teo, K. L. and Goh, C. J. (1991) MISER3: solving optimal control problems – an update, *Advances in Engineering Software*, **13** 190–6.

Martin, R. and Teo, K. K. (1994) *Optimal Control of Drug Administration in Cancer Chemotherapy*, World Scientific, Singapore.

Powell, M. J. D. (1978) A fast algorithm for nonlinearly constrained optimization calculations, in *Numerical Analysis*, Lecture Notes in Mathematics 630, Springer, Berlin, pp. 144–57.

Schittkowski, K. (1985) NLPQL: A Fortran subroutine solving constrained nonlinear programming problems, *Operations Research Annals*, **5** 485–500.

Teo, K. L. and Goh, C. J. (1989) A computational method for combined optimal parameter selection and optimal control problems with general constraints, *J. Austral. Math. Soc., Ser. B*, **30** 350–64.

Teo, K. L. and Jennings, L. S. (1991) Optimal controls with a cost of changing control, *J. Optim. Theor. Appl.*, **68** 335–57.

Teo, K. L., Goh, C. J. and Wong, K. H. (1991) *A Unified Computational Approach to Optimal Control Problems*, Longman Scientific & Technical, London.

Teo, K. L. Wong, K. H. and Clements, D. J. (1984) Optimal control computations for linear time-lag system with linear terminal constraints, *J. Optim. Theor. Appl.*, **44**(3) 509–26.

Zhou, J. L. and Tits, A. L. (1991) *User's Guide for FSQP Version 2.4, A Fortran Code for Solving Optimization Problems, Possibly Minimax, with General Inequality Constraints and Linear Equality Constraints, Generating Feasible Iterates*, Electrical Engineering Dept. and Systems Research Center, University of Maryland.

7

Proof of Pontryagin theory and related results

7.1 INTRODUCTION

This chapter presents proofs of the Pontryagin theory for fixed-time optimal control problems (sections 7.1–7.3). Related results for time-optimal problems can be deduced from the fixed-time case, as in sections 4.6, 5.2 and 6.4.3.

The sensitivity of a control problem to small perturbations is analysed in terms of the stability to small perturbations of a *strict* local minimum constrained optimization problem. Since the computational approximation of a control function by a step function can also be considered as a Lipschitz perturbation, the accuracy of such an approximation can also be discussed (sections 7.6 and 7.8).

Of course, the Pontryagin principle has also been proved in several other ways, often commencing with the time-optimal problem. Some other approaches are described in Macki and Strauss (1982) and in Fleming and Rishel (1975).

7.2 FIXED-TIME OPTIMAL CONTROL PROBLEM

Consider the fixed-time optimal has control problem, formulated as follows.

7.2.1 Formulation

$$\text{Minimize}_{\mathbf{x}\in \mathbf{x}_0+X,\, \mathbf{u}\in U}\ F(\mathbf{x},\mathbf{u}) = \int_0^T f(\mathbf{x}(t),\mathbf{u}(t),t)\,\mathrm{d}t \tag{7.1}$$

subject to

$$\mathrm{d}\mathbf{x}(t)/\mathrm{d}t = m(\mathbf{x}(t),\mathbf{u}(t),t) \quad (0 \le t \le T), \quad \mathbf{x}(0) = \mathbf{x}_0$$

$$(\forall t, 0 \le t \le T) \quad u(t) \in \Delta(t), \quad n(x(t),t) \in V(t)$$

Assume, for now, that U consists of piecewise continuous functions with the uniform norm $\|\mathbf{u}\|_\infty$, and that X consists of piecewise smooth functions with the graph norm $\|\mathbf{x}\| = \|\mathbf{x}\|_{\infty}, + \|\mathrm{D}\mathbf{x}\|_{\infty}$ defined in section 4.1.2, with D representing d/dt. (Other norms may be considered; see section 7.2.3.) As shown in section 4.1.2, the constraints (with initial condition) can be expressed as

$$\mathrm{D}\mathbf{x} = M(\mathbf{x}, \mathbf{u}), \quad N(\mathbf{x}) \in K_x, \quad \mathbf{u} \in \Gamma$$

where

$$\Gamma := \{\mathbf{u} \in U: (\forall t)\, \mathbf{u}(t) \in \Delta(t)\}$$
$$K_x = \{\mathbf{p} \in P: (\forall t \in I)\, \mathbf{p}(t) \in V(t)\}$$

Thus, the control constraint has been expressed by $\mathbf{u}(t) \in \Delta$. Note that Δ and V are given functions of t; in particular, constant sets. The given minimization problem (7.1) is thus expressed as

$$\text{Minimize}_{\mathbf{x} \in \mathbf{x}_0 + X,\, \mathbf{u} \in U}\; F(\mathbf{x}, \mathbf{u}) \quad \text{subject to} \quad \mathrm{D}\mathbf{x} = M(\mathbf{x}, \mathbf{u}),\; G(\mathbf{u}) \in K_u,\; N(\mathbf{x}) \in K_x \tag{7.2}$$

Assume that a local optimum is reached at $(\mathbf{x}, \mathbf{u}) = (\bar{\mathbf{x}}, \bar{\mathbf{u}})$. Assume the representation (section 1.5.10)

$$(\forall \mathbf{w} \in W) \quad \bar{\boldsymbol{\lambda}}\mathbf{w} = \int_I \lambda(t)\mathbf{w}(t)\,\mathrm{d}t$$

for $\bar{\boldsymbol{\lambda}}$, where $\bar{\boldsymbol{\lambda}}$ is the Lagrange multiplier in the Karush–Kuhn–Tucker conditions, associated to the given differential equation. Assume a similar representation for the multiplier $\bar{\mathbf{v}}$ associated to the state constraint (if there is one present). Define the *integrated Hamiltonian* function

$$H(\mathbf{x}, \mathbf{u}; \boldsymbol{\lambda}, \mathbf{v}) = F(\mathbf{x}, \mathbf{u}) + \bar{\boldsymbol{\lambda}} M(\mathbf{x}, \mathbf{u}) - \bar{\mathbf{v}} N(x)$$

Then H is the integral of the Hamiltonian function

$$h(x(t), u(t), t;\; \lambda(t), v(t)) = f(x(t), u(t), t) + \lambda(t)m(x(t), u(t), t) - v(t)n(x(t), t)$$

where $\lambda(.)$ and $v(.)$ are the functions (or generalized functions – see section 1.5.9), representing the Lagrange multipliers $\bar{\boldsymbol{\lambda}}$ and $\bar{\mathbf{v}}$. Let $\bar{\boldsymbol{\rho}} = (\bar{\boldsymbol{\lambda}}, \bar{\mathbf{v}})$, $\boldsymbol{\rho} = (\boldsymbol{\lambda}, \mathbf{v})$. (Thus the function $\boldsymbol{\rho}(.)$ represents both the Lagrange multipliers.)

For simplicity, the terms in $\bar{\mathbf{v}}$ and $\mathbf{v}(.)$ can be omitted on first reading this chapter, thus assuming that there is no state constant.

7.2.2 Theorem (Karush–Kuhn–Tucker conditions)

The necessary Karush–Kuhn–Tucker conditions for a local minimum of (7.2) at $(\bar{\mathbf{x}}, \bar{\mathbf{u}})$ hold if and only if both the adjoint differential equation holds:

$$-\dot{\boldsymbol{\lambda}}(t) = f_{\mathbf{x}}(\bar{\mathbf{x}}(t), \bar{\mathbf{u}}(t), t) + \boldsymbol{\lambda}(t)m_{\mathbf{x}}(\bar{\mathbf{x}}(t), \bar{\mathbf{u}}(t), t) - \mathbf{v}(t)n_{\mathbf{x}}(\bar{\mathbf{x}}(t), t), \quad \boldsymbol{\lambda}(T) = \mathbf{0},$$
$$\mathbf{v}(t) \in V(t), \quad \mathbf{v}(t)\mathbf{n}(x(t), t) = 0 \quad (0 \leq t \leq T) \tag{7.3}$$

and the necessary Karush–Kuhn–Tucker conditions hold for a minimum at $\bar{\mathbf{u}}$ of $H(\bar{\mathbf{x}}, \mathbf{u}, \bar{\boldsymbol{\lambda}}, \bar{\mathbf{v}})$.

Proof
See section 4.3. □

In order to state the next result, the following notation is required. If there is no state constraint, let $P(\mathbf{x}, \mathbf{u}) := -\mathrm{D}\mathbf{x} + M(\mathbf{x}, \mathbf{u})$. If there is a state constraint, let $P(\mathbf{x}, \mathbf{u}) := [-\mathrm{D}\mathbf{x} + M(\mathbf{x}, \mathbf{u}), N(\mathbf{x}) - \mathbf{s}]$, where $\mathbf{s}$ is a nonnegative slack variable (e.g. $\mathbf{s} = (q_1^2, q_2^2, \ldots)$), and adjoin $\mathbf{s}(.)$ to $x(.)$, so that $P(\mathbf{x}, \mathbf{u}) = \mathbf{0}$ combines the given differential equation and the state constraint, with the latter treated as an equation. Define also $Q(.,.; \rho) := F(.,.) + \rho P(.,.)$, where $\boldsymbol{\rho} := [\bar{\boldsymbol{\lambda}}, \bar{\mathbf{v}}]$. (Note that the theorem does *not* assume differentiability with respect to $\mathbf{u}$.)

7.2.3 Theorem (on quasimin)
For the problem (7.2), assume that

(i) the differential equation $\mathrm{D}x = M(x, u)$ (including initial condition) determines a unique $\mathbf{x} = \Phi(u)$ for each $u \in \Gamma$, where the map Φ satisfies a Lipschitz condition, for some constant κ:

$$(\forall \mathbf{u}_1, \mathbf{u}_2 \in \Gamma) \quad \|\Phi(\mathbf{u}_1) - \Phi(\mathbf{u}_2)\| \leq \kappa \|\mathbf{u}_1 - \mathbf{u}_2\|$$

(ii) F and M are partially Fréchet differentiable with respect to $\mathbf{x}$, uniformly in $\mathbf{u}$ near $\bar{\mathbf{u}}$, thus that:

$$F(\mathbf{x}, u) - F(\bar{\mathbf{x}}, u) = F_{\mathbf{x}}(\bar{\mathbf{x}}, \bar{\mathbf{u}})(\mathbf{x} - \bar{\mathbf{x}})\, o(\|\mathbf{x} - \bar{\mathbf{x}}\| + \|\mathbf{u} - \bar{\mathbf{u}}\|)$$

and a similar equation for M.

If $\bar{\boldsymbol{\rho}} = (\bar{\boldsymbol{\lambda}}, \bar{\mathbf{v}})$ satisfies $Q_x(\bar{\mathbf{x}}, \mathbf{u}; \bar{\boldsymbol{\lambda}}, \bar{\mathbf{v}}) = 0$, then $H(\bar{\mathbf{x}}, .; \bar{\boldsymbol{\lambda}}, \bar{\mathbf{v}})$ has a quasimin on Γ at the point $\bar{\mathbf{u}}$.

Proof
Let $\mathbf{x} = \Phi(\mathbf{u})$ and $\bar{\mathbf{x}} = \Phi(\bar{\mathbf{u}})$. From the Lipschitz assumption (i), a term

$$r = o(\|\mathbf{x} - \bar{\mathbf{x}}\|) \Leftrightarrow |r| < \varepsilon\|\mathbf{x} - \bar{\mathbf{x}}\| \quad \text{when } \|\mathbf{x} - \bar{\mathbf{x}}\| < \delta(\varepsilon)$$
$$\Rightarrow |r| < \kappa\varepsilon(\|\mathbf{u} - \bar{\mathbf{u}}\|) \quad \text{when } (\|\mathbf{u} - \bar{\mathbf{u}}\|) < \kappa^{-1}\delta(\varepsilon)$$

Thus $r = o(\|\mathbf{x} - \bar{\mathbf{x}}\|) \Rightarrow r = o(\|\mathbf{u} - \bar{\mathbf{u}}\|)$. Now, since $P(\mathbf{x}, \mathbf{u}) = \mathbf{0} = P(\bar{\mathbf{x}}, \bar{\mathbf{u}})$, and the property (ii) for F and M applies to Q,

$$\begin{aligned}
H(\bar{\mathbf{x}}, \bar{\mathbf{u}}; \bar{\boldsymbol{\rho}}) - H(\bar{\mathbf{x}}, \mathbf{u}; \bar{\boldsymbol{\rho}}) &= F(\bar{\mathbf{x}}, \bar{\mathbf{u}}) - F(\mathbf{x}, \mathbf{u}) + F(\mathbf{x}, \mathbf{u}) \\
&\quad - F(\bar{\mathbf{x}}, \mathbf{u}) + \boldsymbol{\rho}[P(\bar{\mathbf{x}}, \bar{\mathbf{u}}) - P(\bar{\mathbf{x}}, \mathbf{u})] \\
&= F(\bar{\mathbf{x}}, \bar{\mathbf{u}}) - F(\mathbf{x}, \mathbf{u}) + F(\mathbf{x}, \mathbf{u}) - F(\bar{\mathbf{x}}, \mathbf{u}) \\
&\quad + \boldsymbol{\rho}[P(\mathbf{x}, \mathbf{u}) - P(\bar{\mathbf{x}}, \bar{\mathbf{u}})] \\
&= F(\bar{\mathbf{x}}, \bar{\mathbf{u}}) - F(\mathbf{x}, \mathbf{u}) \\
&\quad + Q(\mathbf{x}, \mathbf{u}; \bar{\boldsymbol{\rho}}) - Q(\bar{\mathbf{x}}, \mathbf{u}; \bar{\boldsymbol{\rho}}) \\
&= F(\bar{\mathbf{x}}, \bar{\mathbf{u}}) - F(\mathbf{x}, \mathbf{u}) + Q_x(\bar{\mathbf{x}}, \bar{\mathbf{u}}; \bar{\boldsymbol{\rho}})(\mathbf{x} - \bar{\mathbf{x}}) \\
&\quad + o(\|\mathbf{x} - \bar{\mathbf{x}}\| + \|\mathbf{u} - \bar{\mathbf{u}}\|)
\end{aligned}$$

$$= F(\bar{\mathbf{x}}, \bar{\mathbf{u}}) - F(\mathbf{x}, \mathbf{u}) + Q_x(\bar{\mathbf{x}}, \bar{\mathbf{u}}; \bar{\boldsymbol{\rho}})(\mathbf{x} - \bar{\mathbf{x}}) + o(\|\mathbf{u} - \bar{\mathbf{u}}\|) \tag{7.4}$$

If $(\mathbf{x}^-, \mathbf{u}^-)$ minimizes the control problem, and if $\boldsymbol{\rho}^-$ satisfies $Q_x(\mathbf{x}^-, \mathbf{u}^-; \boldsymbol{\rho}^-) = 0$, then

$$H(\bar{\mathbf{x}}, \mathbf{u}; \bar{\boldsymbol{\rho}}) - H(\bar{\mathbf{x}}, \bar{\mathbf{u}}; \bar{\boldsymbol{\rho}}) \geq o(\|\mathbf{u} - \bar{\mathbf{u}}\|) \quad (\mathbf{u} \in \Gamma, \mathbf{u} \to \bar{\mathbf{u}})$$

Thus the problem

$$\text{Minimize}_{\mathbf{u} \in \Gamma} H(\bar{\mathbf{x}}, \mathbf{u}; \bar{\boldsymbol{\lambda}}, \bar{\mathbf{v}}) \tag{7.5}$$

has a quasimin at $u = \eta$. □

7.2.4 Remarks (associated problem and adjoint equation)

Observe that (7.5) is equivalent to the associated problem:

$$(\forall t \in [0, T]) \quad \text{Minimize}_{\mathbf{u}(t)} h(\bar{\mathbf{x}}(t), \mathbf{u}(t), t, \boldsymbol{\lambda}(t), \mathbf{v}(\mathbf{t})) \quad \text{subject to } \mathbf{u}(t) \in \Delta(t) \tag{7.6}$$

This happens because the constraint on **u** is a constraint on $\mathbf{u}(t)$ for each separate time t; it would be otherwise if, for example, the problem included some term involving $|u(t) - u(t - \delta)|$, for some constant time lag δ.

The equation $Q_x(\bar{\mathbf{x}}, \mathbf{u}; \bar{\boldsymbol{\lambda}}, \bar{\mathbf{v}}) = 0$ is equivalent, by the calculation in section 4.2.7, to the adjoint differential equation:

$$-\dot{\boldsymbol{\lambda}}(t) = h_{\mathbf{x}}(\bar{\mathbf{x}}(t), \bar{\mathbf{u}}(t), t; \boldsymbol{\lambda}(t), \mathbf{v}(t)) \quad (t \in I), \quad \boldsymbol{\lambda}(T) = \mathbf{0} \tag{7.7}$$

in which the function $\boldsymbol{\lambda}(.)$ is usually called the costate function. In expanded form, the adjoint differential equation is:

$$-\dot{\boldsymbol{\lambda}}(t) = f_{\mathbf{x}}(\bar{\mathbf{x}}(t), \bar{\mathbf{u}}(t), t) + \boldsymbol{\lambda}(t) m_{\mathbf{x}}(\bar{\mathbf{x}}(t), \bar{\mathbf{u}}(t), t) - \mathbf{v}(t) n_{\mathbf{x}}(\bar{\mathbf{x}}(t), t), \quad \boldsymbol{\lambda}(T) = \mathbf{0}$$

7.2.5 Chattering property

For the next theorem, it must be assumed that the minimum at $\mathbf{u} = \bar{\mathbf{u}}$ of $J(\mathbf{u}) := F(\Phi(\mathbf{u}), \mathbf{u})$ is either a global minimum, or a local minimum when **u** has the $L^1(I)$ norm, $\|\mathbf{u}\|_1 := \int_I |\mathbf{u}(t)|\, \mathrm{d}t$. The latter requires that $J(\mathbf{u}) \geq J(\bar{\mathbf{u}})$ whenever **u** is feasible and $\|\mathbf{u} - \bar{\mathbf{u}}\| < \delta$, for some $\delta > 0$. From the form of the control constraint, $(\forall t)\, \mathbf{u}(t) \in \Delta(t)$, it follows that if $\mathbf{u} \in \Gamma$, $\mathbf{v} \in \Gamma$ and $A \subset I$, then also $\mathbf{w} \in \Gamma$, where $\mathbf{w}(.)$ is defined by $\mathbf{w}(t) = \mathbf{u}(t)$ for $t \in A \subset I$, and $\mathbf{w}(t) = \mathbf{v}(t)$ for $t \in I \backslash A$. This property of the controls is called the *chattering* (or *interpolable*) property. (The term chattering derives from bang-bang controls with electrical relays.) A property holds for *almost all* $t \in I$ if it holds for all $t \in I$ except a set N for which $\int_N \mathrm{d}t = 0$.

In Figure 7.1, $u(t)$ is shown by a thick line, $v(t)$ by a thin line, and $w(t)$, which jumps back and forth between $u(t)$ and $v(t)$, by a very thick line.

7.2.6 Theorem (Pontryagin property)

Assume that $H(\bar{\mathbf{x}}, .; \bar{\boldsymbol{\lambda}}, \bar{\mathbf{v}})$ has a quasimin on Γ at $\bar{\mathbf{u}}$, when **u** has the $L^1(I)$ norm, and that the constraints on the control have the chattering property. Then

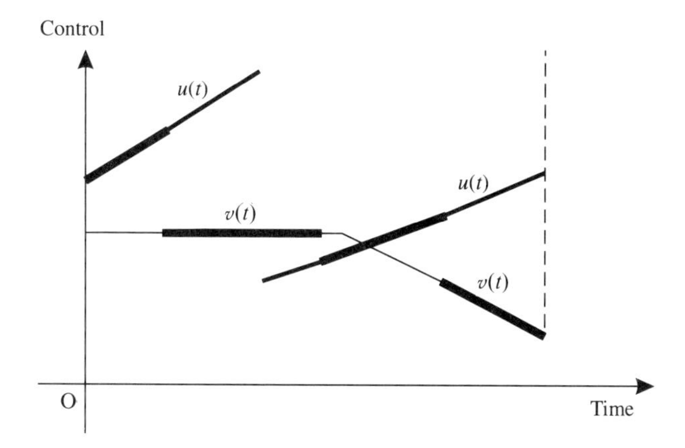

Figure 7.1 The chattering property.

$$\tilde{h}(\boldsymbol{\zeta}, t) \equiv h(\bar{\mathbf{x}}(t), \boldsymbol{\zeta}, t, \bar{\boldsymbol{\lambda}}(t), \bar{\mathbf{v}}(t))$$

is minimized, with respect to $\boldsymbol{\zeta} \in \Delta(t)$, when $\boldsymbol{\zeta} = \bar{\mathbf{u}}(t)$, for *almost all* $t \in I$.

Proof

Suppose, if possible, that the conclusion does not hold. Then there is some $\mathbf{v} \in \Gamma$ and some set $A^{\#} \subset I$ with $\int_{A^{\#}} \mathrm{d}t > 0$, such that

$$(\forall t \in A^{\#}) \quad \tilde{h}(\mathbf{v}(t), t) < \tilde{h}(\bar{\mathbf{u}}(t), t)$$

Define $\varphi : I \to \mathbf{R}$ by

$$\varphi(t) = 0 \quad \text{when } h(\mathbf{v}(t), t) = \tilde{h}(\bar{\mathbf{u}}(t), t)$$

and

$$-\tilde{h}(\mathbf{v}(t), t) + \tilde{h}(\bar{\mathbf{u}}(t), t) = \varphi(t)|\mathbf{v}(t) - \bar{\mathbf{u}}(t)|$$

otherwise. Then $\varphi(t) \geq 0$.

Consider first the case when for some $t_0 \in A^{\#}$; the function $\varphi(.)$ is continuous at t_0. Then there exist $A \subset A^{\#}$ and $\sigma > 0$ for which $\int_A \mathrm{d}t > 0$ and $(\forall t \in A)\ \varphi(t) \geq \sigma$. Define a curve $\{\mathbf{u}_\beta : \beta \geq 0\}$ in U by

$$\mathbf{u}_\beta(t) := \mathbf{v}(t) \quad \text{for } t \in A_\beta := \{t \in A: |t - t_0| \leq \psi(\beta)\}$$

$$\mathbf{u}_\beta(t) := \bar{\mathbf{u}}(t) \quad \text{otherwise}$$

with the function $\psi(\beta)$ chosen as $\omega^{-1}(\beta)$, where $\omega(\theta)$ is the integral of $|\mathbf{v}(.) - \bar{\mathbf{u}}(.)|$ over $\{t \in A: |t - t_0| \leq \theta\}$. By this construction, the L^1-norm $\|\mathbf{u}_\beta - \bar{\mathbf{u}}\|_1 = \beta$. Now $\mathbf{u}_0 = \bar{\mathbf{u}} \in \Gamma$, and $\mathbf{u}_\beta \in \Gamma$ for $\beta > 0$, by the chattering property. Then $\tilde{H}(.) := H(\bar{\mathbf{x}}, .; \bar{\boldsymbol{\lambda}}, \bar{\mathbf{v}})$ satisfies

$$\tilde{H}(\bar{\mathbf{u}}) - \tilde{H}(\mathbf{u}_\beta) \geq \int_{A_\beta} \varphi(t)|\mathbf{u}_\beta(t) - \bar{\mathbf{u}}(t)|\,\mathrm{d}t$$

$$\geq \sigma \int_{A_\beta} |\mathbf{u}_\beta(t) - \bar{\mathbf{u}}(t)| \, dt$$
$$= \sigma \| \mathbf{u}_\beta - \bar{\mathbf{u}} \|_1$$
$$= \sigma\beta$$

by construction of $\mathbf{u}_\beta$. But this contradicts the quasimin of $\tilde{H}(.)$ at $\bar{\mathbf{u}}$ which requires that

$$\tilde{H}(\mathbf{u}_\beta) - \tilde{H}(\bar{\mathbf{u}}) \geq o(\beta) \quad \text{as } \beta \downarrow 0$$

Suppose now that there is no such point t_0 of continuity. A point t_0 is a *point of density* of $A^\# \subset I$ if

$$\lim_{\tau \downarrow 0} (2\tau)^{-1} [m(E \cap [t_0 - \tau, t_0 + \tau]) = 1$$

where $m(E) = \int_E dt$ denotes the Lebesgue measure of a set E. The function φ is approximately continuous at $t_0 \in I$ if, for some measurable set E, t_0 is a point of density of E (hence $m(E) > 0$), and

$$\lim_{t \to t_0,\, t \in E} \varphi(t) = \varphi(t_0)$$

Since $A^\#$ is a measurable set with $m(A^\#) > 0$, a theorem of measure theory (Munroe, 1953, section 4.2) states that φ is approximately continuous at almost all points $t_0 \in I$. Obtain the set $A_0^\#$ by excluding the set N of zero measure on which φ is not approximately continuous, or where t_0 is not a point of density. Then each $t_0 \in A_0^\#$ is a point of density, and there is a set A with $m(A) > 0$ such that

$$\lim_{t \to t_0,\, t \in A} \varphi(t) = \varphi(t_0)$$

Then $\varphi(t) \geq \sigma := \frac{1}{2}\varphi(t_0) > 0$ for all $t \in A$ and $|t - t_0|$ sufficiently small. The remainder of the proof is then valid with this A and σ. □

7.3 PONTRYAGIN THEOREM

These results are summed up in the following theorem.

7.3.1 Pontryagin theorem

Assume that the fixed-time optimal control problem (7.1) reaches a local minimum when $\mathbf{u} = \bar{\mathbf{u}}$ with respect to the L^1-norm for the control $\mathbf{u}$. Assume hypotheses (i) and (ii) of Theorem 7.2.3, and also the chattering property for the constraints on the control. Then necessary conditions for the minimum of (7.1) are that the costate $\boldsymbol{\lambda}(.)$ satisfies the adjoint differential equation (7.7) with boundary condition, and that, for all t except a set of zero measure, the associated problem (7.6) reaches a minimum when $\mathbf{u}(t) = \bar{\mathbf{u}}(t)$.

7.3.2 Remarks

Theorem 7.7 assumes Fréchet derivatives. There is also an analogous result when L^2-norms and Hadamard derivatives (section 4.2.8) are used. Fréchet derivatives are not usually available when the L^2-norm is used, but often Hadamard derivatives can be used then. The formulae appear the same, but the derivatives are Hadamard.

The Pontryagin theorem gives *necessary* conditions, which correspond to the Karush–Kuhn–Tucker necessary conditions for constrained optimization. In both cases, the necessary conditions become also *sufficient* if additional convex or invex (section 3.6) hypotheses are made on the functions in (7.1). Theorems 4.2.3 and 3.6.2(d) apply here, without further proof. For this reason, certain upward arrows in Figure 4.2 are marked *convex*; these arrows indicate valid sufficiency results, assuming that the functions $f(.,.,t)$ and $m(.,.,t)$ of the control problem are convex (or invex properties hold).

7.3.3 No explicit dependence on *t*

Consider now the case where the functions in the control problem do not depend *explicitly* on t; thus one can write $f(\mathbf{x}(t), \mathbf{u}(t))$ and not $f(\mathbf{x}(t), \mathbf{u}(t), t)$, and similarly $m(\mathbf{x}(t), \mathbf{u}(t))$, $n(\mathbf{x}(t))$ and Δ (not $\Delta(t)$). The following theorem then holds.

Theorem

Assume that the functions f, m, Δ in the optimal control problem (7.1) do not depend explicitly on t, the point $\bar{\mathbf{u}}$ is a minimum both when $\mathbf{u}$ has the uniform norm and also when $\mathbf{u}$ has the L^1-norm, and there is no state constraint. Then $h(\bar{\mathbf{x}}(t), \bar{\mathbf{u}}(t), \bar{\boldsymbol{\lambda}}(t))$ is constant in t.

Proof

The Hamiltonian may now be written in the form

$$h(\mathbf{x}(t), \mathbf{u}(t), \boldsymbol{\lambda}(t)) \equiv k(t, \mathbf{u}(t))$$

where the dependence on $\mathbf{x}(t)$ is hidden in the explicit dependence of k on t. Then

$$k_t(t, \bar{\mathbf{u}}(t)) = (f + \boldsymbol{\lambda} m)_x \dot{\mathbf{x}}(t) + \dot{\boldsymbol{\lambda}}(t) m = -\dot{\boldsymbol{\lambda}}(t)\dot{\mathbf{x}}(t) + \dot{\boldsymbol{\lambda}}(t)\dot{\mathbf{x}}(t) = 0$$

using the given and adjoint differential equations. The differential equations in section 4.2.7 show that, for each t, $k(t, \mathbf{u}(t))$ has a Karush–Kuhn–Tucker point at $\mathbf{u}(t) = \bar{\mathbf{u}}(t)$. Since $\bar{\mathbf{u}}$ is a minimum when uniform norms are used, differentiability follows; then Theorem 3.8.3 shows that this Karush–Kuhn–Tucker point is a quasimin.

Since $\bar{\mathbf{u}}$ is a minimum with respect to the L^1-norm, the controls $\bar{\mathbf{u}}(.+\varepsilon)$ and $\bar{\mathbf{u}}(.-\varepsilon)$ are close to $\bar{\mathbf{u}}(.)$ when ε is small positive. Hence

$$k(t, \bar{\mathbf{u}}(t+\varepsilon)) - k(t, \bar{\mathbf{u}}(t)) \geq o(\varepsilon)$$

and

$$k(t, \bar{\mathbf{u}}(t-\varepsilon+\varepsilon)) - k(t, \bar{\mathbf{u}}(t-\varepsilon)) \geq o(\varepsilon)$$

hold as $\varepsilon \downarrow 0$. Thus $k(t, \bar{\mathbf{u}}(t+\varepsilon)) - k(t, \bar{\mathbf{u}}(t)) = o(\varepsilon)$. Hence $\varphi(.) := k(., \bar{\mathbf{u}}(.))$ has zero derivative; moreover, the sum of absolute values of changes in $\varphi(.)$ over a number of intervals of t of total length ε is less than $c\varepsilon$ for some constant c. This means that φ is *absolutely continuous*. (For this property, see any book on measure theory, e.g. Munroe (1953) or Craven (1982).) A consequence of absolutely continuous is the property, for any $[\alpha, \beta] \subset I$, that

$$\varphi(\beta) - \varphi(\alpha) = \int_\alpha^\beta \varphi'(t)\, dt = 0$$

Therefore $\varphi(.)$ is constant. □

Remark

This result is usually stated for a *global* minimum with respect to $\mathbf{u}$; then $\bar{\mathbf{u}}$ is automatically a local minimum with respect to each of the two stated norms.

7.4 SENSITIVITY TO PERTURBATIONS

In section 4.7, various sensitivity results were stated and a theorem proved on the sensitivity of the optimal objective function to small changes of a parameter. To obtain estimates for the sensitivity of the optimal point to parameter changes requires more assumptions and more proofs (e.g. Lempio and Maurer, 1980; Craven, 1983).

These questions have many applications. In computing a constrained minimum, it is often necessary to approximate the given minimization problem. For example, a nonsmooth problem may be approximated by a smooth problem, or a minimization over a space of piecewise continuous functions (as is often required in optimal control) may be approximated by minimization over a prescribed sequence of step functions. In numerical experiments such approximations often work very well (see the references cited in section 7.8). This raises the question: what general properties are required of the functions in a minimization problem in order that such approximation methods will converge to the minimum of the given problem, and at what rate?

7.4.1 Example

As a simple example, let $f(x, q) := x_+^2 + 2qx$ (where $x \in \mathbf{R}$ and $x_+ := x$ if $x \geq 0$, 0 if $x < 0$). If the parameter $q = 0$, then $f(., 0)$ reaches an unconstrained minimum at 0. If $q < 0$, then $f(., q)$ reaches a minimum at $-q$; but if $q > 0$ then $f(., q)$ does not reach any minimum. This function may be compared with

$$g(x, q) := x_+^2 + \alpha(-x)_+^2 + 2qx$$

where α is small positive. Unlike $f(., q)$, $g(., q)$ has a *strict* minimum at 0; thus, for some $\theta(r) = \alpha r^2 > 0$, $g(x, 0) \geq \theta(r)$ whenever $|x - 0| = r$. When $q > 0$, $g(., q)$ is minimized at $-q/(2\alpha)$. As q varies from 0, the minimum of $g(., q_0)$ remains in some neighbourhood of the minimum of $g(., 0)$; the size of this neighbourhood depends on α. From this simple example, stability to perturbations should not be expected if the given minimum is not strict.

7.4.2 Parametric family of problems

Consider a parametric family of constrained minimization problems:

$$\text{Minimize}_x\ f(x, q) \quad \text{subject to} \quad p(x, q) \in K \tag{7.8}$$

in which X and Y are normed spaces, $K \subset Y$ is a closed convex cone, $x \in X$, and the parameter $q \in \mathbf{R}^l$. When $q = 0$, let $f(x, 0) = f_0(x)$ and $p(x, 0) = p_0(x)$, where f_0 and p_0 are given functions. Assume that the functions $f(x, q)$ and $p(x, q)$ are uniformly continuous in (x, q), when x is in a bounded neighbourbood $\mathbf{N}'(q)$ of $\Delta_q := \{x \in X: p(x, q) \in K\}$, and $q \in \mathbf{N}$, a bounded neighbourhood of $0 \in \mathbf{R}^l$. (In finite dimensions, a compact set $\mathbf{E}$ may exist such that $(\forall q \in \mathbf{N})\ \mathbf{N}'(q) \subset \mathbf{E}$, and then it suffices to assume that the functions are continuous on $\mathbf{E} \times \mathbf{N}$. However, other validation must be found in infinite dimensions.) The following results also apply when q is restricted to a discrete set, such as $\{1/n: n = 1, 2, 3, \ldots\}$. Let the point a satisfy $p_0(a) \in K$. Define $\tilde{p}(., .)$ by

$$\tilde{p}(x, q) := p(x, q) - p(a, q) + p(a, 0)$$

then $\tilde{p}(a, q) = \tilde{p}(a, 0) = p_0(a)$. Let $\Gamma_q := \{x \in X: \tilde{p}(x, q) \in K\}$. For $a \in X$, let $B(a, r) := \{x \in X: \|x - a\| \le r\}$. For $e \in E$, the distance $d(x, E) := \inf_{e \in E} \|d - e\|$. Let $A_q(r) := \Gamma_q \cap B(a, r)$ and $A'_q(r) := \Gamma_q \cap \{x \in X: \|x - a\| = r\}$.

7.4.3 Lemma (Craven, 1986, 1994)

Assume $p(., .)$ is uniformly continuous on bounded sets and that the point a satisfies $p_0(a) \in K$. Fix $r > 0$; let $0 < \varepsilon < r$. Then, for sufficiently small $\|q\|$, each point of $A_q(r)$ is distant less than ε from $A_0(r)$.

Proof

By construction, $a \in \Gamma$ and $a \in \Gamma_q$. For $0 < \varepsilon < r$, define

$$\pi(\varepsilon) := \inf \{d(p(x), K) : \|x - a\| \le r, \quad d(x, A_0(r)) \ge \varepsilon)$$

Since K is closed, if $\|x - a\| \le r$ and $x \notin A_0(r)$, then $d(x, A_0(r)) > 0$, $d(p(x), K) > 0$ and $\pi(\varepsilon) > 0$. By the uniform continuity, there is $\delta > 0$ such that, if $\|q\| < \delta$, then $\|\tilde{p}(x, q) - \tilde{p}(x, 0)\| < \pi(\varepsilon)/3$, uniformly in x in a neighbourhood of $A_0(r)$. Since also $\|-\tilde{p}(a, q) - \tilde{p}(a, 0)\| < \pi(\varepsilon)/3$, it follows that $\tilde{p}(x, q)$ lies in a ball with centre $p(x, 0)$ and radius $2\pi(\varepsilon)/3$, which is disjoint from the closed convex set K; hence $\tilde{p}(x, q) \notin K$. Taking the contrapositive,

$$[\|x - a\| \le r \text{ and } \tilde{p}(x, q) \notin K] \Rightarrow [\|x - a\| \le r \text{ and } d(x, A_0(r) \le \varepsilon] \qquad \square$$

7.4.4 Remarks

If a is a boundary point of Γ, then there is a sequence $\{x_j\} \to a$, with each $\{x_j\} \notin \Gamma$; hence $\pi(\varepsilon) \downarrow 0$ as $\varepsilon \downarrow 0$. Otherwise, $\pi(\varepsilon) \downarrow \pi(0) > 0$, and δ does not tend to 0 as $\varepsilon \downarrow 0$ in this proof. It suffices if the uniform continuity holds for x in some bounded neighbourhood of $A_0(r)$.

7.4.5 Lemma

Assume $p(.,.)$ is uniformly continuous on bounded sets and the point a satisfies $p_0(a) \in K$. Fix $r > 0$; let $0 < \varepsilon < r$. Then, for sufficiently small $\|q\|$, each point of $A'_q(r)$ is distant less than ε from $A'_0(r)$.

Proof

The same proof holds with $A'_q(r)$ replacing $A_q(r)$. □

Now suppose that the point a is a local minimum of the problem (7.8) with $q = 0$. For constant $\gamma > 0$, define $F(x, q) := f(x, q) + \gamma \|x - a\|^2$.

7.4.6 Definition

The point a is a strict local minimum of $F(x, 0)$, subject to $p(x, 0) \in K$, if

$$(\forall \delta > 0)\ (\exists \kappa > 0)\ (\forall x, p(x, 0) \in K, \|x - a\| = \delta) \quad F(x, 0) \geq F(a, 0) + \kappa$$

The next lemma shows that, because of the strict minimum, $F(., q)$ has a constrained minimum near to a when $\|q\|$ is small.

7.4.7 Lemma

Let $f(x, q)$ and $p(x, q)$ be uniformly continuous, for x is a bounded neighbourhood of $A_0(r)$ and q is a neighbourhood of 0. When $q \neq 0$, assume that $F(., q)$ reaches a minimum on each closed bounded set. If $\|q\|$ is sufficiently small, then $F(x, q)$ reaches a minimum, subject to the constraint $\tilde{p}(x, q) \in K$, at a point $\bar{x}(q)$, where $\bar{x}(q) \to a$ as $\|q\| \to 0$.

Proof

Let $0 < r < \delta$; let $\varepsilon > 0$. Since $F(., 0)$ has a strict minimum at a, there is $\theta > 0$ such that $F(\xi, 0) \geq F(a, 0) + 4\theta$ for all $\xi \in A'_0(r)$. By the uniform continuity, $|F(x, 0) - F(x', 0)| < \theta$ and $|F(x, q) - F(x, 0)| < \theta$ whenever $\|x - x'\| < \delta_1(\theta)$, say. If $\|q\|$ is sufficiently small, and $x \in A'_q(r)$, then $d(x, A'_0(r)) \leq \delta_1(\theta)$ by Lemma 7.4.5. So, for some $x' \in A'_0(r)$,

$$\begin{aligned} F(x, q) &= F(a, 0) + [F(x, q) - F(x, 0)] + [F(x, 0) - F(x', 0)] + [F(x', 0) - F(a, 0)] \\ &\geq F(a) - \theta - \theta + 4\theta \end{aligned}$$

Similarly, if $\|q\|$ is sufficiently small, then

$$F(a, q) = F(a, 0) + [F(a, q) - F(a, 0)] \leq F(a) + \theta$$

For $q \neq 0$, $F(., q)$ reaches a local minimum on the closed bounded set $A_q(r)$, say at $x = \bar{x}(q)$. Since $(\forall x \in A'_q(r))\ F(x, q) - F(a, q) \geq \theta > 0$, $\bar{x}(q) \notin A'_q(r)$. Hence $\bar{x}(q)$ is a local minimum of $F(., q)$ on Γ_q.

If now $r \downarrow 0$, then the corresponding $\theta \downarrow 0$ and $q \downarrow 0$. Since $\|\bar{x}(q) - a\| < r$, it follows that $\|\bar{x}(q) - a\| \to 0$ as $\|q\| \downarrow 0$. □

7.4.8 Remark

Since the minimum is not on the boundary of the ball, it must lie inside the ball.

7.5 LIPSCHITZ STABILITY

The results in section 7.4 assume uniform continuity, but do not require any differentiability.

7.5.1 Differentiable assumptions

Consider now the case where $p(.,.)$ is continuously (Fréchet) differentiable, and also the Robinson stability condition holds at the point a when $q = 0$, namely that

$$0 \in \text{int}\,[\tilde{p}(a,0) + \tilde{p}_x(a,0)(X) - K]$$

From Robinson (1976), for some constant $\kappa > 0$, and all sufficiently small $\|q\|$,

$$d(a, \{x: \tilde{p}(x,q) \in K\}) \leq \kappa\, d(0, \tilde{p}(a,q) - K) = \kappa \mathbf{O}(\|q\|) \tag{7.9}$$

using differentiability. Hence, for some constant ν, if $\|q\| < \nu\varepsilon$, then $A_q(r)$ lies in an ε-neighbourhood of $A_0(r)$. Otherwise expressed, $A_q(r)$ lies in a $\kappa^{-1}\|q\|$-neighbourhood of $A_0(r)$. A similar result holds also when $A_q(r)$ and $A_0(r)$ are replaced by $A'_q(r)$ and $A'_0(r)$ respectively. It is deduced by adjoining the additional constraint $\|x - a\| = r$; since this constraint is not generally active, it does not change the stability condition.

7.5.2 Lipschitz assumptions

If X and Y are finite-dimensional, p is Lipschitz (no longer continuously differentiable), the mapping $(x, q) \to \partial\tilde{p}(x, q)$ (the Clarke generalized Jacobian with respect to x) is upper semicontinuous at $(a, 0)$, and the generalized stability condition holds:

$$(\forall M \in \partial\tilde{p}(a,0)) \quad 0 \in \text{int}\,[\tilde{p}(a,0) + M(X) - K]$$

then Yen (1990) has shown that (7.9) still holds. Hence, in this case also, it follows that $A'_q(r)$ lies in a $\kappa^{-1}\|q\|$-neighbourhood of $A'_0(r)$ for some constant κ.

7.5.3 Lemma

Assume that $p(.,.)$ is uniformly continuous on bounded sets and Lipschitz; assume either the Robinson stability condition or the generalized stability condition in finite dimensions. Let $F(.,.)$ be Lipschitz, with Lipschitz constant κ_2. Let $F(.,0)$ reach a strict minimum at $x = a$, subject to $p(x, 0) \in K$, satisfying also the *linear growth* condition:

$$F(x,0) - F(a,0) \geq \theta = (\kappa_3/4)r \quad \text{when } x \in A'_0(r)$$

in feasible directions, for some constant κ_3. Then, for sufficiently small $\|q\|$, $F(x,q)$ reaches a minimum, subject to $\tilde{p}(x,q) \in K$, at $x = \bar{x}(q)$, satisfying (for some constant κ_4) the Lipschitz condition

$$\| \bar{x}(q) - a \| \leq \kappa_4 \| q \|$$

Proof

From either stability hypothesis, there is a constant κ_1 so that $A'_q(r)$ lies in a $\kappa_1\|q\|$-neighbourhood of $A'_0(r)$, whenever $\|q\|$ is sufficiently small. This result is used in place of Lemma 7.4.5 in a modification, as follows, of the proof of Lemma 7.4.7.

If $x \in A'_0(r)$ and $\|q\| \leq (\kappa_2\kappa_1)^{-1}\theta$, then

$$d(x, A'_0(r)) \leq d(A'_q(r), A'_0(r)) \leq \kappa_2^{-1}\theta$$

So, for some $x' \in A'_0(r)$, $\kappa_2\|x - x'\| \leq \theta$; hence

$$F(x, q) \geq F(a, 0) - \theta - \theta + 4\theta$$

The rest of the proof is unchanged, showing that there is a minimum $\bar{x}(q)$ of $F(., q)$ on Γ_q, with $\| \bar{x}(q) - a \| < r$, where now

$$\|q\| = (\kappa_2\kappa_1)^{-1}\,\theta = [\kappa_3/(4\kappa_1\kappa_2)]r$$

Thus, setting $\kappa_4 := 4\kappa_1\kappa_2/\kappa_3$, the stated conclusion follows. □

7.5.4 Hölder continuity

If the linear growth condition of Lemma 7.5.3 is weakened to

$$F(x, 0) - F(a, 0) \geq \theta = (\kappa_3/4)r^{\sigma} \quad \text{when } x \in A'_0(r)$$

for some constant $\sigma > 1$ (e.g. $\sigma = 2$), then the proof of Lemma 7.5.3 is modified slightly, concluding with

$$\|\bar{x}(q) - a\| \leq \kappa_4\|q\|^{1/\sigma}$$

When $\sigma > 1$, this property, called Hölder continuity, is weaker than the Lipschitz property.

7.5.5 Exercise

Discuss the sensitivity of the optimum point to a small change of the parameter $\mathbf{p}$ for the quadratic program:

$$\text{Minimize}_{\mathbf{x}}\ \tfrac{1}{2}\mathbf{x}^{\mathrm{T}}P\mathbf{x} + \mathbf{c}^{\mathrm{T}}\mathbf{x} \quad \text{subject to} \quad A\mathbf{x} + K\mathbf{p} = \mathbf{b},\ \mathbf{x} \geq \mathbf{0}$$

Exercise

What happens to the optimum of the control problem in Exercise 5.1 when the coefficient of $x(t)^2$ is changed from ½ to ½ + ε?

7.6 TRUNCATION IN l^2

This is described in Craven (1994).

In section 6.4, an approach to computing optimal control is described where the control function is approximated by a step function. In order to bring such approximations within the scope of sensitivity theory, consider first a con-

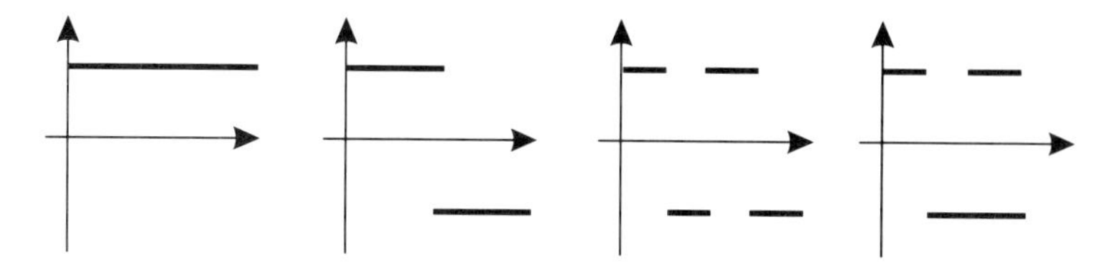

Figure 7.2 Step function components of a Fourier–Walsh series.

strained minimization problem with an objective function $\hat{f}(\xi)$, where the variable $\xi(.) \in L^2[0, 1]$; thus

$$\|\xi\|_2 = \left[\int |\xi|(t)\|^2 \, dt \right]^{1/2}$$

is finite. Let ξ be expanded in a Fourier–Walsh series, analogously to a Fourier series, but with step functions, as illustrated in Figure 7.2, replacing the sines and cosines in a Fourier series. Let the coefficients of the expansion be x_1, $x_2, \ldots$; then the vector $\mathbf{x} := (x_1, x_2, \ldots) \in l^2$, and thus

$$\|\mathbf{x}\|_2 = \left[\sum |x_j|^2 \right]^{1/2}$$

is finite. The same symbol can be used for the two norms, since in fact $\|\xi\|_2 = \|\mathbf{x}\|_2$, by Plancherel's theorem in Fourier theory.

If the Fourier–Walsh series is truncated after the kth term, this gives an approximation to $\xi(.)$ by a step function, where $[0, 1]$ is divided into 2^k subintervals and the step function is constant on each subinterval. Since $\xi = U\mathbf{x}$ for some *unitary* transformation U, meaning that $\|U\mathbf{x}\| = \|\mathbf{x}\|$ for each vector $\mathbf{x}$, the convergence of such approximations to the solution of the original, infinite-dimensional, minimization problem may be studied equivalently in terms of $f(\mathbf{x}) := \hat{f}(Ux)$, where $\mathbf{x} \in l^2$.

Consider then $X = l^2$. If $\mathbf{x} = (x_1, x_2, \ldots) \in l^2$, denote by $\mathbf{x}^{(n)} := (x_1, \ldots, x_n, 0, 0, \ldots)$ the truncation of $\mathbf{x}$ to the first n nonzero components. Similarly define $\mathbf{c}^{(n)}$ from $\mathbf{c} \in l^2$. Consider a minimization problem:

$$\text{Minimize}_{\mathbf{x} \in l^2} f(\mathbf{x}) \quad \text{subject to} \quad p(x) \in K \tag{7.10}$$

Assume (7.10) reaches a local minimum at $\mathbf{x} = \mathbf{a}$. Let $q := 1/n$ $(n = 1, 2, \ldots)$, $f(\mathbf{x}, q) := f(\mathbf{x}^{(n)})$, $p(\mathbf{x}, q) := p(\mathbf{x}^{(n)})$, and $\tilde{p}$ is defined from p and $\mathbf{a}$ as in section 7.4.2. In some sense, the truncated problem:

$$\text{Minimize}_{\mathbf{x} \in l^2} f(\mathbf{x}, 1/n) \quad \text{subject to} \quad \tilde{p}(\mathbf{x}, 1/n) \in K \tag{7.11}$$

approximates the given problem (7.10).

In the particular case of a linear objective function $f(\mathbf{x}) := \langle \mathbf{c}, \mathbf{x} \rangle$ with $\mathbf{c} \in l^2$,

$$|f(\mathbf{x}, 0) - f(\mathbf{x}, 1/n)| = |\langle \mathbf{c}, \mathbf{x} - \mathbf{x}^{(n)} \rangle| = |\langle \mathbf{c} - \mathbf{c}^{(n)}, \mathbf{x} - \mathbf{x}^{(n)} \rangle| \leq \theta_n \|\mathbf{x}\|$$

where $\theta_n := \sup_{j>n} |c_j| \to 0$ as $n \to \infty$. So the desired uniform approximation occurs on norm-bounded sets. For a quadratic function, uniform approximation is attained only by restricting $\mathbf{x}$ to a subspace of l^2. Let $0 < \{b_j\} \to \infty$; let $w_j := b_j^2$; restrict x so that the weighted norm

$$\|\mathbf{x}\|^w := \left[\sum w_j x_j^2\right]^{1/2} < \infty$$

Then

$$\langle \mathbf{x}, \mathbf{x}\rangle - \langle \mathbf{x}^{(n)}, \mathbf{x}^{(n)}\rangle = \langle \mathbf{z}, \mathbf{t}\rangle \quad \text{where } \mathbf{t} := \mathbf{x} - \mathbf{x}^{(n)} \text{ and } \mathbf{z} := 2\mathbf{x} - \mathbf{t}$$

Hence

$$|\langle \mathbf{z}, \mathbf{t}\rangle| \le \left[\sum_{j>n} (b_i^{-2} b_i z_i)^2 \sum_j (b_j t_j)^2\right]^{1/2}$$

$$\le \theta_n \left[\sum (b_i z_i)^2 \sum_j (b_j t_j)^2\right]^{1/2}$$

$$= \theta_n \|\mathbf{z}\|^w \|\mathbf{t}\|^w \le \theta_n (3\|\mathbf{x}\|)\, \|\mathbf{x}\|$$

where $\theta_n := \sup_{j>n} w_j \to 0$ as $n \to \infty$.

Consider now a function $\xi(.)$ in a suitable subspace $\mathbf{S}$ of $L^2[0, 1]$ and $\mathbf{x} \in l^2$ the sequence of the Fourier–Walsh coefficients for ξ. Assume that $\|\mathbf{x}\|^w < \infty$; this restricts high-frequency oscillations of $\mathbf{x}(.)$, but does not restrict $\xi(.)$ to a finite-dimensional subspace. Let $f(\mathbf{x}) := \langle \mathbf{x}, \mathbf{x}\rangle$. From the previous paragraph, $f(\mathbf{x}^{(n)})$ uniformly approximates $f(\mathbf{x})$ on norm-bounded subsets. Now $\mathbf{x}^{(n)}$ is the transform of a step function, $\xi_n(.)$ say, on $[0, 1]$. Let $\varphi(\xi) := \int_0^1 \xi(s)^2\, \mathrm{d}s$; then $\varphi(\xi) = f(\mathbf{x})$. Hence $\varphi(\xi_n)$ uniformly approximates $\varphi(\xi)$ on bounded subsets of $\mathbf{S}$. A similar construction applies to a functional $\varphi(\xi) := \int h(\xi(s), s)\, \mathrm{d}s$ if the transform $f(\mathbf{x})$ satisfies

$$|f(\mathbf{x}) - f(\mathbf{x}^{(n)})| \le \langle \mathbf{c}(\mathbf{x}), \mathbf{x} - \mathbf{x}^{(n)}\rangle$$

where $\|\mathbf{c}(\mathbf{x})\| \le \text{const}\, \|\mathbf{x}\|$.

7.7 APPROXIMATING A CONSTRAINED MINIMUM

Consider a constrained *continuous programming* problem:

$$\min f(x) \quad \text{subject to} \quad p(x) \in K \tag{7.12}$$

where $f(x) := \int_I h(x(t), t)\, \mathrm{d}t$, $I = [0, 1]$, $x(.) \in L^2[0, 1]$, $p(x)(.) = k(x(.), t)$ $(t \in I)$ and K is a cone of nonnegative functions on I. Denote by $x_n(.)$ a step function approximation to $x(.)$, obtained by truncating I into n subintervals (where n is a power of 2). Denote by $(7.12)_n$ the approximated problem, obtained by replacing $x(.)$ by $x_n(.)$ in (7.12).

Theorem 7.7.1

For problem (7.12), let $x(.)$ lie in a subspace $\mathbf{S}$ of $L^2(I)$, such that uniform approximation of (7.12) by step functions $x_n(.)$ holds:

$$(\forall \varepsilon > 0, r > 0)(\exists n)(\forall x \in \mathbf{S}, \|x\| < r) \quad |f(x_n) - f(x)| < \varepsilon, \quad \|p(x_n) - p(x)\| < \varepsilon$$

Assume that (7.12) reaches a minimum at $x(.) = \bar{x}(.)$. Then this approximation for (7.12) reaches a local minimum at a point $\bar{x}_n$, where $\bar{x}_n \to \bar{x}$ as $n \to \infty$.

If, in addition, the functions h and k are Fréchet differentiable and the Robinson stability condition holds, and f satisfies the linear growth condition, then there is a Lipschitz constant κ so that $\|\bar{x}_n - \bar{x}\| \leq \kappa/n$.

Proof

The approximated problem $(7.12)_n$ involves only finitely many (say r) variables, hence $f(x)$, with x restricted to **S**, reaches a minimum on each closed bounded subset. From Lemma 7.4.7, the approximate problem (7.12) reaches a minimum at a point, x_n say, for which $\|\bar{x}_n - \bar{x}\| \to 0$ as $n \to \infty$.

The remaining conclusion follows from Lemma 7.5.3. □

7.7.2 ***Remarks***

A similar result holds for various optimal control problems. If linear growth is weakened to quadratic growth (e.g. Alt, 1989, 1990), then (section 7.5.4) the Lipschitz conclusion of Lemma 7.5.3 is weakened to a Hölder condition: $\|\bar{x}(q) - a\| \leq \kappa_4 \|q\|^{1/2}$, and then in Theorem 7.7.1, $\|\bar{x}_n - \bar{x}\| \leq \kappa/n^{1/2}$.

7.8 APPROXIMATING AN OPTIMAL CONTROL PROBLEM

Consider an optimal control problem:

$$\min J(u) = \int_0^T f_0(x(t), u(t), t)\, \mathrm{d}t + \Phi(x(T)) \quad \text{subject to}$$

$$x(0) = x_0; \quad \mathrm{d}x(t)/\mathrm{d}t = m(x(t), u(t), t) \quad (0 \leq t \leq T) \tag{7.13}$$

$$u(t) \in \Gamma(t) \quad (0 \leq t \leq T)$$

Here, $x(t)$ is the state function, $u(t)$ is the control function, $f(., ., .)$, $\Phi(.)$ and $m(., ., .)$ are continuously differentiable (and hence Lipschitz) functions, and $\Gamma(t)$ is a convex set, depending smoothly on t. The objective function can be written as $J(u)$, since the differential equation for $\mathrm{d}x(t)/\mathrm{d}t$ determines $x(.)$ uniquely as a Lipschitz function of $u(.)$. Assume (as commonly happens in applications) that (7.13) reaches an optimum, with the optimal control $\bar{u}(.)$ a piecewise continuous function.

The conclusions of Theorem 7.7.1 apply also to the optimal control problem (7.13), now replacing $x(.)$ of (7.12) by the control function $u(.)$ of (7.13), taking the subspace **S** in Theorem 7.7.1 to be a subspace of piecewise continuous function such that uniform approximation by step functions holds, and assuming the Robinson stability condition and the linear growth condition. Then Theorem 7.5.3 shows that approximating $u(.)$ by a step function $u_n(.)$ (whereby the interval $[0, T]$ is divided into 2^n equal subintervals, with $u_n(.)$ constant on each), leads to an approximate optimum function $\bar{u}_n(.)$, such that

$\|\bar{u}_n(.) - \bar{u}(.)\| < \kappa/n$ in an appropriate norm as $n \to \infty$. There is an obvious extension when 2^n is replaced by another integer r, with $r \to \infty$.

Such step function approximations to the control function have proved very successful in computational practice – see in particular the references by Jennings *et al.* (1990), Teo, Goh and Wong (1991) and Teo and Womersley (1993). The present results show that such approximations will work well under quite general conditions. An open question remains as to what improvement in convergence might be had by using spline approximations of higher order.

7.9 NONSMOOTH CONTROL PROBLEMS

The analysis of nonsmooth control problems requires two theorems, which will be stated here without proof. Ekeland's theorem (section 7.9.1) is given in a slightly restricted form, appropriate to optimization and control. If a continuous function is bounded below, but does not reach a minimum, then a slightly perturbed function (perturbed by a small norm term) reaches a minimum.

For the second theorem, consider the convex cone

$$C := \{f \in C(I, \mathbf{R}^m) : (\forall t \in I) f(t) \in S\}$$

where $C(I, \mathbf{R}^m)$ is the space of continuous functions from I into $\mathbf{R}^m$ and S is a convex cone in $\mathbf{R}^m$; in particular, $S = \mathbf{R}^m_+$. The dual cone of C is a cone consisting of vector-valued measures on I, taking values in the dual cone S^*. If Λ is such a measure, then its norm is

$$\|\Lambda\| := \sup\left\{\left|\int_I f \, \mathrm{d}\Lambda\right| : f \in C(I, \mathbf{R}^m), \|f\| \leq 1\right\}$$

where $|.|$ denotes the usual (Euclidean) norm in $\mathbf{R}^m$. Given a sequence $\{\Lambda_k, k = 1, 2, \ldots\}$ of such measures, there is a compactness property (section 7.9.2) which provides a convergent subsequence. The case $m = 1$ is well known; for a proof of the vector case, $m > 1$, see Craven (1986, Lemma 6). If $\|\Lambda\| = 1$, the vector measure is called a vector probability measure. The probability distribution corresponding to Λ_k is the function $F_k(.)$ on I, given by $\mathbf{F}_k(t) := \Lambda_k[0, t]$.

7.9.1 Ekeland's theorem (Ekeland, 1974; Clarke, 1983)

Let V be a closed subset of a Banach space (of any dimension); let the function $f: V \to \mathbf{R}$ be continuous and bounded below. For each positive ε' and each $y \in V$ for which $f(y) < \inf_V f(.) + \varepsilon'^2$, there is $v \in V$ such that $\|v - y\| < \varepsilon'$, and the function $f(.) + \varepsilon' \|. - v\|$ reaches a minimum on V at v.

7.9.2 Theorem (compactness of probability measures)

Let I be a compact interval, $S \subset \mathbf{R}^m$ a closed convex cone, and let $\{\Lambda_k: k = 1, 2, \ldots\}$ be a sequence of vector measures on I, taking values in S^*. If

$(\forall k)\ \|\Lambda_k\| \leq 1$, then some subsequence converges weakly to a vector measure $\bar{\Lambda}$; thus

$$(\forall f \in C(I, \mathbf{R}^m)) \quad \int_I f\,\mathrm{d}\Lambda_k \to \int_i f\,\mathrm{d}\bar{\Lambda} \tag{7.14}$$

holds for the subsequence, and also the corresponding subsequence of probability distributions $\{\mathbf{F}_k\}$ converges pointwise to a distribution $\bar{\mathbf{F}}$ on I at each point where $\bar{\mathbf{F}}$ is continuous.

Remark

Since $\bar{\mathbf{F}}$ is a monotonic function it is continuous, except for at most a sequence of points where $\bar{\mathbf{F}}$ has a jump. Since $I = [0, T]$ is a compact interval, the Helly–Bray theorem (Loève, 1955) shows from (7.14) that

$$\Lambda_k(I) = \int_I 1\,\mathrm{d}\Lambda_k \to \int_I 1\,\mathrm{d}\Lambda = \Lambda(I)$$

So $\Lambda \neq 0$ if $\Lambda_k(I) = \|\Lambda_k\| = 1$ for each k.

7.9.3 Continuous programming

Consider first a *continuous programming* problem:

$$\text{Minimize } F(z) := \int_I f(z(t), t)\,\mathrm{d}t \quad \text{subject to} \quad (\forall t \in I := [0, 1])\ g(z(t), t) \leq 0 \tag{7.15}$$

The functions $z(.)$ considered are in $X = C(I)$ (or, alternatively, $X = L^2(I)$). Lipschitz conditions are assumed for the functions f and g, namely

$$(\forall t \in [0, 1], \forall z, z' \in E)\ |f(z, t) - f(z', t) \leq \kappa_1\|z - z'\|;\ \|g(z, t) - g(z', t)\| \leq \kappa_2\|z - z'\|$$

for each bounded subset $E \subset X$ (and the Lipschitz constants κ_1, κ_2 depend on E). Assume that problem (7.15) reaches a minimum at $z(.) = \bar{z}(.)$, and choose E to contain some open ball with centre $\bar{z}(.)$. Assume that $z(t) \in Z := \mathbf{R}^k$, and $g(z(t), t) \in \mathbf{R}^m$.

Smooth the function F by

$$F(z\colon \varepsilon) := \int_I \mathrm{d}t \int_Z f(z(t) - s, t)\varepsilon^{-1}\varphi(\varepsilon^{-1}s)\,\mathrm{d}s \tag{7.16}$$

where $\varepsilon > 0$, and $\varphi(.)$ is a (nonnegative) mollifier (as in section 3.9.1), but now chosen to be continuously differentiable, in particular by

$$\begin{aligned} \varphi(y) &:= 3(y + 1)^2 - 2(y + 1)^3 \quad \text{when } -1 \leq y \leq 0 \\ &\quad\ 3(1 - y)^2 - 2(1 - y)^3 \quad \text{when } 0 < y \leq 1 \end{aligned}$$

and 0 outside $(-1, 1)$; then $\varphi(.) \geq 0$ and $\int_{\mathbf{R}} \varphi(t)\,\mathrm{d}t = 1$. Smooth g by

$$g(z(t), t\colon\varepsilon) := \int_Z g(z(t) - s, t)\varepsilon^{-1}\varphi(\varepsilon^{-1}s)\,\mathrm{d}s \tag{7.17}$$

Then define

$$\tilde{g}(z(t), t{:}\varepsilon) := g(z(t), t{:}\varepsilon) - g(0, t{:}\varepsilon) + g(0, t) \tag{7.18}$$

Then the smoothed problem:

$$\text{Minimize } F(z{:}\varepsilon) \quad \text{subject to} \quad (\forall t \in I)\, \tilde{g}(z(t), t{:}\varepsilon) \leq 0 \tag{7.19}$$

also has $\bar{z}(.)$ as a feasible point.

Define now

$$p(z{:}\varepsilon) := \int_I [\tilde{g}(z(t), t{:}\varepsilon)]_+ \,\mathrm{d}t$$

Then an equivalent problem to (7.19) is

$$\text{Minimize } F(z{:}\varepsilon) \quad \text{subject to} \quad p(z{:}\varepsilon) \leq 0 \tag{7.20}$$

The perturbation theory of section 7.4 (with ε as parameter) requires modification in order to apply to (7.20), since (7.20) need not reach a minimum. The Lipschitz property implies that $\gamma := \inf\{F(z{:}\varepsilon) : p(z;\, \varepsilon) \leq 0\}$ is finite. For sufficiently small ε', choosing feasible z^* such that $F(z^*{:}\varepsilon) < \gamma + \varepsilon'^2$, Ekeland's theorem 7.9.1 shows that there is feasible v with $\|v - z^*\| < \varepsilon'$, such that

$$\hat{F}(.{:}\varepsilon) := F(.{:}\varepsilon) + \varepsilon' \|. - v\| \tag{7.21}$$

is minimized at v over the feasible region of (7.20). The set V is here the (closed) feasible region for the problem (7.15). Then Lemma 7.4.7 can be applied to $\hat{F}(.{:}\varepsilon)$ in place of $F(.{:}\varepsilon)$, choosing ε' small enough that a ball of radius ε' and centre v lies inside the ball of radius r in the construction of section 7.4.2. Since the added term in ε' is another Lipschitz perturbation, it follows from Lemma 7.4.7 that $\hat{F}(.{:}\varepsilon)$ is minimized, subject to the constraint of (7.20), at a point $\hat{z}(.;\, \varepsilon)$, where $\hat{z}(.;\, \varepsilon) \to \bar{z}(.)$ as $\varepsilon \downarrow 0$ and $\varepsilon' \downarrow 0$.

Since $p(.;\, \varepsilon)$ is a function taking values in $\mathbf{R}^m$, the construction of section 3.9.8 then applies, showing that there are Lagrange multipliers $\tau(\varepsilon) \geq 0$ and $\mu(\varepsilon) \geq 0$, not both zero, for which $\|\zeta(\varepsilon)\| \leq 1$ and

$$0 \in \tau(\varepsilon)[F'(\hat{z}(.;\varepsilon){:}\varepsilon)) + \varepsilon'\zeta(\varepsilon)] + \partial[\mu(\varepsilon)\, \partial(\hat{z}(.;\varepsilon){:}\varepsilon)]; \quad \mu(\varepsilon)p(\hat{z}(.;\varepsilon){:}\varepsilon) = 0 \tag{7.22}$$

Now taking $\varepsilon = j^{-1}$ $(j = 1, 2, \ldots)$, letting $\varepsilon' \to 0$ with ε, and choosing successive subsequences as in section 3.9.1, it follows that

$$0 \in \bar{\tau}\partial F(\bar{z}) + \bar{\mu}\partial p(\bar{z}), \quad \bar{\mu}p(\bar{z}) = 0$$

Now $p(.{:}\varepsilon)$ is a composition $\int \circ\, [.]_+ \circ \tilde{g}(.{:}\varepsilon)$, and thus of two smooth functions, and a Lipschitz function $[.]_+$ which is smooth except at 0. Following section 4.2.7, the gradient of $\mu(\varepsilon)p(.{:}\varepsilon)$ at $\bar{z}(.)$, applied to a vector $w(.)$ (which may be taken as a continuous function), is given by

$$p'(\bar{z};\varepsilon)w = \int_I \mathrm{d}\Lambda(t;\varepsilon).[\hat{g}(\bar{z}(t);\, \varepsilon)]_+\, w(t)\, \mathrm{d}t \tag{7.23}$$

where (section 1.10) integration with respect to some measure $\Lambda(.;\varepsilon)$ represents the dual space of a space of continuous functions. The corresponding distribution $\bar{\Lambda}(.;\varepsilon)$ has a derivative $L(.;\varepsilon)$, except on a set of zero Lebesgue measure. Note that, multiplying (7.22) by a suitable positive number, the vector measure $(\tau(\varepsilon), \Lambda(.;\varepsilon))$ may be taken to have norm 1 for each ε.

Expressing the left-hand side of (7.22) as an integral, taking subsequence limits as $\varepsilon \downarrow 0$, and applying the compactness theorem 7.9.2 for probability measures, it follows that

$$\int_I [\bar{\tau}\xi(t) + \lambda(t)\eta(t)]w(t)\,dt = 0, \quad \lambda(t)g(\bar{z}(t), t) = 0 \qquad (7.24)$$

where $\xi(t) \in \partial f(\bar{z}(t), t)$, $\eta(t) \in \partial g(\bar{z}(t), t)$ and an exceptional set of t of zero Lebesgue measure is allowed, with $\Lambda(.)$ a subsequence limit of $\Lambda(.;\varepsilon)$ for a sequence of $\varepsilon \downarrow 0$ and $\lambda(.)$ its derivative (except on a set of zero Lebesgue measure). Applying Theorem 7.9.2 and its following remark to $(\tau(\varepsilon), \Lambda(.;\varepsilon))$ in place of Λ_k, it follows that $(\bar{\tau}, \bar{\Lambda}) \neq (0, 0)$. The ε' term vanishes in the limit, since $\varepsilon' \downarrow 0$ as $\varepsilon \downarrow 0$. Since $w(.)$ is arbitrary, there follows

$$\bar{\tau}\xi(t) + \lambda(t)\eta(t) = 0 \qquad (7.25)$$

except for a set of t of zero measure.

Suppose, if possible, that $\lambda(t)b = \infty$ for $t \in E$, a set of positive measure, and some $b > 0$. Then $\lambda(.)$ is the limit of some sequence $\{\lambda_j(.)\}$, and applying a theorem of integration (Fatou's lemma) shows that

$$\infty = \int_i \liminf[\lambda_j(t)b]\,dt \leq \liminf \int_i [\lambda_j(t)b]\,dt = \Lambda(E)b$$

contradicting the finiteness of $\Lambda(.)$. Hence $\lambda(t)$ is finite except on a set of zero measure.

Each $\eta(t)$ is an $m \times n$ matrix. Assume that $m < n$. Denote by I_0 the set of $t \in I$ for which $g(\bar{z}(t), t) = 0$; note that $\lambda(t) = 0$ for $t \notin I_0$. Consider the *constraint qualification*, that each $M \in g(\bar{z}(t), t)$ has full rank whenever $t \in I_0$. If this holds, and if $\bar{\tau} = 0$, then $\lambda(.)$ is nonzero, but also

$$0 = \int_{I_0} \lambda(t)\eta(t)\,dt \Rightarrow \lambda(t) = 0 \quad \text{on } I_0$$

(except on a set of zero measure), a contradiction. Note that full rank happens exactly when the mapping of $\lambda(t)$ to $\lambda(t)\eta(t)$ is surjective.

These results may be summed up as a theorem.

7.9.4 Theorem (continuous programming with Lipschitz functions)

For the continuous programming problem (7.15), assume that the functions $f(.,t)$ and $g(.,t)$ are Lipschitz, uniformly in $t \in I$; that the problem reaches a local minimum when $z(.) = \bar{z}(.)$; and a constraint qualification holds there. Then, for some costate function $\lambda(.) \geq 0$, the following KKT conditions hold:

$$0 \in \partial(f(\bar{z}(t), t) + \lambda(t)g(\bar{z}(t), t)); \quad \lambda(t)g(\bar{z}(t), t) = 0$$

except for a set of t of zero Lebesgue measure.

Proof

The result follows from (7.25) above, noting that $\bar{\tau} > 0$ from the constraint qualification. □

7.9.5 Optimal control under Lipschitz conditions

Consider now the optimal control problem (with no state constraint):

$$\text{Minimize}_{x,u}\, F(x, u) := \int_I f(x(t), u(t), t)\,\mathrm{d}t \quad \text{subject to} \quad x(0) = 0$$

$$(\forall t \in I) \quad \dot{x}(t) = m(x(t), u(t), t), \quad (\forall t \in I) \quad u(t) \in \Gamma(t) \tag{7.26}$$

where I is the interval $[0, 1]$, $\Gamma(t)$ is a specified set for each t (with all the $\Gamma(t)$ lying in some bounded region), and the functions f and m satisfy Lipschitz conditions, uniformly in $t \in I$. If $z(t) := (x(t), u(t))$ and $z^*(t) := (x^*(t), u^*(t))$, these Lipschitz conditions become

$$(\forall t \in I)\ |f(z(t), t) - f(z^*(t), t)| \le \kappa_1\|z - z^*\|, |m(z(t), t) - m(z^*(t), t)| \le \kappa_2\|z - z^*\|$$

whenever z and z^* lie in some ball centred on a minimum point $\bar{z} := (\bar{x}, \bar{u})$ for the control problem. The differential equation can be expressed, as before, as $\mathrm{D}x = M(x, u)$.

A smoothed problem can be constructed in the same manner as for continuous programming in 7.9.2, now smoothing f and m in relation to $z(.)$. Denote the smoothed version of $F(z)$ by $F(z{:}\varepsilon)$, where $z := (x, u)$. Application of Ekeland's theorem shows that there is a constrained minimum of

$$F(z;\varepsilon) + \varepsilon'\|z - v\|$$

at $z = v$, with v constructed as in section 7.9.3. This gives a problem to which Theorem 7.9.4 can be applied.

Consider now minimization with respect to x, with u held at $\bar{u}$. The constraint qualification of section 7.9.3, applied to the constraint $\mathrm{D}x = M(x, \bar{u})$, requires that whenever $t \in I$ and $\eta(t) \in \partial_x m(x(t), \bar{u}(t), t)$, the differential equation $\dot{x}(t) - \eta(t) = w(t)$, $x(0) = 0$, has a solution $x(.)$ for each continuous $w(.)$. This means that the expression on the left defines a surjective mapping from (piecewise smooth) functions $x(.)$ onto continuous functions.

Assuming this constraint qualification, $\bar{\tau}$ may be taken as 1 in Theorem 7.9.4. Treating the term $\lambda\mathrm{D}$ by integration by parts, as in section 4.2.7, there results the following *adjoint differential inclusion*:

$$-\dot{\lambda}(t) \in \partial_x h(\bar{x}(t), \bar{u}(t), t, \lambda(t)) \quad (t \in I); \quad \lambda(1) = 0 \tag{7.27}$$

in terms of the Hamiltonian

$$h(x(t), u(t), t, \lambda(t)) = f(x(t), u(t), t) + \lambda(t)m(x(t), u(t), t)$$

The Pontryagin theory is now obtained by a generalization of section 7.2.3, now applied to $F(z;\varepsilon) + \varepsilon'\|z - v\|$ in place of $F(z : \varepsilon)$. Define the corresponding

Hamiltonian as

$$\tilde{H}(z) := F(z;\varepsilon) + \varepsilon'\|z - v\| + \bar{\lambda}(\varepsilon)M(z;\varepsilon)$$

and let $v = (\tilde{x}, \tilde{u})$. The construction of section 7.2.3 then shows that, if the multiplier $\bar{\lambda}(\varepsilon)$ satisfies an adjoint differential equation, then

$$\tilde{H}(\tilde{x}, \tilde{u}) - \tilde{H}(\tilde{x}, u) \geq o(\|u - \tilde{u}\|) + O(\varepsilon')$$

the last term coming from $\varepsilon'\|z - v\|$. The hypotheses required for this are that x has Lipschitz dependence on u, and that $F(z;\varepsilon) + \bar{\lambda}(\varepsilon)M(z;\varepsilon)$ is partially differentiable with respect to x, uniformly in u in some ball with centre $\bar{u}$. The latter property holds (section 4.3.2) since the function is twice continuously differentiable by construction of the smoothing.

Now subsequence limits, such as $\varepsilon \downarrow 0$ and $\varepsilon' \downarrow 0$, may be applied to deduce that the Hamiltonian h for the nonsmooth control problem is such that $H(\bar{x}, .)$ has a quasimin, subject to the control constraints, at $u = \bar{u}$.

The remainder of the proof of the Pontryagin principle consists of an application of Theorem 7.2.6, which uses the quasimin, but makes no use of differentiability. The following theorem has thus been proved.

7.9.6 Theorem (Pontryagin theorem for Lipschitz control problem)
Assume that the fixed-time optimal control problem (7.26), with Lipschitz functions, reaches a local minimum when $\mathbf{u} = \bar{\mathbf{u}}$ with respect to the L^1-norm for the control $\mathbf{u}$. Assume that the chattering property holds for the constraints on the control, and that the objective function satisfies the linear growth condition. Then necessary conditions for the minimum of (7.26) are that the costate $\boldsymbol{\lambda}(.)$ satisfies the adjoint differential equation (7.27) with boundary condition, and that, for all t except a set of zero measure, the associated problem (7.6) reaches a minimum when $\mathbf{u}(t) = \bar{\mathbf{u}}(t)$.

7.10 REFERENCES

Alt, W. (1989) Stability of solutions for a class of nonlinear cone constrained optimization problems, *Numer. Funct. Anal. Appl.*, **10** 1053–64; 1064–76.

Alt, W. (1990) Stability of solutions and the Lagrange–Newton method for nonlinear optimization and optimal control problems, Habilitationsschrift, Universität Bayreuth.

Clarke, F. H. (1983) *Optimization and Nonsmooth Analysis*, Wiley-Interscience, New York.

Craven, B. D. (1982) *Lebesgue Measure and Integral*, Pitman, Boston.

Craven, B. D. (1983) Perturbed minimization, with constraints adjoined or deleted, *Optimization*, **14** 23–6.

Craven, B. D. (1986) Nondifferentiable optimization by smooth approximations, *Optimization*, **17** 3–17.

Craven, B. D. (1994) Convergence of discrete approximations for constrained minimization, *J. Austral. Math. Soc., Ser. B*, **35** 1–12.

Ekeland, I. (1974) On the variational principle, *J. Math. Anal. Appl.*, **43** 324–53.

Fleming, W. H. and Rishel, R. W. (1975) *Deterministic and Stochastic Optimal Control*, Springer, Berlin.

Jennings, L. S., Fisher, M. E., Teo, K. L. and Goh, C. J. (1990) *MISER3, Optimal Control Software, Theory and User Manual.*

Lempio, F. and Maurer, H. (1980) Differential stability in infinite-dimensional nonlinear programming, *Appl. Math. Optim.*, **6** 139–52.

Loève, M. (1955) *Probability Theory*, Van Nostrand, New York.

Macki, J. and Strauss, A. (1982) *Introduction to Optimal Control Theory*, Springer, New York.

Munroe, M. E. (1953) *Measure and Integration*, Addison-Wesley, Reading MA.

Robinson, S. M. (1976) Stability theory for systems of inequalities, Part II: Differentiable nonlinear systems, *SIAM J. Numer. Anal.*, **13** 497–513.

Teo, K. L. and Goh, C. J. (1989) A computational method for combined optimal parameter selection and optimal control problems with general constraints, *J. Austral. Math. Soc., Ser. B*, **30** 350–64.

Teo, K. L., Goh, C. J. and Wong, K. H. (1991) *A Unified Computational Approach to Optimal Control Problems*, Longman Scientific & Technical, London.

Teo, K. L. and Womersley, R. S. (1983) A control parametrization algorithm for optimal control problems involving linear systems and linear terminal inequality constraints, *Numer. Funct. Anal. Opt.*, **6** 291–313.

Yen, N. D. (1990) *Stability of the Solution Set of Perturbed Nonsmooth Inequality Systems*, International Centre for Theoretical Physics, International Atomic Energy Agency, UNESCO, Trieste, Italy; to appear in *J. Optim. Theor. Appl.*

Index